Geomorphology

B. W. Sparks

University Lecturer in Geography,
Fellow and Senior Tutor, Jesus College, Cambridge

Longman

LONGMAN GROUP LIMITED
London
*Associated companies, branches and representatives
throughout the world*

First edition © B. W. Sparks 1960
This edition © Longman Group Ltd. 1972

First published 1960
Second edition first published 1972

ISBN 0 582 48147 3

*Printed in Great Britain by William Clowes & Sons, Limited
London, Beccles and Colchester*

Contents

Contents

List of text figures

List of text figures

List of plates

List of acknowledgements

We are indebted to the following for permission to base diagrams on existing material:

Edward Arnold (Publishers), Ltd for Fig. 13.22 from de Filippi, *Himalaya, Karakoram and Eastern Turkestan*; Stephen Austin & Sons, Ltd for Fig. 9.14 from J. Hanson-Lowe, *Geol. Mag.*, **72**; the author and William Collins Sons & Co., Ltd for Figs. 8.19, 8.21, 8.26 from Steers, *The Sea Coast*; Columbia University Press for Fig. 6.11 from Howard, *J. Geomorph.*, **1**; the author and the Council of the Croydon Natural History Society for Fig. 7.22 from Fagg, *Proc. and Trans. Croydon Natural Hist. Soc.*, **9**; the editor for Figs. 6.12, 6.15 from Sparks, *Geography*, **36**; The author and editor for Fig. 4.10 from Gifford, *Geography*, **38**; the author and the Council of the Geological Society for Fig. 14.12 from Wills, *Quart. J. geol. Soc. Lond.*, **68**, and the Council for Figs. 14.8–9 from Kendall, *Quart. J. geol. Soc. Lond.*, **58**; the authors and the Council of the Geologists Association for Fig. 4.11 from Grove, *Proc. Geol. Ass., Lond.*, **64**, Fig. 8.5 from Wilson, *Proc. Geol. Ass., Lond.*, **63**, Figs. 4.14, 4.16 from Wood, *Proc. Geol. Ass., Lond.*, **53**, and the Council for Fig. 4.8 from Gossling, *Proc. Geol. Ass., Lond.*, **59**, Fig. 4.9 from Gossling & Bull, *Proc. Geol. Ass., Lond.*, **59**, Fig. 7.24 from Bull, *Proc. Geol. Ass., Lond.*, **51**; The Geological Society of America for Fig. 5.1 from Horton, *Bull. geol. Soc. Amer.*, **56**; the Controller of H.M. Stationery Office for Fig. 14.6 from T. H. Whitehead and others, *The Country Between Wolverhampton and Oakengates*; Hutchinson & Co. (Publishers), Ltd for Fig. 8.1 from Russell and Macmillan, *Waves and Tides*; the Council of the Institute of British Geographers for Fig. 6.26 from Wooldridge & Linton, *I.B.G. Publication, No. 9*; Macmillan & Co., Ltd for Figs. 4.17, 4.18 from Penck, *Morphological Analysis of Landforms* and Figs. 13.2, 13.3 from Wright, *The Quaternary Ice Age*; Methuen & Co., Ltd for Figs. 3.3, 3.4 from Tyrrell, *The Principles of Petrology*; the author and Methuen & Co., Ltd for Figs. 11.1–4, 11.6, 11.7 from Bagnold, *Physics of Blown Sand and Desert Dunes*; the author for Fig. 13.14 from Nye, *Journal of Glaciology*, **2**; the author and Oliver & Boyd, Ltd for Figs. 7.14, 7.18 from Charlesworth, *The Geology of Ireland*; the authors and George Philip & Son, Ltd for Figs. 14.11, 16.6, 16.7 from Wooldridge & Linton, *Structure, Surface and Drainage in South-Eastern England*; the authors and the Royal Geographical Society for Figs. 8.2, 8.30 from Lewis, *Geogr. J.*, **78** and **80** and for Figs. 6.20, 6.21 from Wager, *Geogr. J.*, **89**; the author and the Royal Society for Fig. 14.10 from Shotton, *Phil. Trans. B*, **237**; the author and the editor for Fig. 13.20 from Linton, *Scot. Geogr. Mag.*, **65**; the author for Figs. 4.4, 4.5, 4.6 from Sharpe, *Landslides and Related Phenomena*; the University of Chicago Press for Fig. 5.9 from Kesseli, *J. geol.*, **49**; the author and Whitcombe

Acknowledgements

& Tombs Ltd for Fig. 7.20 from Cotton, *Volcanoes as Landscape Forms*; John Wiley & Sons, Inc. for Fig. 8.17 from Johnson, *Shore Processes and Shoreline Development*, 1919; Yale University Press for Figs. 8.33, 8.34, 13.7 from Daly, *The Changing World in the Ice Age*; Prof. J. A. Steers for Fig. 8.22; Fig. 7.19 is modified from Gilbert, *The Geology of the Henry Mountains*, U.S. Department of the Interior; Fig. 13.1 is reproduced from Stamp and Beaver, *The British Isles*, Longmans.

For the second edition we are indebted also to the following for permission to base diagrams on existing material:

the author and the American Meteorological Society for Fig. 11.9 from Hanna, *J. appl. Met.*, **8**; the authors and the British Association for the Advancement of Science for Figs. 16.9–11 from Sparks and West, 'The relief and drift deposits' in *The Cambridge Region* ed. Steers; the author and Cambridge University Press for Fig. 13.6 from Donner, 'Land sea level changes in Scotland' in *Studies in the Vegetational History of the British Isles* eds. Walker and West; the author and the former Geographical Branch of the Canadian Department of Mines and Technical Surveys for Fig. 15.2 from Mackay, *The Mackenzie Delta Area, N.W.T.*; the author and the University of Chicago Press for Figs. 4.12 and 4.13 from Kirkby, *J. Geol.*, **75**; the authors and W. H. Freeman and Company, Publishers, for Fig. 6.33 from Leopold, Wolman and Miller, *Fluvial Processes in Geomorphology*; the authors and the Institute of British Geographers for Figs. 11.8 and 13.4 from Peel, *Trans. Inst. Br. Geogr.*, **38**, and Cullingford and Smith, *Trans. Inst. Br. Geogr.*, **39**, respectively; the author and Litton Educational Publishing, Inc., for Fig. 8.36 from an article by Coleman in *Encyclopedia of Geomorphology* ed. Fairbridge, copyright 1968; the authors and Oliver and Boyd, Edinburgh, for Figs. 3.1 and 14.7 from Ollier, *Weathering*, and Sissons, *The Evolution of Scotland's Scenery*, respectively; the author and the Paleontological Institute, Upsala, for Fig. 5.4 from Hjulström, *Bull. Geol. Inst., Upsala*, **25**. Plate 33 is reproduced from a photograph by Graf von Castell in Machatschek, *Geomorphologie*, published by Teubner, Stuttgart.

Preface to the first edition

In writing this book I have had in mind mainly the requirements of under-graduates reading Part I of the Geographical Tripos or its equivalent, the compulsory geomorphology needed for the Final Honours examination in Geography. At the same time, I hope that it may provide an introduction, mainly through its references, to the more advanced geomorphology, which may be taken as an optional subject in the final year at most universities. The references given are, wherever possible, to papers or books which summarise present knowledge on certain aspects of geomorphology, be-cause these usually contain large bibliographies of original sources. Where summary accounts are not available, for example on rivers and slopes, my treatment of the subjects is deliberately a little more extended and more references to original papers are given.

In spite of an enormous and ever-increasing literature, there are few, if any, landforms the origins of which are known with certainty. The generally accepted theory is usually only accepted because it explains a larger proportion of observed facts than its rivals. But with most theories there are facts which are difficult to explain and assumptions which seem improbable, while, if one cares to enquire into the compatibility of ex-planatory theories of different landforms or to analyse the full consequences of any theory, one will probably find inconsistencies and defects. The stu-dent of geomorphology should realise this. By training himself to detect the limitations of explanations, he may become sufficiently interested to study some particular problem himself and ultimately, perhaps, to provide a better theory. Accordingly, I have tried to stress the deficiencies of theories, even of those which seem the best available at present.

Because geomorphology involves the study of processes and the study of the evolution of landscapes, the barriers between it and other natural sciences are almost non-existent. Thus, the geomorphologist requires an appreciation of physics and chemistry in the consideration of process, and an acquaintance with the numerous techniques of Pleistocene geology and the validity of the conclusions drawn from them in studying problems of landscape evolution. In this merging of subjects lies one of the fascinations of geomorphology.

Preface to the first edition

An interest in scenery may easily be aroused in anyone interested in any aspect of natural history, but such an interest may as readily evaporate when confronted with the technical jargon which seems inseparable from many natural sciences today. For concepts outside the range of everyday language geomorphology obviously needs a technical vocabulary in the interests of brevity. But there is no need for the hordes of synonyms from many different languages and dialects, which have been applied to features readily described in ordinary English words, for example in connection with limestone relief. I have, therefore, tried to limit the number of such words, but some jargon is so deeply rooted that it is impossible to eradicate it.

Finally, as I have personal knowledge of only a small fraction of the field of geomorphology, I must express my thanks to colleagues and friends for advice, opinions and references, especially to Mr A. T. Grove and Mr W. V. Lewis for reading and criticising chapters on subjects on which their knowledge is much greater than mine. I am also indebted to Mr H. Brammer for the photographs forming Plates 31 and 32, to Dr J. K. S. St Joseph for photographs from the collection of the University of Cambridge forming Plates 18, 22, 24, 26, 30 and 46, and to Mr L. R. Thurston for undertaking the drawing of the line diagrams. The remainder of the photographs are my own.

Cambridge, 1959

Preface to the second edition

The first edition of this book was an attempt to review the field of geo-morphology in a general and critical way. As such it seems to have interes-ted a rather wider audience than I had then anticipated. In the twelve years since it was written there has been a geometric increase in the number and length of publications on geomorphology, especially on the philosophy and methodology of the subject, so much so that the general geomorpholo-gist has almost become extinct and replaced by a variety of geomorpholo-gical races who hardly understood each other's language. Nevertheless, as geomorphology is basically a field subject about landforms and of interest to a wide variety of people, I thought it worth while to try to bring the book up to date without changing its essential character. This has involved two new chapters, the redistribution of the material of one chapter, and additions of varying length and complexity to all other chapters. If it was difficult to cover the field twelve years ago, it is even more difficult now. Hence, in my efforts to reduce the number of my errors to a minimum, I have repeatedly consulted my colleagues, Mr R. J. Chorley, Mr A. T. Grove, Dr B. A. Kennedy, Dr D. R. Stoddart and Dr R. G. West, and I am very grateful to them for their help. The line diagrams for this new edition have been redrawn throughout by Mr R. D. Hipps. Finally, I must thank Miss Elizabeth Sparks for the provision of a much fuller index for this edition.

Jesus College
Cambridge
June 1970

METRICATION

Metric units are the primary units in this revised text: English equivalents follow in brackets, the accuracy or otherwise of the conversion reflecting the precision of the original. Diagrams are entirely metric except where accurate contours in standard intervals (i.e. 50 or 100 feet) have been con-verted to unlikely looking metric equivalents: in this case both are included on the diagrams. In a few places where a range of say 25 feet has been men-tioned as a class interval this has been equated with 10 m, simply because one is dealing here not with conversions but with equivalents.

1

The aims and position of geomorphology

An English word may come, through slowly changing usage, to have a meaning different from that which a study of its derivation would indicate. Thus, although geomorphology means the study of the form of the earth, the subject in practice covers a narrower field. The study of the form of the earth is shared by a number of sciences, among them geography, geology, geodesy and geophysics, in addition to geomorphology. In practice these sciences overlap, so that hard-and-fast definitions of their scope are neither desirable nor practicable. An attempt will be made, therefore, to indicate the scope of geomorphology without giving it a precise definition, as it merges with geography on one side, with geology on another, with engineering on a third, with physics on a fourth, with Pleistocene studies on a fifth, and, in its methods, with statistics on a sixth.

Essentially it is the study of the evolution of landforms, especially landforms produced by the processes of erosion. But such a statement is both too broad and too narrow to act as a definition. While erosional landforms are the main interest, the study of certain types of constructional landforms is just as legitimately a part of geomorphology. The study of the formation of coastal features, for example, has always attracted the geomorphologist, who has not limited his attention to the erosional features of the cliffs, but has rightly considered also the constructional forms, such as beaches, spits and bars, produced by the action of waves. Similarly, it would be ridiculous, in the study of landforms produced by glaciation, to concentrate attention on corries, hanging valleys and similar features of erosion to the complete exclusion of constructional landforms such as kames, eskers and moraines. Yet the study of deposition as a whole is not within the scope of geomorphology, but only those aspects which produce features on the surface of the land.

The definition is too broad, in that the geomorphologist does not study all aspects of the evolution of landforms. The origin of the major features of the earth's crust is usually not considered to be within the sphere of geomorphology. An understanding of the forces and processes responsible for folds and faults, earthquakes and volcanoes, lands and oceans, was attempted in the past by the geologist, but is becoming more and more within the

I

field of the geophysicist. Such studies require more knowledge of mathematics and physics than the average geomorphologist, usually a geologist or geographer by initial training, possesses. Many years ago W. M. Davis defined the position quite clearly when he relegated the study of the evolution of structure to geology and the study of the wearing down of that structure to geography.

Even within geomorphology there are recognisably different aspects of study. In its simplest form geomorphology is the study of the relations between landforms and the underlying rocks. We are all familiar with this type of geomorphology in its crudest form from early lessons in physical geography. Yet, in many ways, it is a very neglected branch, as the intimate relations between landforms and rocks are often insufficiently appreciated. It is true that the broad relationships are known: for example, most of us were taught at an early stage that the Lower Greensand forms an important cuesta within the Weald, but the detailed relations between the different beds of the Lower Greensand and the relief are too often ignored at a later stage. The basic knowledge of the effect of rocks on relief is fundamental in any regional study, especially when systems, such as the Carboniferous or Jurassic, which undergo great lateral changes in lithology, are being considered.

The study of the effects of rocks must eventually go well beyond this. Ultimately geomorphology is concerned with the interactions between denudational process and rock strength. Hence, in the precise investigation of the resistance of rocks to denudation, detailed experimental work on rocks must be done. Questions such as the relative resistance of different rocks to processes such as freeze-thaw may only be decided by detailed laboratory experiments, which may be difficult not only to set up but also to interpret. The greatest steps in the understanding of landforms may well depend on the successful development of such experiments, on which a start has been made.

In its second sense, geomorphology is the study of the evolution of landscapes. Such study is often termed denudation chronology, a phrase so charged now with emotive overtones that the simpler and less pretentious term, evolution of relief, may well be preferred. Essentially any study of the evolution of landforms attempts to reconstruct a succession of pictures of the relief at different periods. Alterations of relief are usually caused by changes of base level and climate. Thus, the reconstruction of former base levels and of the landforms related to those base levels, by means of marine and river terraces and long profiles of rivers, is one part of the study, while the study of landforms produced, for example, by glacial and periglacial conditions is the other. The dividing line between this branch of geomorphology and the geological science of stratigraphy is not at all clear. Landforms dating from the Miocene and Pliocene periods have been widely recognised, so that many geomorphological studies of the evolution of relief commence at those periods, and overlap studies of stratigraphy. The latter extend to include the study of glacial and Post-glacial deposits, which are of interest to the geomorphologist in a somewhat different way. In so far as stratigraphy interprets the past primarily by reconstructions based upon deposits and geomorphology interprets the past mainly from the study of landforms, this aspect of geomorphology might almost be called the extension of stratigraphy by geomorphological rather than by palaeontological methods.

In recent years more and more geomorphology is being integrated with Pleistocene studies to produce a synthetic reconstruction of the evolution of past landscapes using as many lines of convergent evidence as possible. These studies have usually involved the collaboration of specialists, as many of the techniques of study require long experience before they can be used satisfactorily.

The third aspect of geomorphology is the study of the actual processes of erosion which give rise to landforms. Unlike the first two aspects, which are essentially regional in approach, this third and last aspect is systematic. It endeavours to understand the action of waste movement, of water movement, of ice action and wind action, as well as the processes of weathering. It is the most difficult of the branches of geomorphology, as all natural processes are the results of the interaction of many factors. It is the most fundamental, because geomorphological studies of the first two types rest on the assumption that the processes operating are known. While it is probably true that we know what happens in general terms and can observe the results of the processes, the actual mechanism of most processes of erosion is still the subject of intensive research. The problem of process can be approached in a number of different ways.

A method which is becoming increasingly popular is the highly deductive argument. It is usually known that any given process involves a number of factors, but the exact effect of each cannot be assessed. In this instance, the assumption can be made that all but one of the factors are constant. With this assumption made, the effects of variations in the unknown factor may then be analysed. While such arguments may lead occasionally to the understanding of natural phenomena, they are sometimes so artificial as to be divorced from nature which is essentially complex.

In the past this approach was largely by verbal argument, but it is now possible by the use of the mechanical aid of computers to quantify results provided that the various factors can be quantified. In fact, the variations of more than one of the variables can be fed in to produce estimates of the combined effects. The precision of the results depends on the accuracy of the observations originally used as a basis for the quantification of the variables. Thus, the method ultimately depends, like all geomorphology, on meticulous observation and field work.

On the other hand, processes may be inferred in precise or general terms from field observations. Regional geomorphological studies may lead to the conclusion that certain forms are probably the results of certain processes, the precise actions of which cannot be clearly defined. The value of such inferences or hunches is that they are the ideas which can be tested by mechanical or experimental methods. No computer in the middle of its operations says to its programmer that the theory behind the program should be scrapped and another adopted: no laboratory experiment really suggests different theories of origin. Landforms alone suggest theories, some of which are undoubtedly absurd. These can be tested by machine and laboratory techniques.

Finally, there is the laboratory method, in which natural processes are reproduced as closely as possible in miniature under controlled conditions. Undoubtedly geomorphological laboratories help greatly in understanding some processes and in a qualitative appreciation of others, although it is doubtful whether they can lead to a complete understanding of the quantitative nature of all processes. The limitations imposed by scale are very great: it is not at all easy to make scale replicas of different rock types, to quote one example. Laboratory rocks are usually various sand, gravel and mud mixtures, and probably correspond to unconsolidated grits and conglomerates in nature. Any attempt to provide greater resistance seems to lead to an enormous retardation of erosion and to offset one of the objects of the laboratory, which is to speed up natural processes on a reduced scale.

All methods of studying processes involve difficulties, but together they lead to a better understanding of both systematic and regional geomorphology.

Relations with other subjects

It is pointless to argue whether geomorphology should be a part of geology, a part of geography or a science in its own right, as the three points of view can probably be equally justified. In this country geomorphology is traditionally a part of geography, in other countries a part of geology, its exact affiliation probably reflecting the interest shown by geographers and geologists in landforms in the infancy of geomorphology.

In most courses of geography, geomorphology stands a little apart, usually as a specialist study for the final year. As described above, it is not the physical basis of human geography, which consists of the relevant parts of many sciences designed to aid the geographer in the study of human beings and their activities. As the study of geography is usually understood to include the study of the earth's surface whether inhabited or not, geomorphology is an essential part of geography, but its results may or may not be of interest to the human geographer. Geomorphology is concerned with the understanding of landforms rather than with their description and distribution alone, but the distribution of landforms is usually of more significance to the human geographer than their origin. It matters little to him that much of Cambridge is built upon a terrace which dates from the last interglacial period: the location, the elevation and the nature of the deposits of the terrace are of greater interest because of their effects on settlement site and water supply. Many of the results of geomorphological studies relate only to knowledge about landforms and have little or no bearing on human studies.

Geomorphology is probably more closely related to geography than to geology in so far as it studies surface forms rather than deposits. Its main contribution to geology is in its ability to help in the unravelling of geological successions when the deposits are thin and scattered and the landforms plentiful, such conditions being most characteristic of the Pliocene and Pleistocene periods in certain areas. But geomorphologists must possess a good working knowledge of geology if they are to have a full appreciation of landforms, as the importance of the lithology of the rocks in the formation of relief can hardly be overemphasised. The lithological characteristics and variations of rocks is the aspect of geology of most importance to the geomorphologist, and to acquire this information he must be capable of digesting and selecting from geological accounts, few of which have lithology as the main object of study.

The student of process often finds a greater need of mathematical and physical knowledge, especially when he is concerned with the movement of different types of material, whether by water, ice or wind. Other scientists have made great contributions to our knowledge of some of these processes, physicists having studied ice movement and hydraulic engineers water movement, while the deformation of waste mantles on slopes concerns constructional engineers in a wide variety of fields including road and building construction.

It is, however, in the unravelling of the history of erosion that the geomorphologist has come into closest contact with other workers, especially in those countries where some of the landforms are the relics of the glaciations of the Pleistocene period. The study of the Pleistocene period, especially the elucidation of the succession of events, is one of the greatest problems in natural science. It cannot readily be solved by geological methods alone, as there are no fossils of sufficiently wide range and sufficiently restricted to certain horizons to act as zone fossils. Both the archaeologist, with his study of human cultures, and the botanist, with a study of plant successions, have contributions to make to the subject, while at times the geomorphological study of landforms and distribution of relief seems to provide the only way out of some of the difficulties.

Thus, it can be seen that the study of geomorphology, while clearly defined in its central aspects, merges with several other sciences towards its margins. In the latter aspects the geomorphologist must be prepared to work with other scientists, to take their conclusions and to test their application to the phenomena which are his own special study, and, in return, to offer his own conclusions for testing by other techniques.

Obviously there will not be space in the following chapters to cover fully and systematically all aspects of geomorphology. The emphasis will be

mainly upon the effects of the most important factors which control geomorphological processes, illustrated as far as possible by regional examples.
Some of the more important regional studies will be noted in the selection
of references given at the end of each chapter.

2
The Davisian geographical cycle

The integration of geomorphology into a coherent subject was largely the result of the work of the American geomorphologist, W. M. Davis, in the last decade of the last century. His main concept was the arrangement of the phenomena of erosion into a cycle of the development of landforms, which he called the geographical cycle. This concept and some of Davis's other ideas have been criticised severely, as later work has proved that his ideas were not always correct. It must be stated, however, that much of the criticism appears to have arisen through an insufficiently thorough acquaintance with Davis's work, which is not as rigid and limited as some of the criticism would suggest.

One should endeavour to put Davis into perspective in the history of the development of geomorphological thought. Some of the geomorphological opinions current in the second half of the last century appear very antique today: the idea of continental glaciation by an ice sheet was not accepted without difficulty; the idea that Pleistocene deposits were due to a vast flood was only recently forgotten; the concept of peneplanation had made little headway, at least in England. The main geomorphological concepts had been stated largely by American geologists working in the western part of their country in the decade 1870 to 1880 and had been admirably summarised by Gilbert. Davis attempted little less than the organisation of a discipline. The amazing fact is not that he has been shown to be wrong on some points, but that his ideas on many things are still accepted today.

His aims were not quite the same as those of modern geomorphology. He called the subject physical geography, and occasionally geomorphogeny. Its end was to enable the trained reader to understand the descriptions of the trained observer of landforms, as both would speak the same language and have the same concepts. The aim was to effect a great economy in description.

His weakness, at least in the eyes of many geomorphologists, lay in looking at landforms on too large a scale and in too descriptive a manner. Undoubtedly this led to the neglect of the detailed mechanics of process operation, which must be studied to enhance our basic knowledge of landscape formation. Yet it is possible for interests to swing too far in the opposite direction and the detailed analyses of a few slopes or drainage basins must be supplemented by attempts to interpret the physical landscape on a regional scale. The latter type of study is not often cast within the Davisian framework, but its object is one which Davis would have appreciated. Both types of study, the one detailed and short-term and the other general and long-term, are required for the full development of geomorphology.

The geographical cycle or the cycle of erosion

Any landform was, according to Davis, a function of structure, process and time. By structure he meant something more than is usually meant today: the word includes not only the attitude of the rocks, the nature of the dip of the beds, their folds and faults, but also what would now be termed the lithology of the beds, the nature of the rocks, their relative hardness and relative permeability. Process includes the different agents of weathering and of erosion: water, wind, ice and gravity. Finally, the length of the time during which the processes have acted obviously must have a profound effect on the landforms. For the purposes of his cycle, Davis divided the landforms produced during a cycle of erosion into three stages: youth, maturity and old age.

The ideal geographical cycle

In order to simplify the teaching of the elements of his geographical cycle, Davis made certain assumptions, which he himself knew to be not justifiable in nature. The chief of these was the simplification of the uplift of an area, so that there should not be appreciable erosion during uplift, as the discussion of such erosion would complicate the working out of the cycle considerably. Such an assumption is not fatal to the scheme of the cycle provided that the scheme is sufficiently flexible to allow for complicated uplifts to be considered in more advanced discussions, as, indeed, it is. Davis has been criticised on the grounds that he misinterpreted uplift, whereas he categorically stated that he was using only a simplifying assumption. Indeed, in his German essays, he stressed the amount of erosion which is sometimes performed during uplift, and even analysed the case where uplift is so slow that the landscape, 'Minerva-like', passes straight into the stage of maturity. The following analysis of his cycle also makes use of a simple uplift merely for the sake of convenience.

A mass of land, the structure of which is complicated by harder and softer beds and by faults, may be imagined recently to have emerged from beneath the sea. It is very unlikely that its surface will be absolutely smooth, as there are bound to be initial irregularities. Obviously rain falling on to the surface will be concentrated into the depressions, which may or may not reach the sea. If they do reach the sea, the land may be drained directly, but, if they lead into closed depressions, lakes will gradually develop until, finally, they spill over and the drainage continues on its way to the sea. Streams of this type, which follow initial irregularities in the land's surface, dominate the drainage in the early stages of the cycle, and are known as consequent streams. At the beginning, their beds, or, in more geomorphological terminology, their long profiles, will be irregular, due to initial irregularities in the depressions and to the presence of harder and softer rocks. Slowly, however, the courses of the rivers will become smoothed out, first at the mouths and then progressively and slowly upstream (Fig. 2.1). At its mouth the river cannot cut down below the level of the sea, so that an imaginary projection of the sea surface beneath the land represents a level below which the river cannot erode: it is known as base level, or, as some later writers would have it, the general base level. Except at its mouth no river can ever reach base level, as it must always preserve sufficient gradient to allow water to flow down its course.

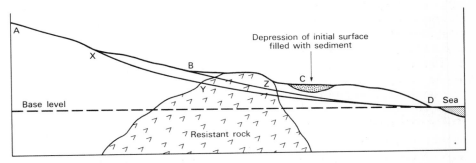

FIG. 2.1. Development of a smooth graded profile. ABCD is the irregular course of early youth with resistant rock outcrops causing rapids and an infilled depression still present. ABZD and AXYD are successive stages in grading: only stretch AX remains ungraded to the end

The development of the consequent stream has been followed regardless of other drainage developments that take place. Obviously the consequent stream will receive tributaries, and, if these are also guided by initial depressions, they, too, will be termed consequent streams. But other classes of streams are included among the tributaries. An area of soft rocks between two areas of harder rocks will provide the ideal course for the development of a stream. Such a stream is termed a subsequent stream. Although it must develop subsequently to the consequent stream, it is defined as a stream which owes its course to some weakness in the structure of the area subject to erosion. From this definition it can be seen that there is another important class of subsequent streams: those which are guided by faults and shatter belts in the rocks. A fault does not necessarily provide an especially favourable course for a river. If, however, it throws harder rocks against softer, or if it causes considerable dislocation of the rocks, it will probably provide an easier course. Examples of the latter type are provided by the

9

rivers Neath and Tawe in South Wales. If, on the other hand, there is a great thickness of homogeneous rock, with an extensive system of faults without much shattering, the stream pattern may show little relation to the faults. Such conditions would obtain in a landscape composed of a considerable thickness of clay for example the Weald Clay in southeast England. Subsequent streams, like consequent streams, slowly smooth out their profiles until they lead progressively and gradually to the sea, usually via the consequent streams to which they are tributary.

Yet another class of streams may be formed, streams which do not appear to depend upon either initial depressions or upon weaknesses in the rocks. Such streams are termed insequent streams. The term really covers two possibilities: either the streams owe their courses completely to chance, or they are guided by lithological differences too small to be detected by man. It will be seen in a later chapter how very minute lithological differences can cause great differences in relief. Insequent streams are characteristic of areas of homogeneous rocks, especially areas where thick clays predominate.

Finally, the last class of streams is termed obsequent: these were defined by Davis as streams which flow in a direction opposite to that of the consequent streams, but they are usually defined today as streams which flow in a direction opposite to that of the geological dip of the beds.

By the slow extension of the drainage network the land is assured of an efficient drainage pattern, by means of which the water and debris supplied to the rivers are transported to the sea. When the rivers have such gradients that everywhere along their courses the slope is sufficient to discharge the water and the debris, and yet allow little energy for further erosion, the rivers are said to be at grade. The concept of grade is a difficult and complicated one and will be discussed more fully in Chapter 5. Davis himself defined grade as that state when there was a balance between erosion and deposition, but he did not maintain, as Gilbert had done at an earlier date, that further erosion was inhibited when a river reached a state of grade.

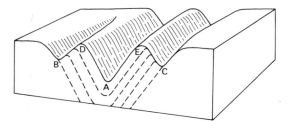

Fig. 2.2 Stream abstraction. Deepening of valley A, as shown by pecked lines could lead to the breakdown of divide D and capture of valley B, and later the breakdown of divide E and capture of valley C

But he did stress that, once grade is reached, erosion by the stream is greatly reduced in intensity.

Besides adjusting themselves to carry away the water and waste supplied by the valley side slopes, various other modifications occur within the drainage pattern. In the initial stages a process of stream abstraction may take place, by means of which one of a closely spaced set of consequent streams actually breaches its valley wall to capture an adjacent consequent stream (Fig. 2.2).

Far more important is the process of stream capture, which is especially favoured when the rivers are flowing at an angle to each other. The stream which has the more readily eroded course will capture the stream with the more resistant course (Fig. 2.3). The way in which the succession of rocks

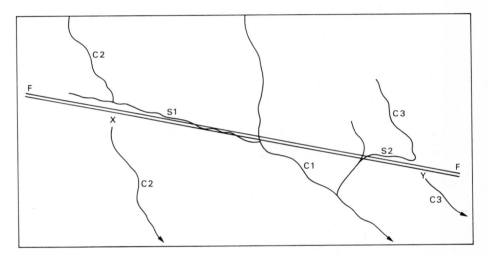

FIG. 2.3. Capture by subsequent streams. Consequent stream C1 has developed subsequent tributaries S1 and S2 along the fault FF to capture the heads of consequents C2 and C3, leaving the lower parts of those streams as misfits and wind gaps at X and Y

is arranged may lead to one stream being favoured at an early stage in the erosion of the area, while at a later stage and at a lower level the disposition of the strata is so changed that a different stream is favoured. The arrangement of the rocks may thus lead to a complicated series of captures taking place in the course of a cycle of erosion. Yet the captures generally lead slowly to the domination by the subsequent streams of all the other streams. There are simple reasons for this state of affairs: the subsequents are those streams following lines of geological weakness and, therefore, the lines of least resistance. Streams in such courses would be expected to be able to cut down more readily than the other streams and hence to capture them. Therefore, towards the end of the cycle of erosion the subsequent drainage predominates. The only advantage of the consequent streams lies in the initial depressions which they follow, and this advantage is soon lost once subsequent streams begin to develop. Insequent and obsequent drainage may survive as small streams, but rarely as the major rivers of an area.

River capture may also be caused by the relative length of the streams: one which has a quick and direct course to the sea possesses an advantage over one following a long and circuitous course to the sea. By this means one consequent stream may capture another (Fig. 2.4).

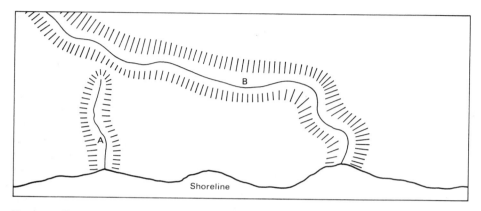

FIG. 2.4. Capture caused by differences in stream gradients. Streams A and B both have excavated valleys along consequent depressions, but A is likely to capture B as it can cut its valley down faster. If the rocks are permeable, the capture may be facilitated by underground seepage from B into the head of A

Finally, the courses of the streams in plan will not remain straight. Any chance irregularities, such as slight bends in the initial consequent depressions or hard projections of rock, may serve to divert the river first one way and then another. The stream will tend to erode on the outside of the bends, or meanders, so formed and to deposit on the inside of the bends. Thus it will slowly cut a level surface or floodplain, which will first appear in small patches on the inside of the meanders and slowly extend as the meanders move downstream until a continuous floodplain results (Fig. 2.5). This phase of valley widening as distinct from valley deepening becomes important, according to Davis, as soon as the stream reaches the state of grade, as it then ceases to be an effective agent of downcutting and spends its energy in lateral cutting. There is a limit to the width of the meanders, depending on the size of the stream, a larger stream producing larger meanders. The limit to the width of the meander belt is probably caused by two main factors. As the meanders become larger, there is always the tendency for one meander to break through into the next with the formation of a cutoff (Fig. 2.6). Secondly, as meandering increases the length of a stream, it reduces its gradient and decreases its energy. The stream, therefore, ceases to be so effective an eroding agent and both downward and

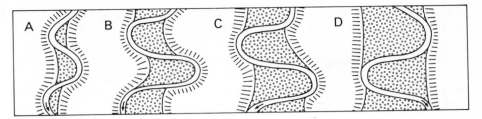

FIG. 2.5. Successive stages in development of a floodplain
A. Floodplain fragments form on inner sides of bends
B. Fragments enlarge and meanders move downstream
C. Movement of meanders downstream leads to continuous narrow floodplain
D. Floodplain as wide as the meander belt develops

lateral erosion diminish. In fact, meandering, by decreasing the gradient of a stream, tends to limit the amount of its erosion.

To sum up the changes which occur in the drainage during the course of a cycle, it may be said that, with the progress of the cycle from youth to old age, there is a gradual development of the drainage pattern to form an effective system for the whole of the area, a progressive increase of the graded condition starting at the mouths of the streams and proceeding headwards, a gradual adaptation of the drainage to the structure of the region until the subsequent streams come to have a dominance over all the others, usually by the process of river capture, and finally the development of meanders and floodplains until the streams occupy the wide flatfloored valleys of old age instead of the narrow steepsided valleys of youth.

So far the processes which have been acting on the sides of the valleys have been ignored. By the action of weathering, a mantle of rock waste is

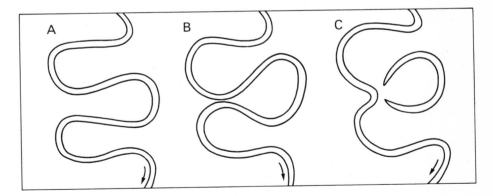

FIG. 2.6. Formation of a cutoff

formed and this, under the action of gravity, slowly migrates down the slopes into the streams, by which it is removed as part of their load. Just as

13

stream courses, in the youthful stage, may have irregularities due to the presence of harder and softer beds of rock, so may the valley side slopes be similarly irregular. In the early stages wholesale slipping may occur when the gradients of the slopes are steep, but in time a moving waste mantle will cover all irregularities and in it the products of subaerial weathering are transported to the rivers for removal. Davis visualised a complete series, ranging from a river, in which water greatly exceeds waste, to a waste mantle, in which the reverse is true. As he considered them to be the end members of a continuous series, he spoke of them in the same terms. Thus, a waste mantle is graded when the slope is so arranged that the products of weathering are transported systematically to the rivers. After that stage a progressive comminution of the debris leads to a progressive reduction in the angle of slope of the valley sides and to the development of a thick mantle of waste, which, in the last stages of the cycle, covers the whole of the land-scape and effectively prevents the weathering of the solid rock below. It can be seen then that, after a certain stage, not only the rivers, but also the slopes are graded for the efficient evacuation of debris over the whole land-scape. As regards the actual processes which act upon slopes, Davis was not explicit, and the subject must be left for discussion in Chapter 4.

The combination of graded water streams and graded waste streams re-moves the products of weathering and allows the whole of the land surface to be eroded slowly down, faster in the early stages and progressively more slowly later, until the area is reduced to a lowland of faint relief or peneplain. In this process the slow flattening of slopes under the agents of weathering is more important than the action of rivers, which, although active agents of downcutting in the early stages, later become mere transporting agents for the waste delivered to them down the slopes. The development of the landscape under these processes is divided into the stages of youth, maturity and old age, and the visible signs of each stage may now be defined.

YOUTH

The stage of youth begins with the very first action of erosion on the land-scape: its end is defined differently according to whether the landscape or the rivers are the main objects of consideration. The land surface itself passes from youth to maturity when all traces of the initial surface dis-appear, a criterion very difficult to apply in practice in most areas. The drainage passes from youth to maturity when the main streams and their main tributaries are graded. The attainment of grade means that the streams commence to form floodplains, so that the formation of these features is a good sign that youth has passed and maturity commenced.

The destruction of the initial surface of the land may not, as clearly stated by Davis, occur simultaneously with the development of grade in the main rivers. Thus, an early mature landscape may have a river pattern which has only reached the state of late youth. The definition of the stage is also further complicated by the fact that Davis mentioned in his German essays that maturity starts when the slopes are mantled by graded waste sheets. According to Davis, areas of very high relief will show maturity of the surface long before the slopes and streams have attained the same state (Fig. 2.7A). In areas of low relief, on the other hand, the streams and slopes will reach maturity before the initial surface has finally disappeared (Fig. 2.7B). In areas of moderate relief there is the greatest likelihood of the maturity of the surface, the slopes and the streams occurring at the same time (Fig. 2.7C). Furthermore, in any of these types of relief the texture of the drainage will play an important part, as a system of closely spaced streams will obviously tend to destroy the initial surface before a system of widely spaced streams (Fig. 2.8). On theoretical grounds, therefore, it may well be impossible to define the stage in the cycle of erosion reached in any area without separate consideration of surface, streams and slopes.

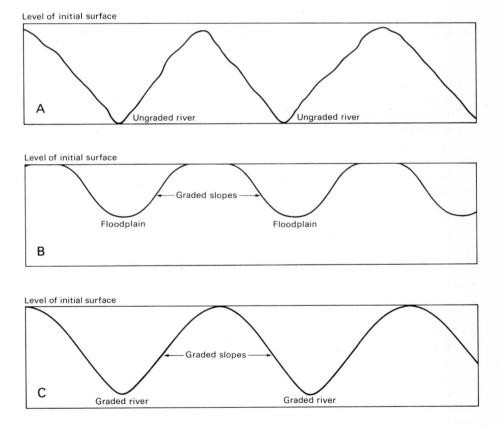

FIG. 2.7. Effect of depth of relief on attainment of maturity (for explanation see text)

15

Level of initial surface

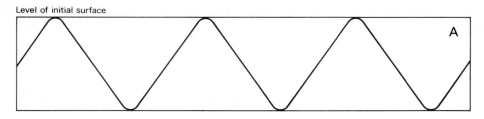

Level of initial surface

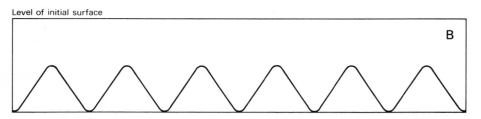

FIG. 2.8. Effect of spacing of streams on attainment of maturity. The streams are presumed to have attained grade in each example

MATURITY

Youth is a period of increasing relief as the streams cut down until they reach grade, while the initial surface of the land in the ideal case is not destroyed. Maturity, at least in its later parts, is a phase of decreasing relief, as, although the streams no longer cut down rapidly because they have reached grade, the initial surface of the land has been destroyed and the interfluves become progressively reduced in elevation.

Maturity is the period when the drainage is most closely adapted to structure, as the development of subsequent streams proceeds apace. Further, the rivers develop floodplains and adopt typically meandering courses. The grading of the streams is extended and by the end of maturity not only the main streams and tributaries but even the wet weather rills are graded. The whole land surface is covered by a graded waste mantle and slopes become progressively less steep due to the comminution of the debris forming the waste mantle (Fig. 2.9). At this stage the landscape passes into the third and last phase of the cycle.

OLD AGE

During the stage of old age the landforms become progressively more subdued, the slopes are covered with thick waste mantles which effectively prevent any further weathering of the rock below, the debris in the waste sheets becomes finer and finer and consequently the angle of slope of the waste becomes less and less. The lithological differences which were important in the formation of relief in youth and maturity are now concealed beneath the waste mantle. The streams begin to wander over their flood-

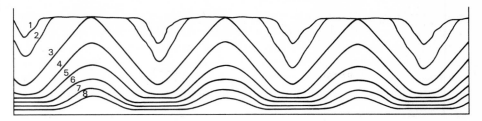

FIG. 2.9. Sections to illustrate the cycle of erosion
1 & 2 Stage of youth
    3 Early maturity
4 & 5 Middle maturity
    6 Late maturity
7 & 8 Old age leading to peneplain

plains on very gentle gradients and to lose their adaptation to structure, as they are insulated from the effects of that structure by the thickness of alluvium flooring their valleys. The length of some of the streams is reduced by the loss of headwaters, as the lowering of the surface means a reduction in rainfall and the thickening waste mantle means an increase in permeability and, therefore, in the proportion of the rain which percolates into the soil as compared with the proportion that runs off on the surface. On the major divides somewhat higher residual hills may be left: these were termed monadnocks by Davis, a name which is not altogether a good one as the type example, Mt Monadnock in New Hampshire, is by no means the gentle eminence that should be formed at this stage. The lowland of faint relief, characteristic of an advanced stage of old age, was termed a peneplain by Davis. The final stage would be a perfect plain, but this, of course, cannot be realised by the processes of subaerial erosion, as slopes are always necessary to evacuate the rain falling on the land and the debris being worn from it. In later literature the peneplain is sometimes spelled peneplane, but it seems better to retain the word as used by its author. Similarly, the whole cycle has been called the geomorphic cycle, but the phrase was not used by Davis to describe his own concept.

Interruptions of the cycle

Two main types of event can upset the orderly progress of the cycle of erosion: substantial alterations in the level of base level and climatic changes.

Small movements of base level may have little effect on the general relief of a great land area developing under the action of a long and uninterrupted cycle of erosion, but large alterations may have pronounced effects on the relief and lead to the development of what may be termed a polycyclic landscape. These movements of base level may best be called positive and negative, depending on whether the land sinks or rises relative to the sea.

Thus, the awkward question whether such movements are caused by uplift or subsidence of the land or by changes in the level of the sea may be avoided. Any large movement of such type initiates a new cycle, the relief forms of which are superimposed on those of the earlier cycle. It is not necessary for the number of cycles to be limited to two, and the relief of any area may have been formed under the influence of several cycles or partial cycles of erosion.

Climatic change leads to an interruption in the cycle because it causes a marked change in the processes of erosion operating on the landscape. Consequently, the relief forms developed under the new climate may be markedly different from those formed under the old climate without there being any change in base level. The most common climatic changes are caused by the onset of glacial or arid conditions. Because geomorphology developed mainly in the humid temperate areas of Europe and North America, glacial and arid conditions are often called climatic accidents. From our point of view they certainly are, as glaciation intervened for a part of the million years assigned to the Pleistocene period and arid conditions were characteristic of certain geological periods in the past. Yet, in regions which are deserts today the climatic accident was probably a brief interlude of humid conditions during the Pleistocene when the climatic belts seem to have shifted towards the equator for a time. Nevertheless, climatic changes have caused important modifications of many landscapes, short periods of different climate having had, in some areas, more effect than long periods of 'normal' climate.

Finally, another interruption may be caused by the outbreak of vulcanicity on a large scale, but this constitutes a much less profound interruption of the cycle of erosion. It results in the addition to the landscape of a series of constructional forms, which, because of their later origin, will not necessarily be in the same stage of the cycle of erosion as the landscape as a whole.

The use of the cycle

It must be admitted that the cycle of erosion has not, as a concept, found the wide acceptance among geomorphologists that Davis envisaged. In very few later geomorphological studies does one find the use of the terms youthful, mature and old to describe the landscape. There are several reasons for this. Many of the later studies have been mainly concerned with the elaboration of the mechanism of the processes of erosion, and, in such studies, there is little place for the Davisian terminology of the cycle of erosion. The regional studies that have accumulated since the concept was introduced have emphasised the complexity of the history of the evolution of relief, and have shown that relief usually has been formed under the influence of a series of partial cycles, often with the intervention of changes of climate. For these reasons the Davisian terms have not and are never likely to receive widespread currency.

It is also doubtful whether the term cycle is used properly in the context. Davis himself considered this aspect and defined cycle in the sense of a period of time in which a succession of events is completed, and then returns again and again. The succession of events, however, cannot return without the intervention of uplift and, even with uplift, it is doubtful whether they would be exactly repeated. Further, it is somewhat doubtful whether climate is sufficiently stable over long periods to ensure that a similar cycle of erosion can ever succeed the first. But, however unsatisfactory the word cycle, it is so established as the term for Davis's concept that any attempt to change it would be both pointless and unsuccessful.

It might well be asked why one should bother with the geographical cycle if so much of it consists of general inference and if its language has never been taken up seriously by later geomorphologists. The answer is that before one has doubt one must have dogma. The Davisian cycle provides the dogma, the comprehensive theoretical arrangement of all the aspects of denudation. There are plenty of doubts about the exact modes of development of many of the features contained therein, but imagine trying to learn, for example, the inconclusive, mutually contradictory, often differently orientated mass of research on slopes if one had no idea how it might fit into a broader pattern.

Perhaps we could make do with an even vaguer generalisation about landscape development; for that old wartime truism, 'what goes up must come down', we might substitute 'what endogenetic (tectonic) forces lift up, exogenetic (denudational) forces must wear down'. But this, although true, is altogether too slight. Again, it has been said that we ought to take the pediplain cycle as the norm and that our thinking is conditioned to regard humid temperate climatic conditions as normal. This may be true. But the application of the pediplain cycle can be assessed when, in the light of field evidence, the probability of such features as parallel slope retreat or the formation of pediplains rather than peneplains and so on can be discussed. It does not matter whether we start from the Davisian cycle dogma or the pediplain cycle dogma, so long as we realise the place in the general scheme of denudation of the subjects under discussion.

At the outset certain characteristics of the Davisian view must be stressed. Essentially it is a long-term view of landscape, a view geological in its scale: so too are other cycle concepts such as the pediplain. Objections have been lodged that, in the cycle, uplift is quickly completed and that denudation then runs its course until the potential energy provided by the uplift is used up, rather like charging the batteries of an electric car, then watching it run until it slowly comes to a halt. Davis, of course, made the perhaps rash assumption that users of his scheme would have enough intelligence to deduce the effects of complicated uplift patterns. Yet, whether the uplift is intermittent, smoothly accelerated and then retarded, whether it is mainly early or mainly later in the cycle, however much the restoration of isostatic equilibrium offsets the results of denudation, in the end denudation wins and the land is worn down to low relief, whether or not in the form of a Davisian peneplain. This is incontrovertible: it is a lesson taught by the study of the geological science of stratigraphy.

Many modern geomorphologists, mainly interested in the physics of process, see, for example, slopes and river profiles as equilibrium states between a number of interacting factors. A change in one tends to cause a series of linked changes in the others. Time is not seen as a major factor in landform development, as it may well not be in the study of equilibrium states where the forces are powerful and the rocks weak. Usually the interest of such geomorphologists is concentrated on one facet of the landscape and how it may be formed, and not on the way in which all the facets combine in the general landscape.

An analogy with human geography may be made. Cambridge may be regarded as a town that has slowly evolved and now includes legacies of varying importance from a number of past periods going back at least to Roman times. These are all reflected in various features of the town's physiognomy. Alternatively one might study the effects of the journey to work and property values in the town and its surroundings. These two studies are both aspects of the geography of towns, just as the two considered above are different, not superior and inferior, aspects of geomorphology.

But there is much more in the Davisian cycle than the presentation of a long-term view of general landscape development. It stresses the following points:

(*a*) The development of slope forms is fundamental in geomorphology. It is true that little attention was paid to processes, and that the model of slopes flattening with age may be incorrect or only correct in certain conditions, but changes in slope forms are prominent in the cycle.

(*b*) Rivers and their profiles and plans: the grade concept may not be of much use and may even be an impediment to progress, while the treatment of meanders is very descriptive, but attention is again focused on a vital point of the landscape.

(*c*) The development of drainage systems is treated as one of the main themes. It is true that the system is largely discussed in relation to geological structures, whereas later work has been largely in terms of patterns very often in areas of homogeneous rocks.

(*d*) Rock type exerts effects of different degree at different stages in the cycle. In the very beginning there may be little effect, but lithologically induced irregularities appear in both slopes and streams and are later ironed out. A similar point of view was adopted by Johnson in his suggested cycles of marine erosion. On the other hand, in the development of drainage systems structural control increases as the cycle progresses except possibly at the very end. These ideas may not be fully correct: they were an early assessment of the role of rocks in the development of landscape.

(*e*) Climatic changes, or accidents as Davis termed them, are very important as they control the whole complex of denudational forces acting on the
landscape. The school of climatic geomorphology later emphasised this,
and, although some authorities try to see unity of process in all climatic
regions, the climatic approach is still profoundly ingrained. One can see it in
the writing of books dealing with glacial, periglacial, arid and equatorial
landforms. The emphasis on climate also drew attention to the possibility
of finding relics from past processes among presentday landforms.

Thus the cycle is not negligible as a dogma to be dissected and discussed
in its various parts in later chapters. Some of its aspects will be more relevant
than others to particular geomorphologists. For example, if one studies
stream and slope development on waste heaps, or on their nearest natural
equivalents, soft rocks in semi-arid environments, landform change is seen
to accompany process change very closely. Thus one would naturally
minimise the importance of legacies from past phases of erosion. On the
other hand a worker in an area of glaciated Pre-Cambrian crystalline
rocks may be excused for minimising the effects of present processes and
stressing the legacy of past glacial erosion.

The importance of Davis is not whether he was right or wrong. It would
be a sad reflection on our subject if, after three-quarters of a century, we
found nothing to criticise or disbelieve. But Davis drew attention to the
main features of landform development and suggested an overall scheme
to embrace them, so providing the dogma to accommodate the expression
of our doubts.

3
Weathering

Weathering may be defined as the mechanical fracturing or chemical decomposition of rocks *in situ* by natural agents at the surface of the earth. It is the initial phase in the denudation of any landscape, as the rocks must be weathered before the debris can be transported and erosion is very limited unless the agents of transport are carrying a load of debris.

Although it is convenient to differentiate between physical, or mechanical, weathering and chemical weathering, it is improbable that either of the processes ever operates completely alone. The presence of a certain amount of water vapour in the whole of the earth's atmosphere ensures that chemical weathering is likely to occur everywhere. It is true that in different climates the relative importance of the two may be somewhat different and there is a general tendency to characterise the weathering by the name of the dominant process. Thus, although it is convenient in the ensuing discussion to consider the various aspects of physical and chemical weathering in isolation, it must be constantly remembered that they act together.

The agents of weathering

Physical weathering

Physical weathering is brought about by two main processes, temperature change and crystallisation. To these might be added, as a lesser effect, the action of plant roots in the purely mechanical sense. Doubt has been cast on the efficacy of temperature change, which will be discussed first.

As rock is a poor conductor of heat the effects of diurnal heating are confined to the surface layers and not transmitted freely through the rock. With the concentration of heat in the upper layers, the surface of the rock tends to expand more than the rock at depth and this should lead to the formation of stresses within the rock. If such stresses exceed a certain critical amount, there should be fracturing parallel to the surface. It has been calculated, for instance, that a rise of temperature of 83 degrees C (150 degrees F) in a sheet of granite 30 m (100 ft) in diameter would lead to an expansion of 25 mm (1 in) in the granite: and a fall in temperature of the same amount would lead to a corresponding contraction. The assumption that a process such as this operates has been held to explain the flaking off of sheets parallel to the surface of the rock, a process which is termed exfoliation. However, such sheeting is more likely to be due to the operation of the process of pressure release (see below).

Most igneous and metamorphic rocks are composed of different minerals, which, because of different specific heats, expand at different rates. Such expansions should lead to minute internal fracturing within the crystals composing the rock and at their boundaries. The fact that minerals often have different colours also leads to differential absorption of heat, to differential expansions and, hence, to minute internal fracturing.

Processes such as these have been held to operate mainly in desert regions, where the diurnal range of temperature is very great and where the humidity of the atmosphere is low. Travellers' reports mention sounds like pistol shots, which are thought to be the sounds of rock fracturing under the effect of temperature changes. The process has been heard rather than seen.

Considerable doubt has been thrown on the efficacy of these processes by a number of American investigators, who cited both field and laboratory evidence against the possibility of the processes occurring in nature. Blackwelder in 1925 described an experiment performed with a block of rock, which he subjected to a temperature range of from 15 to 210 degrees C by immersion in boiling oil. In spite of this somewhat drastic treatment, the rock showed no signs of spalling, cracking or visible weakening, even though it was subject to temperature changes of 195 degrees C, i.e. approximately three times the maximum known diurnal temperature range of deserts. He later (1933) described basalts and granites as resisting sudden heating to at least 200 degrees C, and usually to 300 degrees C, before fracture took place. Even obsidian, an acid volcanic glass, withstood sudden chilling to the extent of 250 degrees C without fracturing. Finally, a more comprehensive experiment was made in 1936 by Griggs, who devised an apparatus capable of subjecting a mass of rock to very many rapid temperature changes. He placed a block of coarse-grained granite, with a polished surface, in front of a red-hot electric heater and arranged for it to be alternately heated and then cooled by a stream of cool dry air. The temperature change involved was 110 degrees C and the rate of heating and cooling very sudden. Sufficient alternations of heating and cooling were performed to represent the diurnal temperature changes of 244 years. At the end of the experiment there had been no exfoliation, and even under a moderate power microscope examination, there was no detectable change within the rock: minute cracks within individual crystals appeared to be no more pronounced than at the beginning of the experiment.

However, after the alternate dry heating and cooling for '244 years', Griggs then subjected the rock to a mere '2½ years' weathering, the heating being the same but the cooling performed by a spray of tap water, which naturally contained oxygen and carbon dioxide in solution. At the end of the second experiment the block showed surface cracking of exfoliation type, the cracks within the crystals had widened and the surface of the rock had lost its polish. His conclusion was, naturally, that chemical weathering played a more important part than the presumed processes of mechanical weathering.

It is always difficult to be sure that laboratory experiments of this type are true reproductions of natural conditions, but Griggs went into that question to some extent. The significant factor in exfoliation should be the temperature gradient in the surface layers of the rock, and it was easily shown that the temperature gradient in the rock in the experiment was considerably greater than that which can occur in rocks in nature. There are, however, as pointed out by Ollier (1969), two aspects in which the laboratory experiments do not repeat natural conditions. The first lies in the fact that the surface of the laboratory slab was free to expand as it was not confined in any way. This may not be true in nature: in large masses of unjointed rock small sections of the surface are not only confined but may be considered to be subject to a compressive effect from the expansion of adjacent sections. Hence, there may be a much greater tendency for the lifting of surface layers. The same should apply less to a rock with a fairly closely spaced and open joint system. It might make an interesting field study to see whether there was any detectable difference in superficial sheeting between closely jointed and widely jointed rocks of the same type. The second difference is that fatigue is difficult to simulate in a laboratory. In metals it is a feature only too familiar to aircraft operators and owners of vintage motor cars. It results from a vast number of repetitions of stress, not of themselves large enough to cause quick failure of the material. It may well affect rocks under natural conditions.

The suggestion, derived from the laboratory experiments, that chemical weathering may be of importance even in deserts would help to explain some of the natural phenomena observed. Studies of granite monuments in Egypt (Barton, 1916) have shown that certain situations appeared to favour weathering. The granite statues near Cairo, where there is some slight rainfall, are more weathered than the statues in other, drier parts of Egypt. Furthermore, the bases of some of the statues appeared to be more affected by weathering than the upper parts, presumably because they are more often wetted where they are in contact with the soil. As Blackwelder pointed out, in humid climates rounded boulders are known to form beneath ground level, where they are more subject to chemical attack and to a considerable extent protected from the effects of insolation. In addition, exfoliation landforms are typical of many areas of the humid tropics, such as eastern Brazil and eastern China. Blackwelder concluded that chemical decay and hydration (see below) were necessary factors for weathering even in arid regions.

Another possible explanation of the process of exfoliation must be considered. Sheet structures have been observed down to a depth of 75 m (250 ft), which is far deeper than diurnal ranges of temperature penetrate (Blackwelder, 1925). In addition, the layers are sometimes arranged parallel to the present surface of the rock. From this it may be concluded that exfoliation has something to do with the formation of the present surface. It may be that Blackwelder's hypothesis of chemical decay and hydration of minerals is correct, but there is another possible explanation.

Many rocks, especially the coarse plutonic igneous rocks, such as granite, and many of the metamorphic rocks are crystallised under temperature and pressure conditions very different from those found at the surface of the earth. Consequently the minerals may not be stable at surface temperatures and pressures. As the rocks are slowly exposed by the forces of erosion, stresses are set up within the rock and the result is an exfoliated sheet structure parallel to the surface relief. Such a process seems more plausible in the case of sheet structure extending to a depth of 75 m (250 ft) than does the chemical hypothesis. But care must be taken not to apply this pressure release hypothesis too widely. Undoubtedly, jointing due to this mechanism may be expected in all deep-seated rocks from which hundreds, if not thousands, of metres of rock have been removed by erosion, and a tendency for the same thing to happen may be present in some sedimentary rocks which have been fairly deeply buried, but, if joints parallel to the surface are found in lavas and comparatively young sediments, which are known not to have been deeply buried, they cannot be attributed to pressure release.

It is also possible that the release of strain energy may assist in the granular disintegration of rocks (Durrance, 1970). When polycrystalline rocks are buried deformation may occur within the grains, at the grain interfaces, and, in the case of sedimentary rocks, in the cement between grains. When the load is taken off such rocks the first release of energy may result in the formation of joints as suggested in the paragraph above. But with the formation of joints all the strain energy within the joint-bounded blocks may not be released. However, when weathering starts and the rock is weakened, the strain energy may then be released and materially assist the process of granular disintegration. Durrance makes the interesting suggestion that tors may be developed in parts of granite where the initial strain energy was low and hence jointing is coarse and granular breakdown relatively slight, while the zones between are areas of high strain energy and hence massive breakdown at both joint and granular levels.

It is probably best to conclude that chemical weathering and pressure release are allied with temperature changes to produce the typical disintegration structures that have been observed in hard rocks.

Ollier (1969) has mentioned other features which could well be due to physical weathering. He stresses the reports of shattered, chemically inert rocks, such as quartzites and cherts, in desert regions. He also mentions a way in which boulders may be split when forming a part of a conglomeratic material (Fig. 3.1): a boulder half embedded in its conglomerate is free to expand in its upper half much more than in its lower because of the constraint of the matrix there. Not only is it free to expand, but it will also receive more heating in its exposed position. This should lead to a crack developing between portions A and B of the boulder parallel to the general surface: such cracks have been observed in the field.

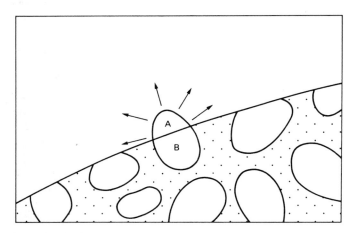

FIG. 3.1. Differential expansion of a boulder in conglomerate (*after Ollier*)

Again, it seems that even the most unlikely rocks suffer volume changes when saturated with moisture. Percentage changes in linear dimensions of the order of 0·015 in basalt and 0·005 in flint consequent on shrinking with reduced moisture content have been reported. On the whole the changes are of the same order of magnitude as those which result from thermal expansion.

Wetting and drying weathering is another feature mentioned by Ollier. Even consolidated rocks, such as Silurian shales, if alternately soaked for a day and then dried for a day, eventually disintegrated. This action is familiar to Pleistocene research workers, for one of the best ways of disintegrating silts so that small fossils may be washed out of them is to leave them to dry out completely and then soak them in water. The suggested mechanism is a complex one involving an ordered arrangement of water molecules, which has certainly not yet been proved though there is suggestive evidence (Ollier, 1969).

Finally a feature which is almost intermediate between weathering and erosion is colloid plucking. It has been shown that a thin film of drying gelatine can pluck glass flakes from a tumbler. Ollier mentions that films of drying clay have been seen to lift grains out of an arkose, a felspathic sandstone, but suggests that the process is of minor importance.

The other main aspect of physical weathering is caused by crystallisation. In cold regions the most effective action is caused by the crystallisation of water into ice and the expansion accompanying this. Although water reaches its greatest density at 4°C (39°F) and then expands with further cooling, this expansion is insignificant compared with the approximately 10 per cent increase in volume when it is changed into ice.

For the operation of freeze-thaw there have to be joints and/or small-scale, intergranular fractures present in the rocks. Fortunately almost all rocks, whether hard or soft, are fractured: without fractures freeze-thaw could not operate as it cannot produce fractures but only widen them. The other condition required is frequent change from above-freezing to below-freezing temperatures. This is most likely to occur in cold climates in the transitional seasons of spring and autumn, while in more temperate climates winter is the most likely season. Even when air temperatures are continually below freezing, surface rock temperatures, especially where the exposure is towards the sun, may be high by day and very low by night because of the low heat conductivity of most rocks, so that freeze-thaw may continue to operate.

A number of conditions have to be fulfilled for freeze-thaw to operate to full effect (Grawe, 1936). It must take place ideally in enclosed cavities. Obviously in completely enclosed spaces there is no means by which water can enter, but certain rocks, such as cellular or carious limestones and dolomites or vesicular lavas, must provide very suitable conditions for freeze-thaw rupturing. The fuller the cavity the less easy it is for the water to migrate when freezing starts. It is true that in open cavities freezing will take place on the surface at first, so that the lower part of the water will already be contained when it freezes in turn, but the force so produced is most likely to be used mainly in displacing the frozen surface layers. Natural water contains dissolved gases and these reduce the efficacy of freeze-thaw, because in such a two component system much of the force exerted by the crystallisation to ice can be absorbed by the gas phase. In hard rocks the blocks are probably moved only minute fractions of a millimetre at a time, but ultimately the structures must be weakened. The abundance of screes in glaciated mountains such as the Alps, even in their southern sections, leaves little doubt about the general effectiveness of the process, while in Britain screes in varying degrees of stabilisation can be found in the highland and upland districts, such as the Lake District, the Highlands of Scotland and the Pennines.

The action on soft rocks is rather different. The rocks most affected are porous rocks, i.e. those with considerable pore or interstitial space in which water can be retained. Good examples are Pleistocene sands and silts, chalk, which besides being very permeable is also very porous except in a few areas where it is cemented to an unusual degree, and clay. If these rocks are soaked in water and then frozen the development of ice in them causes a large degree of disintegration. Faces of gravel and sand pits are

27

strongly affected: it is not unusual to find a talus over 0·5 m (2 ft) high at the foot of a 6 m (20 ft) face after a few days hard frost and a thaw. In chalk pits thin surface layers are sometimes prised off by the development of ice crystals at right angles to the face. Blocks of chalk left on the floor of the pits rapidly disintegrate into flaky rubble after a winter or two, especially when the chalk involved comes from the more marly horizons of the Lower Chalk. The shattering effects of frost on clay is a well-known phenomenon, which is used in good farming to allow the development of a good tilth on clay soils. Freeze-thaw also affects the bricks or stones used in buildings below damp courses, especially when they are porous, and causes them to flake: hence the tendency to use the hard, relatively non-porous engineering brick below the damp course of many buildings.

Laboratory experiments on the effects of freeze-thaw on rocks such as chalk have been made (Williams, 1969, summarised in Sparks, 1971), but there are formidable difficulties in getting the refrigerating apparatus to work with the absolute reliability necessary for the simulation of large numbers of freeze-thaw cycles. In addition, there are difficulties in the selection of samples, for example in the elimination of those affected in some degree by weathering, a very difficult feature to detect, and those which have potential planes of weakness, such as concealed sections of fossils included within them. Inside the refrigerator care has to be taken to ensure that the rocks remain continually and equally saturated and that they are frozen throughout. Finally there are problems of recording and measuring the degree of shatter. It must be done quantitatively because usually the experimenter will be trying to decide whether there are significant differences between materials which are broadly similar. The ideal way would be to measure the total surface area of all the fragments, but this is virtually impossible and one has to make do with a second best, such as determining the percentages of fragments in different size grades by sieving. Because of sample to sample variation and difficulties in assessing results, one would ideally like to perform innumerable experiments and test the statistical significance of the results, but there are rarely enough equipment and man-hours available. Even then, the final assessment of the meaning of this in terms of natural weathering is far from easy.

Similar to freeze-thaw is the weathering induced by the crystallisation of salts in rocks. There are a number of salts which might be involved, especially sodium chloride (common salt), calcium sulphate (gypsum), magnesium sulphate and sodium carbonate. If any of these salts are dissolved in the moisture penetrating rocks, subsequent drying and crystallisation can cause a disruptive mechanical effect. Examples of this sort have been noted in buildings where the sulphates may have been introduced with plaster or formed from the interactions of sulphate ions from the atmosphere with the carbonates present in the rocks. For example, the Houses of Parliament are built of Magnesian Limestone, which contains both calcium and magnesium carbonates. The latter reacts with the sulphurous and sulphuric acids of the industrial atmosphere to produce magnesium sulphate—hence the delightful letter quoted by Ollier which describes the building as being reduced slowly to some sort of a heap of Epsom salts. The magnesium sulphate is very soluble and normally washed away, but it has been known to crystallise in dry weather with disintegrating effects on the stonework. Birot (1960) quotes laboratory experiments in which slabs of granite and sandstone have been destroyed in four months by coating them with salt and moistening them each day.

Such action is possible in nature in different environments: in deserts where sodium chloride and sodium carbonate may be blown about in the general desert dust, settle in crevices and there be moistened by dew or occasional light showers and so crystallise out; in marine environments where dried sea spray may behave in a similar way; in porous rocks in industrial areas where soluble sulphates may be formed in the same way as that described for the Houses of Parliament above. As with building stones granular disintegration seems to be the usual result.

Perhaps the greatest contribution of salt crystallisation is in cavernous weathering. On a small scale a honeycombed rock surface is produced, the cavities in which are known to Continental geomorphologists as alvéoles, which may range up to 0·5 m (nearly 2 ft) but are usually 20–100 mm (approx 4 in) in diameter (Cailleux, 1962). Larger cavities are known as taffonis, although there is no clear distinction between them and alvéoles. Taffonis are of the order of 1 m (3 ft) in diameter but may be double this in extreme examples. Both taffonis and alvéoles are common in igneous rocks. The former were first described from Corsica where they are very well developed, but they are also found in many other parts of the world including Antarctica, of which some splendid photographs are given by Cailleux (1962).

Many authorities would refer alvéoles and taffonis to the mechanical disintegrating effects of salt crystallisation working their way in from original slight hollows in the rock. Others, including Cailleux, hesitate between salt crystallisation and freeze-thaw, especially for the Antarctic forms. The two processes are, of course, mechanically similar. What does seem to be generally agreed is that one or the other of these two mechanisms must be responsible for cavernous weathering and that older explanations, such as wind abrasion or chemical weathering, are not applicable.

The whole literature on salt crystallisation effects has been reviewed by Evans (1970) and the possible operative mechanical effects discussed. The question is not quite as simple as freeze-thaw in which freezing involves the whole liquid and a simple volume change takes place. When a solution crystallises only the salt crystallises out and the liquid remains. Hence the effects cannot be attributed to an overall pressure caused by expansion as in freeze-thaw. There seems to be a number of possible mechanical effects:

(*a*) Crystallisation involves growth in certain directions. A considerable force is involved for crystallisation occurs under retaining pressures that exceed the tensile strength of some rocks: it has been observed to occur against a pressure of 47 bars whereas the tensile strengths of many rocks lie in the range 20–200 bars. Thus crystal growth tends to split rocks.

(*b*) Hydration, which is discussed below under chemical weathering, can affect crystals and has been known to occur against pressures as high as 63 bars. Thus moistening and drying of rocks containing salt crystals may exert considerable disruptive force. This action probably takes place with some carbonates and sulphates but not with common salt.

(*c*) Salts have also usually higher coefficients of thermal expansion than rocks so that simple heating of rocks containing salt crystals may tend to produce a mechanical breakdown. But, as differential expansion of minerals has been questioned as an adequate explanation of rock shattering, this hypothesis awaits laboratory testing.

Some interesting preliminary experiments on the effects of different salts have been made (Goudie and others 1970). The material used was Arden Sandstone of Keuper (Triassic) age. Considerable disintegration was effected by sodium and magnesium sulphate but practically none with calcium chloride, sodium carbonate and sodium chloride in 40 wetting and drying cycles. Different rocks were affected differently by the same salt. However it is too early yet to generalise safely and it must not be concluded that certain salts are necessarily ineffective agents of disintegration until many more experiments have been made under a variety of conditions.

Finally, among the agents of physical weathering may be mentioned the activities of plants and animals. Tree roots can penetrate deep into a well-jointed or fractured rock such as chalk, and animals may help in the excavation of partially weathered fragments of rock. Rabbits are the obvious example.

Chemical weathering

Whereas there are only two main processes of physical weathering, temperature change and crystallisation effects, chemical weathering involves a number of distinct processes.

The first of these is the process of hydration, whereby certain types of minerals take up water and expand, thus causing additional stresses within the rock. A common natural mineral which occurs in both hydrated and unhydrated states is calcium sulphate, which is known as anhydrite when unhydrated and as gypsum when hydrated. Gypsum, however, is not a common rock-forming mineral, although it occurs in beds at certain horizons, especially in the Trias in this country. But the decomposition products of certain rock-forming minerals are subject to hydration and the disintegration of the rock is thereby hastened. In itself hydration produces what is strictly a mechanical effect, but it so often combines with hydrolysis that the whole process illustrates well what was said at the beginning of the chapter about the difficulty of separating physical from chemical weathering.

Oxidation, the taking up of additional oxygen by any mineral, is also a common feature, especially of iron compounds in the zone above the water-table. The most common example is to be found in the sedimentary rocks, where many clays, as long as they are deprived of air by being saturated with water, are blue or grey due to the presence of ferrous compounds, which on exposure to the air are oxidised to red or brown ferric compounds. Gault Clay districts and boulder clay districts display this phenomenon very well: the soil and uppermost layers of the clay are usually brown, but the material below is blue-grey. Road cuttings and pit faces change colour often in the comparatively short space of a few months.

Hydrolysis, a process of chemical reactions with water involving the action of H and OH ions, is very important in the decomposition of the felspars, which are an important mineral group in the composition of igneous rocks. The reaction may be represented in words by saying that orthoclase (potash felspar) plus water leads to the formation of potassium hydroxide and an alumino-silicic acid. As carbon dioxide is almost invariably present in the atmosphere, this reacts to produce potassium carbonate and water from the potassium hydroxide. The unstable alumino-silicic acid breaks down with the formation of clay minerals and colloidal silica, the latter going off in solution. Although the process of hydrolysis occurs through the action of pure water it is accelerated by the presence of carbon dioxide. From the present point of view its chief interest lies in the fact that it is the process whereby felspars are altered to clay minerals. It must be emphasised that it is a weathering process only and that the thick deposits of kaolin found on such granites as that near St Austell in Cornwall are the results of metamorphism.

Simple solution is not a common process of weathering in nature, as very few rock-forming minerals are soluble in water, the few exceptions, rock salt for example, not being common constituents of rocks. Solution may, however, be of importance in the removal of the products of other types of chemical weathering.

Carbonation, a process often termed solution, is the reaction of carbonate ions with minerals. It is vital in the chemical decomposition of limestones and dolomites, which are very common rocks at the surface of the earth. The carbon dioxide in rainwater, and especially within the soil, acts as a weak acid which converts the calcium carbonate into calcium bicarbonate, a soluble product which is taken away in the groundwater. As it is unstable it is usually redeposited later as calc-tufa, a cellular deposit of calcium carbonate.

The acid action of dissolved carbon dioxide is probably supplemented by the action of other acids, especially the obscure group derived from the decomposition of vegetable matter and known as humus acids. Both living plants and the humus derived from their decomposition are capable of extracting metal ions from otherwise insoluble solids by the process known as chelation. Living plants use this process to draw nutrients from minerals.

31

Humus, acting in this way, can cause rapid and serious leaching, i.e. weathering. In a later chapter the question of its probable effect on chalk will be considered, but the action on other rocks may be briefly noted here. Humus acids are often present in greater strength than carbonic acid and, consequently, must have a great effect upon the course of weathering. In the granites of part of the Southern Piedmont region of the United States circular and elliptical pits, 3–12 m (10–40 ft) in diameter and 0·3–1 m (1–3 ft) deep, have been observed (Smith, 1941). They have been attributed to the action of organic acids, as abundant rotting vegetation and acid water have been found in them. The organic acids attack the felspar in the wetter parts of the year, the products of weathering being partly washed away in very wet seasons and partly deflated in dry seasons. What remains provides a basis for the growth of vegetation, the decomposition of which leads to the production of more acid with consequent later deepening of the pit.

Biological weathering

Biological weathering can really be divided into physical and chemical effects, but it is convenient to consider it together because at the present time it is recognised to be very important though its actions are not fully understood. We have already mentioned the simple physical side in talking of the action of plant roots and burrowing animals and the chemical side in considering chelation.

The main effect of both flora and fauna is to increase the carbon dioxide content of the soil through respiration to many times the atmospheric content, so that the major contribution to weathering potential comes from the biosphere and not from the atmosphere.

In addition various organisms may cause reactions with the minerals of rocks. Among these are chemotrophic bacteria which oxidise minerals such as sulphur and iron, and possibly algae, fungi and lichens, the last being symbiotic associations of the other two. Furthermore, bird droppings (guano) are capable of weathering limestones. These and other effects mentioned by Ollier may prove to be far more important in weathering than has been generally realised.

The processes of weathering described above under the headings of physical and chemical are the main factors responsible for the initial stages of the denudation of landscape. It must be emphasised that no rock is subject merely to one single process: the various chemical and physical processes act in cooperation, although their relative importance may vary from area to area.

Factors affecting weathering

The importance of the mineralogical composition of rocks, their textures and structures, as well as the effect of different climates, on the rate and type of weathering must now be considered. Only after all these considerations have been taken into account will the full complexity of the action of weathering be realised.

Rocks fall into three main classes: igneous, sedimentary and metamorphic. Within each group various classifications may be made, and those classifications will be adopted which are of the greatest use in understanding the weathering and erosion of rocks, although such classifications may not be the best from the point of view of rock genesis.

The minerals present in igneous rocks, which affect the course of weathering, fall into several main groups. Broadly they may be divided into light-coloured and dark-coloured minerals.

Among the light-coloured minerals the most important group is the felspars, which are alumino-silicates of potassium, sodium and calcium. The potash felspar is known as orthoclase, while the soda-lime felspars are known collectively as plagioclase felspar. The end members of the plagioclase series are albite, the soda felspar, and anorthite, the calcium felspar. These two end members are mixable in any proportions so that any plagioclase felspar consists of a certain ratio of albite to anorthite. The recognised combinations are albite, oligoclase, andesine, labradorite, bytownite and anorthite, the calcium content increasing through the series and the sodium content correspondingly decreasing. Minerals greatly resembling felspars but containing less silica are known as felspathoids. Finally, the most important light-coloured mineral is quartz, the crystalline form of silica.

The dark-coloured minerals are mainly silicates of iron, magnesium and calcium, and are often known as ferromagnesian minerals. The principal groups of minerals are amphiboles and pyroxenes, the most common amphibole being hornblende and the most common pyroxenes being augite and hypersthene. In general, amphiboles are characteristic of more acid rocks than are pyroxenes. In addition to these two groups, the ferromagnesian minerals include olivine, a very characteristic mineral of basic rocks. The last common minerals are the micas, one of them, biotite, being a silicate of iron and magnesium and belonging to the ferromagnesian group and the other, muscovite, being a silicate of potassium and really belonging to the light-coloured group.

If a very elementary rock classification on the basis of the acidity of the rock, by which is meant the silica content of the rock, is adopted, it can be seen that the mineral composition varies considerably from class to class and consequently the chemical weathering is different (Fig. 3.2). At the same time the mode of intrusion can be used as the other basis for classification and, thus, distinguish between major intrusions, minor intrusions and extrusions of magma. The mode of formation affects the rate of cooling of the magma and, hence, the size of the constituent crystals of the rocks, major intrusions being usually coarsely crystalline, minor intrusions usually more finely crystalline, and extrusions or lavas very finely crystalline or even glassy. Thus, the classification in Fig. 3.2, which is a very elementary

33

	ACID	INTERMEDIATE		BASIC	ULTRABASIC
PLUTONIC (Major Intrusions) Coarse grain	GRANITE GRANODIORITE	SYENITE	DIORITE	GABBRO NORITE	PERIDOTITE
HYPABYSSAL (Minor Intrusions) Medium grain	MICRO-GRANITE —PEGMATITES→ ←VARIOUS PORPHYRIES & APLITES→	MICRO-SYENITE→	MICRO-DIORITE→	DOLERITE	
VOLCANIC (Extrusive) Fine grain	RHYOLITE	TRACHYTE PHONOLITE	ANDESITE	BASALT SPILITE	OLIVINE-RICH BASALTS
VOLCANIC Glassy Rocks	←OBSIDIAN→	PITCHSTONE FELSITE		TACHYLYTE	
PRINCIPAL MINERAL CONSTITUENTS	Quartz Orthoclase Micas	Orthoclase Na-Plagioclase Some Ferromagnesians Little Quartz	Na-Plagioclase Ferromagnesians	Ca-Plagioclase Ferromagnesians	Ferromagnesians predominant

Fig. 3.2. Elementary classification of igneous rocks

one, although basically useful in geomorphology, is derived. A somewhat more complex classification applied to landform development has been described elsewhere (Sparks, 1971).

The common minerals belonging to each class of rock are noted below each class, but a few notes on some of the rocks are required, as it is not immediately obvious why certain categories are represented by two rocks. Granodiorite is a granitic rock in which the felspar is usually the soda plagioclase and not the orthoclase typical of granite. Aplites and pegmatites are fine- and coarse-grained dyke rocks given off by major intrusions and may, therefore, vary in composition with the parent intrusion. Porphyries are fine-grained rocks in which one mineral is present as considerably larger crystals: they may also be called porphyritic microgranites, microsyenites and microdiorites. Phonolite is a lava in which the felspathoids attain notable proportions. Felsite is a devitrified glassy rock usually of considerable age, while pitchstone is predominantly glassy and obsidian and tachylyte completely glassy. Norite has hypersthene as its main ferromagnesian mineral instead of augite, which is characteristic of gabbro. Spilite is characterised by soda plagioclase, as distinct from basalt, in which lime plagioclase is dominant.

The effect the mode of intrusion or extrusion has upon the relief will be considered more fully later, in discussing the effects of lithology on relief (Chapter 7), and attention for the moment must be confined to the effects of weathering on these different rocks.

The effect of mineral composition on weathering

It has been found possible to arrange the common minerals of igneous rocks in order of susceptibility to chemical weathering (Goldich, 1938). In the following table, the most susceptible minerals are at the top of the table and the least susceptible at the bottom, the dark-coloured minerals being on the left and the light-coloured ones on the right.

MOST SUSCEPTIBLE   Olivine —
                     — — — — — — — — — — — — Lime plagioclase
                     Augite — — — — — — — — — — — — — — — — — — —
                     — — — — — — — — — — — Lime-soda plagioclase
                     Hornblende — — — — — — — Soda-lime plagioclase
                     — — — — — — — — — — — Soda plagioclase
                     Biotite — — — — — — — — — — — — — — — — — —
                     — — — — — — — — — — — Orthoclase
                     — — — — — — — — — — — Muscovite
LEAST SUSCEPTIBLE   — — — — — — — — — — — — Quartz

It can be seen that on the whole the dark-coloured minerals are more susceptible to chemical weathering than the light-coloured minerals, and when one studies the detailed mineral composition of the rocks on the table it becomes obvious that the acid rocks are far less susceptible to chemical weathering than the basic rocks. One may take two common plutonic rocks as examples, a muscovite-biotite granite and a gabbro. The main rock-forming minerals in the granite are quartz, orthoclase, muscovite and biotite, which are the four least susceptible minerals on the list. On the other hand, the gabbro is composed primarily of augite and lime-soda plagioclase, two of the most susceptible minerals. Thus, other things being equal, a gabbro should be more susceptible to chemical weathering than a granite. It must be noted carefully that this statement does not imply that a gabbro is more susceptible to weathering than a granite, for other things are rarely equal in nature. The total susceptibility of rocks to weathering depends on other factors, such as the texture and structure of the rock, the nature of the type of weathering and the length of the period of weathering suffered by the rock.

Another point of interest emerges from the table. Quartz and muscovite are not susceptible to chemical weathering at all: hence, although they become detached from the rock as grains and flakes, they are not converted to clay minerals as are the felspars and the ferromagnesian minerals. The weathered debris derived from a granite is, consequently, coarser and more sandy than the debris from a gabbro. It forms a poorer soil, the poorness being accentuated by the fact that granite contains little or no calcium, and gabbro a much higher proportion in its constituent minerals.

It is also possible to derive a weathering series of minerals empirically by studying their changes through soil profiles. Such series include the accessory minerals as well as the main rock-forming minerals mentioned above. The order may be somewhat different from that listed because of different weathering environments and the way in which reaction products may sometimes coat a mineral grain and inhibit further weathering of the kernel. Soil profiles in different regions may yield slightly different weathering series, as can be seen from some of the examples quoted by Ollier, but in general terms they agree with each other and with the table above.

The effect of texture on weathering

For present purposes, texture may be defined as the crystalline state of the rock, whether coarse-grained, fine-grained, porphyritic or glassy. Over a long period, glasses are unstable as they devitrify to a compact stony rock, usually known as felsite. Under certain conditions coarse-grained rocks (Fig. 3.3) should weather more rapidly than fine-grained (Fig. 3.4) rocks of the same mineralogical composition. In the process of weathering it is rare to find all constituents of the rock weathering at an equal rate. Usually one mineral in particular is weathered and the weathering of that mineral loosens the whole fabric of the surface layers of the rock. In a coarse-grained rock the weathering of one constituent may have a proportionately larger effect than the weathering of the same constituent in a fine-grained rock. In addition, fine-grained rocks sometimes possess an interlocking structure in their crystals, and this must retard the decomposition of the rock.

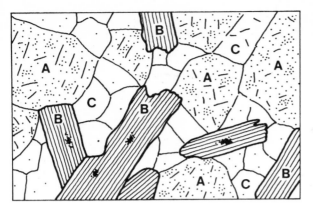

FIG. 3.3. Thin section of granite. A coarse-grained rock with large crystals of orthoclase, A, biotite, B, and quartz, C. Weathering of one constituent would cause considerable weakening of the whole. Magnified × 30 approx. (*after Tyrrell*)

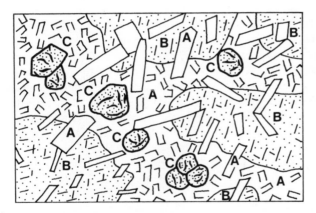

FIG. 3.4. Thin section of olivine dolerite. A medium-grained rock, showing plagioclase (labradorite) crystals, A, interlocking with augite, B. Olivine, C, is the other main mineral. Magnified × 30 approx. (*after Tyrrell*)

Theoretically the rate of weathering could be faster in the case of a finer-grained rock because the smaller the grains the greater the potential surface area. Similarly, strongly cleaved minerals have a greater possible weathering surface area than uncleaved minerals. But these total surface areas are not usually open to the agents of weathering, and weathering susceptibility of certain constituent minerals is probably usually more important than total surface area of mineral grains.

37

The effect of minor structures on weathering

Minor structures found in rocks are principally joints, although the bedding of sedimentary rocks and the foliation of some metamorphic rocks come in the same category and have the same effects. These minor structures allow the ingress of the agents of weathering especially those of chemical weathering. The presence of joints greatly increases the surface available for weathering and accelerates the general rate of weathering of the rock.

Plutonic rocks, of which granites are the most common, develop joints during cooling and probably also as they are relieved of the weight of the rocks above by erosion. There are usually three sets of joints at right angles to each other (Plate 1), thus dividing the rock up into a number of rectangular blocks (Fig. 3.5). Chemical weathering of these blocks causes a slow

FIG. 3.5. Weathered jointed granite, Great Staple Tor, Dartmoor

rounding of the corners of the blocks, and the development of piles of partially rounded boulders (Plate 2). Such features, well developed on Dartmoor and Bodmin Moor, are known as tors. As mentioned above, a considerable amount of this weathering may take place underground, and the tors may be formed before they are exposed by erosion (Linton, 1955). A similar tendency towards the production of rounded boulders by chemical erosion can be observed in many other rocks, and is occasionally well displayed in basalts. A useful term for the process is spheroidal weathering. It

PLATE 1. Jointed granite precipice, Beinn Nuis, Arran

must be distinguished from exfoliation, a process taking place on a much larger scale and not always attributed to chemical weathering.

Of the volcanic rocks, basalt is by far the most common and often displays a very well defined, polygonal, vertical joint pattern (Fig. 3.6), the columns

FIG. 3.6. Polygonal jointing in basalt

usually but not always being hexagonal in plan, a feature also shown by some sills (Plate 3). These features are very well displayed in the Tertiary basalts of the Giant's Causeway, Northern Ireland, and at Fingal's Cave in the Isle of Staffa. As the rocks are usually jointed horizontally as well, a large surface is presented to the agents of weathering.

The effect of climate on weathering

Climate affects the relative importance of the different types of weathering. Generally speaking it is said that physical weathering predominates in cold and dry regions and chemical weathering in humid climates whether they be temperate or tropical, but these opinions may need some qualification.

Undoubtedly in Arctic regions the dominant process is freeze-thaw acting in cracks of the rocks and producing great piles of shattered, angular debris. But there have been suggestions that even here chemical weathering is more important than is generally realised. The basis of these suggestions

PLATE 2. Jointed mass of granite, Cir Mhòr, Arran. Note the varying degrees of weathering of the joints

is that carbon dioxide, the presumed principal agent of chemical weathering, is, like most gases, more soluble at low than at high temperatures. Indeed, carbon dioxide is about twice as soluble at 0°C (32°F) as at 20°C (68°F). This does not mean that its rate of chemical action is twice as fast at freezing point as at 20°C (68°F), as the rate of chemical reactions of this type is reduced as temperatures fall. But the decrease in the rate of its action does not quite offset the effect of the increased solubility. Hence, any solution by carbon dioxide is at least as effective at around freezing point as at considerably higher temperatures.

The hypothesis of carbon dioxide weathering at low temperatures has been used by at least two investigators. Tamm (1924) used it to suggest that much of the rock flour derived from glaciers may not be material ground up very finely by the action of the glacier, but the product of chemical weathering. He observed that, in the Greenland ice cap, air bubbles with a pressure of 10 atmospheres had been noted at a depth of about 7·5 m (25 ft). Any meltwater percolating through the ice and releasing the air from these bubbles would gain a content of gases corresponding to a pressure of 10 atmospheres, which would mean a much higher carbon dioxide content than that corresponding to the partial pressure of that gas in the normal atmosphere. Thus, the possibility of chemical weathering below the glacier could not be excluded.

Williams (1949) suggested the possibility of chemical weathering to explain snow patch erosion in the north-western part of the United States. He measured the concentration of carbon dioxide in certain snow banks and found it to be higher than that found in a certain hall of the University of Washington. Any meltwater percolating through these snow banks would have a high concentration of carbon dioxide and might, therefore, be expected to have a considerable chemical effect on the rock below. Snow banks might, thus, cause the production of hollows by chemical means.

There are, however, other factors which are probably more important than the solubility of carbon dioxide in deciding the effect of solution under cold conditions. In the first place the vegetation cover is much reduced so that the available carbon dioxide in the soil atmosphere is very much lower. In the second place the time during which water with dissolved carbon dioxide is moving over rock surfaces is much less, partly because the water is frozen for much of the year and partly because the precipitation is less. Thirdly, if one looks at the problem from the opposite end and asks what field evidence there is of accelerated carbonation in the Pleistocene, the answer is precious little.

In our own climate, in which the rainfall is not very great but is offset by low evaporation, chemical weathering is probably of greater importance than physical weathering, as the rocks are almost always moist at ground

PLATE 3. Columnar-jointed quartz-porphyry sill at Drumadoon Point, Arran. In the foreground is the '25-foot' (7·5 m) raised beach

43

level. Care must be taken in the interpretation of certain phenomena, especially the presence of large screes in some of the glaciated highlands, as these may be relics of freeze-thaw weathering in the Pleistocene ice age and may not be forming at the present time.

The possibility that chemical weathering may be more important in deserts than had been generally realised has been mentioned earlier in this chapter in connection with exfoliation. However, exfoliation is not the only action of weathering in deserts. Many granites weather into coarse sandy debris corresponding roughly with the crystal sizes of the individual minerals within the rock, while other rocks split up into angular fragments. Blackwelder thought that these products would be more typical of temperature changes than would exfoliation sheets, but doubts have been cast by Griggs's experiments on the efficacy of temperature changes in producing granular disintegration. As mentioned earlier, these experiments showed no signs of such disintegration when the process was carried on completely in the dry, but only when the cooling was effected by water. However dry the air in deserts, it is never completely dry and there is always a certain amount of water vapour present, which may be deposited on rapidly cooling rocks at night as dew. It seems likely that chemical processes, especially hydration with its resulting expansion of minerals, must cooperate with the purely physical processes of temperature change.

Finally in equatorial regions there is a combination of considerable heat and moisture, the ideal conditions for pronounced chemical weathering. Undoubtedly chemical weathering is more pronounced here than physical weathering. The descriptions of many areas near the equator mention intense chemical rotting of the rocks to depths of 30 or 60 m (100 or 200 ft) and occasionally more. To the normal atmospheric carbon dioxide must be added a plentiful supply of organic acids derived from the decaying vegetation and, probably as a minor factor, a certain amount of nitric acid derived from the atmosphere by lightning discharges. The rate of weathering of certain types of rocks is more rapid in these regions than anywhere else on the earth's surface, but there are only rare indications of the actual rate of the process. This process of rotting causes the decay of the rock without any great changes in its texture: one mineral has been completely altered and, although the rock still appears to be sound, it is ready to crumble as soon as it is exposed.

The effect of time on weathering

This factor hardly needs elaboration, as it is obvious that the longer the period the deeper the weathering. There may, however, be a limit to the action of weathering unless the products of weathering are continuously removed. Some authors, including Davis, have spoken of the protective effect of the waste mantle on the rock beneath. While this would appear to be true as far as physical weathering is concerned, it is probably not completely true for chemical weathering. The probability that spheroidal weathering may take place before the rocks are exposed and while they are still buried beneath waste has already been mentioned. In addition, permeable debris of sandy type may nourish an acid vegetation, such as heather moor or coniferous forest, and be saturated with humus acids derived from the decay of that vegetation. It may, then, act like a sponge soaked in weak acid on the rocks beneath.

The weathering of sedimentary rocks

So far the effects of the different factors have been considered mainly in connection with igneous rocks. They apply equally to sedimentary rocks which, however, react in a somewhat different way to weathering. Sedimentary rocks are the products of the weathering of pre-existing igneous, metamorphic and sedimentary rocks. There are really three products of such weathering: coarse debris which is not susceptible to chemical weathering, clays which are the products of chemical weathering, and soluble products carried off in solution. Sedimentary rocks may owe their origin to the deposition and compaction of any of these types. The coarse debris and clays are deposited mechanically, and the soluble products are precipitated either by being incorporated in the skeletons of organisms which accumulate after death or directly due to salt water. Several main classes of sedimentary rock may, therefore, be recognised: conglomerates and sandstones represent the bulk of the debris not affected by chemical weathering; clays and shales are the accumulations of the products of chemical weathering; calcareous and siliceous rocks, as well as a few chemical deposits of no great geomorphological importance, result from the precipitation of dissolved material. The classes react to weathering in different ways.

The conglomerates and sandstones usually consist of pebbles or smaller particles of quartz, a mineral which is not affected by chemical weathering except when strong alkalis are present, a very rare occurrence in nature. The great durability of all forms of silica can be well illustrated by flint, a form of silica occurring as nodules in the Chalk. Flint nodules have been worn from the Chalk by the waves of the seas of Eocene times, battered to roundness on the Eocene beaches, and sometimes suffered further marine hammering in the Pliocene period. They have then, in some instances, been incorporated into ice sheets, survived that violence and been redeposited as outwash gravels. Later the outwash gravels have been subjected to atmospheric weathering, resorting by rivers, and occasionally, to solifluxion in periglacial conditions. Yet the flint has survived. As silica is so durable and resistant to both chemical and physical weathering, conglomerates and sandstones are only susceptible to weathering in the material which cements the grains together (Fig. 3.7).

Four cements are most common: silica, calcium carbonate, iron oxides and clay. If silica is the cement, the rock is most resistant as the cement and

45

FIG. 3.7. Thin section of sandstone, showing grains of quartz and their cement (in black)

the grains are equally resistant to weathering. Silica-cemented sandstones are often called quartzites, although this term is sometimes restricted to metamorphosed sandstones in which the silica is fused together. Quartzite is probably the most resistant rock found at the earth's surface. If the cement is calcium carbonate, the action of acid water dissolves the cement and thus effects the weathering of the rock. If the cement consists of iron oxides it may be susceptible to the processes of oxidation, and a weakening of the rock may be effected as a consequence. Clay cements are, of course, not resistant, and physical processes, such as freeze-thaw, may cause rapid weathering of the rock.

Clays and shales are not generally susceptible to chemical weathering, except through the impurities they may contain. Depending on the position of the water table, oxidation may affect ferrous compounds within the rock. Iron sulphide, a common mineral in some clays, may lead to the production of acids which assist in weathering. On the other hand, clays and shales are usually bedded or laminated and contain some water, thus allowing physical processes such as freeze-thaw to have a considerable effect.

Calcareous rocks, essentially the limestones, consist of either chemically precipitated calcium carbonate or the remains of invertebrate organisms composed of the same material. They are very susceptible to the action of carbon dioxide and humus acids and weather into a variety of forms, which will be discussed more fully in Chapter 7. As the rocks are permeable, either because of a major joint system or because of the presence of a multitude of minor cracks, there is a very large surface available for weathering.

Siliceous rocks are hardly affected by weathering, but they are of far less importance than calcareous rocks, only forming continuous beds of chert or cherty sandstone at a few geological horizons, for example in the Hythe Beds of the Lower Greensand.

The weathering of metamorphic rocks

It is very difficult to generalise about metamorphic rocks as they are the products of complicated processes of change due to temperature, pressure of different types and chemical soaking. They may be formed from any other type of rock, and the nature of the metamorphic rock formed from any one parent rock will depend on the nature of the metamorphism as well as on the lithology of the parent material. On the other hand, almost identical metamorphic rocks may be formed from different parent rocks by different metamorphic processes. Thus, there are a host of metamorphic rocks and a simple classification cannot be made. No more can be done than to indicate the main types and the way in which they may be expected to weather.

The most important are the gneisses, granulites and schists. Schists are rocks which have been subjected to intense directed pressure with the formation of mica and similar minerals on a large scale. The mica crystals are all orientated in the same direction and consequently the rock has a marked texture, usually called schistosity. Granulites are coarse quartz-felspar rocks of equigranular texture. Gneisses (Fig. 3.8) may be regarded

FIG. 3.8. Thin section of a gneiss. Alternate bands of schistose, S, and granulose, G, texture

as alternations of schistose and granulose texture. The granulites and gneisses weather in much the same way as granite, as their composition is somewhat similar. The schists are more susceptible to weathering, as the layered nature of the mica allows the penetration of the agents of weathering much more readily.

47

Slates are the product of a lower grade of metamorphism than schists. They have a marked cleavage developed in them but mica is not usually present. The cleavage is set up at right angles to the direction of the pressure causing the metamorphism (Fig. 3.9), and may have nothing whatever to do with the original bedding of the rocks. The cleavage facilitates the weathering of the rock, which is usually accompanied by oxidation. Weathering leads to the flaking of the rock along the cleavage planes.

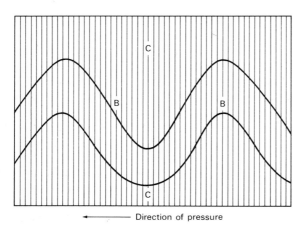

← ———— Direction of pressure

FIG. 3.9. Cleavage in slate. Diagrammatic representation of relations between direction of pressure, original bedding, B, and slaty cleavage, C

Sandstone and limestone fused by heat form quartzite and marble respectively. Marble is susceptible to the attack of acids, as is its parent material, limestone. Quartzite, a mass of silica, with no planes of weakness unless there are joints, is extremely durable.

This does not exhaust the types of metamorphic rock but merely indicates the more important ones. Many different types of very hard rocks, which resist weathering, may be formed by local baking of sediments at the margins of igneous intrusions, but they rarely attain regional importance.

Weathering, then, is a complex phenomenon, involving a variety of processes and being influenced by a number of factors, the most important of which are lithological and climatic. No hard-and-fast rules can be laid down about exactly how different rocks will weather, as so much depends upon minute differences in lithology and process. One can really only tell how a rock will weather when one has seen it weathered. Doubts and differences of opinion exist about the efficacy of the processes of weathering and it will be many years before even some of the doubts are resolved, as weathering is such a difficult process to study.

4
The development of slopes

In this chapter and the next the development of slopes and streams, which are primarily responsible for the removal of the waste produced by weathering from the land, will be considered. The main concern will be with slopes and streams in areas of humid, temperate climate, as the changes in slope and stream development which take place in different climates are considered in later chapters.

The mass of literature on slopes has increased enormously since this book was first published, and it is no easier to present an overall picture than it was then. The problems concern civil engineers in such things as road construction, general building, the stability of waste heaps and so on; they concern agriculturalists, soil conservationists and hydrologists; and, of course, geomorphologists. If anything there has been a welcome tendency for specialists to get together or at least to understand and use each other's methods. Geomorphologists like Carson and Kirkby have used the detailed methods of soil mechanics to try to understand mass movements and slope failure. The school of geomorphology largely inspired by Strahler and including people such as Schumm, Melton and Chorley has made use of statistical methods designed to eliminate observational bias in selection and description and to sort out from a mass of data measured more or less accurately in the field those variables which seem to affect slope form most. Others have continued to look at and speculate on the overall evolution of slope form with time, and in this they have been greatly aided by some of the evidence provided by the physical and analytical methods mentioned above. There seems to have been a swing away from the elaboration of over-simple mathematical models towards a refinement and intensification of field observations and an elaborate processing of the data so obtained to extract from it the last ounce of significance.

The aims of the various approaches are very different. The statistical approach often treats slopes as equilibrium states, exchanging energy and material with their surroundings; they are, in Strahler's phrase, open systems in steady states. On the other hand, the geologically-minded geomorphologist wants to know how slope profiles evolve with time and whether this evolution is lithologically or climatically determined. The return to field studies has raised warnings from workers such as Melton (1960) and Kirkby (1967) that their results only apply to the field areas they study or to very closely comparable areas. This follows from the way in which slopes seem to vary with rock type, soil moisture and climate, which are all capable of wide variation. Once one stresses the importance of rock type it is difficult to see how anything but very simple or basic generalisations may be made. The geometric model approach and naive preconceptions, for example that slope forms in all climatic zones are similar or that certain rocks, as though rocks were not continuously variable, produce certain slope forms, would possibly have added unity to the treatment of slopes at the expense of truth.

The inclusion of material in this review of slopes must, therefore, be the result of personal choice and bias is bound to creep in. If possible, other general reviews of the field of slope geomorphology should be studied e.g. Leopold, Wolman and Miller (1964), Schumm (1966) or Carson (1969A).

The study of slopes may be split up approximately as follows:

1. The nature and form of mass movements on slopes.
   (*a*) Descriptive classification of mass movements.
   (*b*) The mechanics of mass movements.
2. The evolution of slope profiles.
   (*a*) Geometric treatment of slope forms.
   (*b*) Inductive studies of slope forms.
   (*c*) Statistical studies of slope forms.
   (*d*) The development of slope profiles through time.

Mass movement on slopes

The movement of rock debris and soil on slopes is subject to a large variety of processes, some of which act very slowly but continuously, while others cause sudden movements of large masses of material followed by long periods of quiescence. The main factors moulding slopes may not be the present processes acting slowly over long periods, but the large and intermittent movements. The most important may have operated in a recent period of somewhat different climate, such as the Pleistocene glaciations in the British Isles, while the present processes may have had little effect in the Post-glacial period.

Classification of mass movements

A comprehensive study of the forms of mass movement has been made by Sharpe (1938), who, like Davis, visualised a continuous series ranging from a river, in which there is a great dominance of water over debris, through slopewash, sheetflow, mudflow, earthflow and debris avalanche to the landslide, in which there is a great abundance of debris and very little water. Just as there is a progressive increase in the proportion of debris to water through the series, so does the angle required for the processes to operate increase. There is, however, a significant difference between debris avalanches and landslides on the one hand, and all the remaining forms of movement on the other: it is to be found in the nature of the movement, all the more lubricated forms of mass movement being subject to flowing, while the drier types undergo sliding. The difference between sliding and flowing is that sliding involves equal velocity from the top to the bottom of the mass in motion, or in some cases a slight increase towards the bottom, while in flow the velocity is greatest at the surface and decreases to nil at the bottom of the mass. These differences are shown diagrammatically in Fig. 4.1. Thus, mass movements may be classed

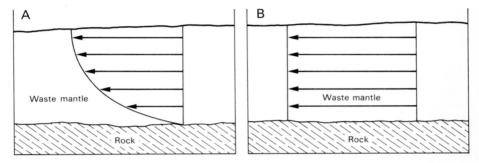

Fig. 4.1. Velocity distribution in a flow, A, and a slide, B

fundamentally into flows and slides, while a further distinction may be made between slow and rapid flow movements.

Mass movements involving slow flow

The most widespread movement falling within this category is that of soil creep. It is the slow downhill movement of debris and soil under the influence of gravity and is ubiquitous in both temperate and tropical climates. Its presence can be detected by a variety of phenomena: fence posts and telegraph posts leaning downhill, walls bulging due to the accumulation of debris on the upslope side, turf rolls below creeping boulders, stone lines in the subsoil traceable to a distinctive outcrop further upslope, and the superficial downslope bending of highly inclined strata (Fig. 4.2). A number of processes, each capable of producing only very slight movements, combine to cause soil creep. Rainwash and raindrop impact may move small particles downslope, and may remove fine material from the downslope side of large stones, thus facilitating the movement of the stones. Any frost heaving, due to the growth of ice crystals beneath particles, lifts those particles up at right angles to the slope but they fall back perpendicularly under the influence of gravity. Sharpe quoted the fact that uplifts of

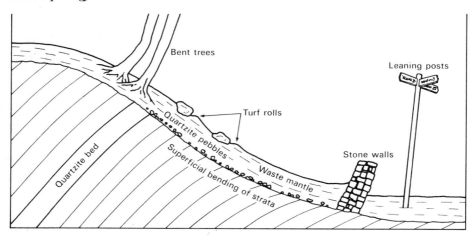

FIG. 4.2. Types of evidence indicating soil creep

150 mm (6 in) have been observed in parts of New York State. But the thawing of the ice crystal has, if sudden, another important effect, as it may start the particle tumbling a short distance downslope. On any slope a stone expands with heating more on the downslope side than on the upslope side as gravity acts with expansion downslope. The contraction during cooling is greater on the upslope side, as gravity acts with the contraction on this side. The net result of heating and cooling is a slow downhill movement of stones. Cracks, caused by drying, animal burrows and plant roots, are filled with material from the upslope side, and consequently assist in the slow operation of soil creep. Other forces such as trees swaying and animals treading also tend to give a slight downslope movement to the weathered material at the surface. Ploughing on slopes, especially when the soil is turned always in one direction, causes appreciable downhill movement of material. The combined effect of all these processes, some of which are illustrated in Fig. 4.3, causes a steady, if very slow, downhill movement of soil.

Among other slow flowage movements, Sharpe distinguished such types as talus creep, rock creep, rock glacier creep and solifluxion. The first three, as their names indicate, are primarily comparatively dry movements of coarse debris under somewhat different conditions: talus creep is the downhill movement of material in screes; rock glaciers consist of streams of boulders with little soil and only interstitial ice; rock creep is a movement of jointed blocks, partly as the result of soil creep and partly as a result of sliding. Sharpe defined solifluxion as a comparatively slow flowage of soil and included boulders, which took place mainly under Arctic conditions and on slopes of angles as low as 2–3 degrees. It depends on an absence of vegetation, a permanently frozen subsoil and melting surface snow to lubricate the moving mass of waste. This process will be considered in the

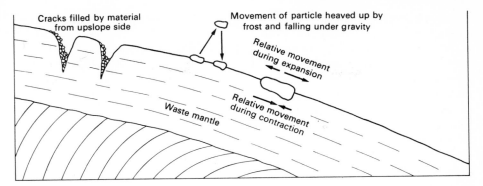

FIG. 4.3. Some of the processes causing soil creep (for explanation see text)

discussion of landforms developed under periglacial conditions (Chapter 15), as solifluxion is of no great importance in our climate at the present time.

Mass movements involving rapid flow

Unlike the slow flows described above, the rapid flow movements all depend to a large degree on thorough lubrication by water of the moving mass. Earthflow and mudflow are two terms used by Sharpe to denote two of the principal types. They both involve flow of wet material, but mudflows usually take place on steeper slopes, involve more thoroughly lubricated material and have higher velocities than earthflows. Earthflows are apparently important on the terraces bordering the St Lawrence river and its tributaries. A layer of clay below the sands of the terrace becomes thoroughly waterlogged and the whole mass of material may then flow through gaps in the banks into the river (Fig. 4.4). As the angle of slope

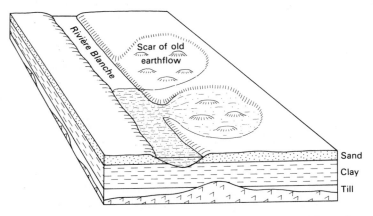

FIG. 4.4. Earthflow of 1898, St Thuribe, Quebec. Material from the terrace blocked the Rivière Blanche (*after Sharpe*)

53

becomes greater this type of movement may involve an increasing degree of sliding.

Mudflows are much more characteristic of steeper slopes, on which heavy rainfall initiates movement in a thick layer of weathered material in the absence of a substantial vegetation cover. One of the classic examples is the Slumgullion flow in the San Juan Mts, Colorado, where a fall of deeply weathered, saturated volcanic rocks started a flow, which moved for a distance of nearly 10 km (6 miles) and descended 800 m (2 600 ft) on a 5-degree slope.

A third type, the debris avalanche, is more characteristic of humid regions with a greater vegetation cover and occurs on steep slopes. Sliding as well as flow may be involved, and then the mass movement becomes one of the landslide features noted below.

The classification of these features is, of course, to some extent arbitrary, as are all classifications of natural phenomena. There is probably no abrupt change between any two types, and, in the case of actual rather than theoretical mass movements, it may be very difficult to place them easily in Sharpe's classification. As an example may be quoted the type of earthflow, mentioned by Sharpe, in which, although the lower part is predominantly a flow movement, the upper part involves sliding (Fig. 4.5).

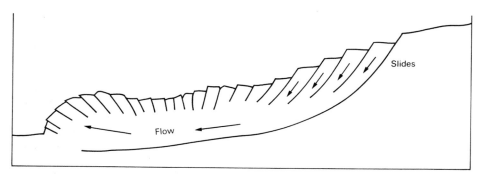

Fig. 4.5. Idealised earthflow. Minor slides at the beginning and flow lower down (*after Sharpe*)

There is, therefore, no clear break between flow and slide movements, just as there is no clear break between the various types of flow.

Mass movements involving sliding

Sliding is said to involve mainly dry material and usually takes place rapidly. Various falls and slides of rock and rubble come within this category, but the most significant movements are the landslides. As the velocity of the movement does not decrease downwards, there must be a surface along which shearing takes place. Very often this surface is a thoroughly lubricated layer. These movements, which are the most significant mass movements of the type, at least in lowlands, are slow and intermittent, as compared with the rock and debris falls and slides. Some of the forms assumed by these movements are illustrated in Fig. 4.6. A useful term for

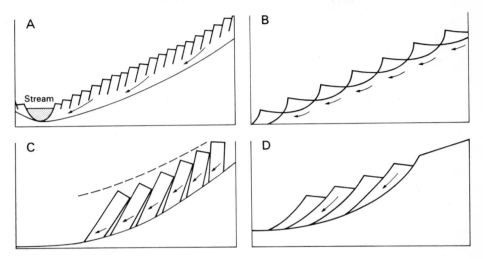

FIG. 4.6. Some of the forms assumed by miniature landslides.
A. Movement of underlying material causing superficial features.
B. Rotation of superficial blocks
C. Slips on major slip surface or lubricated zone
D. Rotational slips on curved surface (*after Sharpe*)

one of the common forms of movement is rotational slip, which describes very clearly the process involved.

The significance of mass movements on slopes in Great Britain

Most of the forms described above seem to involve such catastrophes that one at first doubts their significance with regard to slopes in this country. Soil creep, it is true, is a process which seems to be acting in most places and the evidence for it can usually be observed without much difficulty. Few people, however, have seen the rapid movements in operation, and yet they undoubtedly occur, although at rather rare intervals. But a very occasional movement of catastrophic type can cause far more change than centuries of slow movements of the creep type. They are comparable with the exceptional floods of rivers and storm-surges in the sea, which also perform far more work than centuries of gentle action.

Landslips of a rotational kind have been especially studied along the Hythe Beds escarpment of the north and west of the Weald. Two that have been investigated in great detail occur at Bower Hill, Nutfield (Gossling, 1935) and at Tilburstow Hill (Gossling and Bull, 1948), both in Surrey. It is

always possible to interpret the movement of large masses—that at Tilburstow Hill is about 300 m (1 000 yd) long and up to 60 m (200 yd) in
width—by ordinary tectonic faulting. There should, however, be two significant differences between faulted and slipped masses: the dislocation at
the back of a slipped mass should not continue far into the underlying incompetent rocks, usually clays, the saturation of which contributes largely
to the slipping; a slumped mass, because of the rotational movement,
should possess high angles of dip as compared with the unaffected parts of
the beds. These differences may be seen in idealised form in Fig. 4.7.

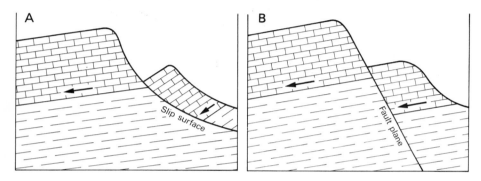

Fig. 4.7. Differences between slipped mass, A, and faulted mass, B; dip angles are
emphasised by black arrows

Theoretically, then, it should be comparatively easy to distinguish between
slipped and faulted masses and it may be wondered why there is ever any
cause of difference of opinion. The rock underlying slipped masses is almost always clay, in which it is very difficult to trace faults, especially faults
of the size which might be responsible for the slipped masses. Not only is
there a lack of exposures, but downwash from the beds above may obscure
the true nature of the ground at the foot of the slipped masses. A similar
lack of exposures may also make it very difficult to demonstrate any backward inclination in the slipped mass. Indeed, the nature of Bower Hill and
Tilburstow Hill was only established after very intensive research, which
involved the observation of sections in sunken lanes and temporary sections, as well as augering and boring, and a detailed subdivision of the beds
involved. The appearance of slipped masses in the field can be seen in both
plan and in section on Figs. 4.8 and 4.9, which represent conditions at
Bower Hill and Tilburstow Hill respectively.

Although slips have been studied so thoroughly in the northern part of
the Weald, it appears that they are of greater regional significance in the
western part of the Weald, where the nature of the Hythe Beds escarpment
has been studied over a considerable area (Wooldridge, 1950). The escarpment at Hindhead, at Blackdown and in various places in the Vale of
Fernhurst shows a number of dislocations which are, in all probability,

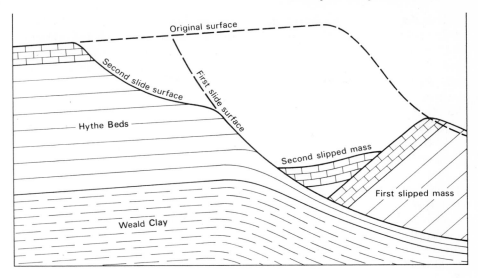

FIG. 4.8. Slipped masses at Bower Hill, Surrey (*after Gossling*)

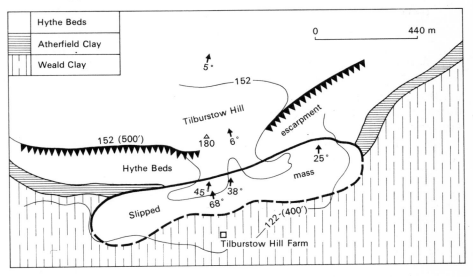

FIG. 4.9. Slipped mass at Tilburstow Hill, Surrey (*after Gossling and Bull*)

slipped masses of the same type as Bower Hill. Some of them are very large, that on the eastern side of Blackdown Hill being 3 km (2 miles) long and 400–800 m (between ¼ and ½ mile) wide and displaced 30–60 m (100–200 ft) downwards. Comparable slips have been reported from the Upper Greensand escarpment of the western Weald (Linton, 1930).

All the landslides noted above have occurred in resistant permeable beds underlain by weak, incompetent, impermeable clays, the Gault

57

underlying the Upper Greensand and the Atherfield and Weald Clays underlying the Hythe Beds. Coastal slips of comparable type at Folkestone occurred in Chalk overlying Gault. It should not be too readily assumed that such slips are still occurring on the Hythe Beds escarpment, as a few slips observed within historical times have not been on a scale comparable with the features described above. Wooldridge has offered the suggestion that they may be early Pleistocene features, caused to a large extent by freeze-thaw processes. There must be, at the end of a glacial period, a period when the landscape is utterly saturated and when the underlying clays are particularly susceptible to deformation.

The ideal conditions for mass movement occurred during the disastrous Exmoor storm of August 1952, when great damage was done to the village of Lynmouth. The first half of August of that year had rainfall considerably above the average, and the storm of August 15th brought a fall of anything up to 225 mm (9 in) in the space of a few hours (Kidson, 1953). This tremendous fall descended on a soil already thoroughly wet. It caused torrential floods on the rivers, the effect of which will be considered in more detail in the next chapter, but is also provoked dozens of mass movements on the slopes of the streams draining Exmoor (Gifford, 1953) (Fig. 4.10). Most of the sites of the slides occurred where the natural vegetation showed a tendency for the soil to be very wet, and, significantly, there were signs in some places of the scars of old movements of the same type. The exceptionally heavy rainfall of the storm caused such saturation that masses of earth and rock debris slipped down into the valleys. The slides are of some additional interest in showing the difficulty of fitting natural phenomena into arbitrary classifications. They were obviously provoked by abnormal saturation, and, taking place on comparatively steep slopes, they should

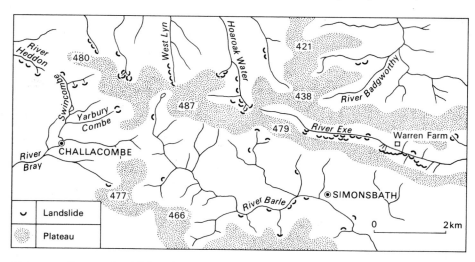

Fig. 4.10. Exmoor landslides after storm of 15 August 1952 (*after Gifford*)

have been movements of rapid flowage, yet observations demonstrated that the movements were slides, which, according to the classification, take place in a fairly dry state.

Somewhat similar to these Exmoor movements are the debris slides which occurred widely in the Fort William area at the southern end of the Great Glen in Scotland in May 1953 (Common, 1954). The rainfall involved was much less than in the Exmoor storm, but it fell on an area in which the slopes are still to a considerable extent probably in equilibrium with the conditions of the last glaciation. A severe night thunderstorm resulted in 40 mm (1·53 in) of rain in two hours at Fort William and 65 mm (2·6 in) in seven hours at Kinlochleven. Debris slides originated from tension cracks developed above wet flushes at middle and lower altitudes, that is in situations similar to the Exmoor movements. They no doubt represent one of the ways in which the landscape is changing into equilibrium with present conditions as modified by man.

Another interesting but slower movement has been in operation on Bredon Hill, one of the main outliers of the Cotswold escarpment (Grove, 1953). It corresponded almost exactly to the earthflow as envisaged by Sharpe and illustrated on p. 54 as Fig. 4.5. In this particular example (Fig. 4.11) water percolating from the Inferior Oolite Limestone caused a

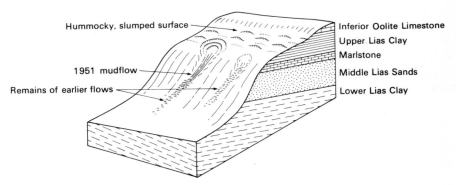

FIG. 4.11. Mudflows on Bredon Hill (*after Grove*)

thorough waterlogging of the underlying Upper Lias Clay, in which small rotational slips were developed, especially at the heads of small valleys notching the Marlstone ledge. The debris moved down through these notches on to the slopes below. In April 1951 a mass about 20 m (70 ft) long, 4·5 m (15 ft) wide and 1·5 m (5 ft) thick was observed to move 15 m (50 ft) in half an hour. The total movement in 1951 was of the order of 180 m (200 yd), but an earlier flow, possibly in 1930, had extended at least 135 m (150 yd) farther down the slope.

The interest of these phenomena lies in the fact that, even in our climate, the possibility of sudden mass movements having an effect on the moulding of slopes cannot be excluded. The examples quoted by no means exhaust the examples known, but merely illustrate them. Some features, such as the Hythe Beds slips, may be entirely related to the past existence of different climates, but others are known to occur in exceptional conditions under the present climatic regime.

Mechanisms of slope movements

To attempt a complete discussion of the mechanisms of mass movements would be impossible, for there is neither the knowledge available nor the necessary space in a general book such as this. One must illustrate.

In Fig. 4.3 some oversimplified versions of stone movements were shown. The precise path of a particle heaved up by frost and falling back under the action of gravity will be mentioned below. Let us concentrate on the stone which is alleged to expand and contract differentially because of gravity, so leading to a downslope displacement. A correspondent, D. T. Radburn of Leicestershire, took this point up (*in litt.*) and offered a number of interesting suggestions. He doubted whether a stone half buried in a waste mantle would behave in the simple manner suggested by the diagram, but thought that the expansion and contraction would be virtually equal upslope and downslope. The gentler the slope the more likely this is to be true. Nevertheless the expansion of the stone would create a cavity partly through the thermal effect on the stone and partly because the heat of the stone would dry out the surrounding soil, so causing it to contract. On cooling the stone would be isolated in a hole slightly larger than itself and so tend to slide downslope within that hole to rest against its lower edge. Repetition of this cycle could lead to the downslope migration of the stone. However, this assumes that the upper part of the hole is filled in by a process such as rainwash or frost action and that the infill is as consolidated as the rest of the soil. Thus we might expect a maximum rate of stone movement in changeable weather, an effect which should be capable of field testing. Mr Radburn also suggested a more general test of the idea, which would involve putting stones of different coefficients of expansion on a homogeneous waste mantle and observing the results. Such testing would be very interesting, though it might be extremely difficult to ensure a waste mantle sufficiently homogeneous for the experiment.

In writing of the various mass movements included in Sharpe's classification, which, as Common (1966) rightly points out, needs some amplification to include the special case of organic deposits, composite movements, and human disturbances of slope equilibrium, we have merely referred movement to lubrication. The mechanism is a lot more complex than this.

Deformation of a waste mantle, for example a clayey soil, takes place in two stages. When stress is applied to a soil, at a certain threshold it begins to yield and creep. This movement is largely intergranular and occurs where the stress is large enough to cause minute shearing of the asperities of grains which are in contact with each other. In this process of creep the pressure of the water in the pores of the material will, if it is saturated, tend to force the grains apart and so reduce its frictional resistance. With increased stress a point is reached at which the soil fails and rupture occurs, which can readily lead to mass movements.

Imagine a potential slip surface in a mass of clay; several opposed forces operate on it. The main force promoting slip is the weight of the vertical column of soil resting on the slip surface. With waste mantles of equal thickness this will obviously increase as the angle of slope increases; hence the reason for the fact that slope is fundamental in determining slip. Opposing this are frictional forces and the cohesion inherent in the soil. In the standard treatment of slope failure the effective stress mobilising frictional resistance is measured normally to the slip surface. In elementary language this is a pressing-together effect. The frictional resistance will be larger in the case of wet, but not saturated, soil than in dry soil, because the weight will be greater. On the other hand, once the soil is saturated and the pores filled with water, the pore water pressure, which depends on depth below ground water level and on water movement, will tend to force the particles apart so decreasing friction. Therefore, dry soils and saturated soils tend to fail more readily than wet soils (Carson, 1969a and b). Cohesion is probably a property of molecular attraction among closely spaced, small particles: it tends to disappear with time.

So far we have considered a homogeneous or isotropic material. But natural clays are full of fissures, cracks, joints and probably old slip surfaces. As Skempton (1964) points out, it is well known that flaws greatly decrease the strength of solid materials and act virtually as stress concentrators. Thus failure first occurs at some such weak point, immediately throwing increased stress on some other point which in turn fails, so leading to a progressive failure of the material; in other words, it shears. It has been observed that most failures in clay slopes occur in the zone affected by cracks and fissures, thus indicating the latters' importance.

As soon as failure starts there may be alterations one way or the other in the pore space and in the pore water pressure, this action acting as a positive or negative feedback mechanism. These two borrowings from the science of electronics are useful pieces of jargon to denote the cases where respectively:

(*a*) one change produces another change which accelerates the first: positive feedback.
(*b*) one change produces another change which tends to neutralise the first: negative feedback.

Geomorphology is full of examples which should spring readily to mind.

If we turn back to the 'lubrication' idea, we can now see that the great saturation of the rocks at the end of periglacial conditions could lead to high pore water pressures especially well below the ground water level. These pressures reduce friction and allow failure, for example, in saturated clays beneath sandstones and limestones, the set-up of many of the Wealden rotational slips attributed by Wooldridge to the Pleistocene period.

Clays are complicated materials and the treatment given above is a very simple case couched in probably oversimplified language. Skempton (1964) has drawn attention to the abnormal conditions present in what he terms overconsolidated clays. These are clays which have been consolidated under a much greater overburden than that which they now possess. For example, any Jurassic clay must have been compacted under a great weight of Mesozoic and Tertiary sediments. In such clays, when the overlying rocks are removed by erosion, the water content is much below that of normally consolidated clays, even allowing for a slight rise which occurs when the overburden is removed. The particles are more tightly packed and hence the shear strength is greater. If these rocks are subjected to shear tests, the peak strength (i.e. the point at which shear occurs) is found to be high. Yet there are known cases of failure well below these peak strength levels. However, if the rock continues to be strained after the peak strength is passed, there seems to be a marked decrease in strength so that the residual strength is lower. The reduction in strength may be due to an increase in water content, as consolidated clays expand when their peak strength is passed, but it seems to be most likely due to a regular orientation in mineral grains. Thus tests of shear strength of unflawed samples may give strengths much above that of the material as a whole, where the residual strength and, as noted above, the flaws are significant.

In a more general treatment of slopes, Carson (1969a) considered variations in some of the controls of slope failure through time. An implicit assumption must be stated at the outset: that slope form is controlled by failure in the waste mantle, an opinion with which all geomorphologists would not agree, at least in all cases. Cohesion may be reduced by weathering changes either along joints or between minerals: it also tends to disappear with time in clays, as stated above. The frictional resistance (in more technical terms, the angle of internal friction) will vary as the material is weathered to finer products. With finer material the friction is likely to be less, partly because the pore spaces are smaller and more likely to be saturated, giving rise to pore water pressures capable of reducing resistance to stress. Mineralogical changes may also help in the process. It may be that these changes cause progressive changes in the frictional resistance of waste mantles and so lead to progressive slope flattening, or that weathering proceeds until a threshold is exceeded and the slope adjusts itself rapidly to a new angle of repose (Carson and Petley, 1970). The number of such phases of instability will probably reflect the number of major steps in the weathering of the material, which could well vary with the rock and the climate. Thus, in a semi-arid region there might be a change on jointed rocks from a rock slope (i.e. a cliff) to a talus slope depending on the size of the fragments, and no further change beyond that because finer fragments are washed away. In humid areas there may be changes from talus slope, through taluvial slope (i.e. a slope developed on a mixture of talus and more weathered material) to soil slope. Carson sees the trimodal distribution of slope angles, observed in areas with thick waste mantles, as probably reflecting the effects of these phases: with thin waste mantles the effect does not appear, possibly because of the binding effects of vegetation. On strong rocks in Exmoor and the Pennines bimodal slope distributions with peaks at 33–35 degrees and 25–27 degrees may well represent limiting angles of stability for scree and taluvium mantles.

The importance of all this work, which is only beginning to open up an understanding of slopes, is that form is related to a detailed understanding of process, so that one can, to use modern terms, begin to construct process-response models.

Another recent trend is an attempt to measure and understand creep measurements and, in this connection, the work of Kirkby (1967) must be quoted. Kirkby set out to measure the rates of creep and to evaluate its causative factors. The process he studied was seasonal creep, because continuous creep would involve so delicate a balance in material as unhomogeneous as soil that strength limits must be locally exceeded with the result that landslips might occur. This had to be avoided, so that the areas studied, which had unbroken slopes, were probably only subject to seasonal creep. A theoretical consideration of creep would suggest that a progression downslope occurred under the influence of the sort of forces outlined earlier in this chapter. In these the downhill effect of gravity dominates. It is not strictly true to say that a particle is heaved up normally to a slope and then falls back vertically under the influence of gravity. This ignores the cohesiveness of the material, which would probably cause the particle to fall back at an angle between the two directions (Fig. 4.12A). If one assumes that any cyclic stress which acts in this fashion, such as freeze-thaw, decreases exponentially with depth, the movement profile in the waste mantle should be as shown in Fig. 4.12B. This, however, ignores the fact that the main force tending to move the waste mantle is the vertical thickness of the overburden at any point, i.e. AB at point B in Fig. 12B. So, at the surface particle movement is at a maximum and shearing stress at a minimum; at intermediate depths both reach medium values; at greater depths, provided that the shear strength is not exceeded, shear stress may be great but particle movement is at a minimum. This could lead to the sort of velocity profile shown in Fig. 4.12C.

This model of soil creep was tested both in the laboratory and the field. In the laboratory a block of boulder clay was put in a glass-sided trough and tilted. The movements of wires driven through the block were measured against the glass sides as the block was submitted to cyclic wetting and drying. On the whole the results agreed with the theoretical model deduced above.

It was also tested in the field in a small drainage basin, that of the Water

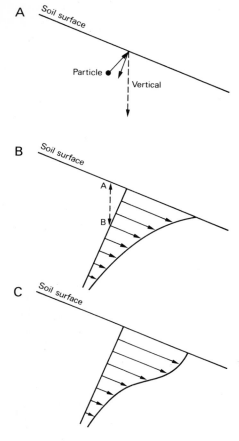

Fig. 4.12. Movement of waste mantle (*after Kirkby*) (for explanation see text)

of Deugh, in the Southern Uplands of Scotland. This was instrumented with Young pits (Young, 1960) and T-pegs to measure creep. A Young pit is formed by digging a hole with sides parallel to the direction of maximum slope and inserting a series of wires in the face as near vertically as possible with the aid of a plumbline. Then, if the pit is re-excavated later, the movements of the wires can be determined by reference to a plumbline and the movement of the soil profile obtained (Fig. 4.13A). With this method a measurement accuracy of ±0·5 mm may be obtained. An alternative method, suggested by Rudberg, is to auger a hole, insert a metal cylinder down which a column of metal or plastic discs can be dropped (Fig. 4.13B), and then remove the cylinder. Because of the thickness of the metal wall of the cylinder one is liable to have errors of ±3 mm when the column of discs is excavated and its deformation used to assess creep. Kirkby also used T-pegs (Fig. 4.13c) which consist of a flat strip of metal bent back parallel to itself, the two limbs connected with an adjusting screw: the whole is attached to a 6 mm square steel rod. It is driven into the ground as near vertically as possible; then levelled by using a level attached to an accurately ground cylinder in the V-shaped level rests and making use of the adjusting screw. At intervals the tilt accumulated in the T-peg can be measured and the waste mantle movement calculated. If different length T-pegs are used profile movement to different depths can be obtained. T-pegs are easier to use than Young pits, in which there is also more disturbance of the ground, but give annual measures of tilt some thirteen times less than those obtained from Young pits. Thus, although they can be used for evaluating factors and assessing relative movements, they do not give long-term measures of creep. One wonders whether a blade, say 50 mm by 4 mm in cross-section, might not give better results than the 6 mm square rod. Rainwash was also measured by means of the quantity of material deposited in small troughs parallel to the contours.

The results of these extremely detailed investigations led Kirkby to conclude that the most important factors in soil creep were soil moisture changes, freeze-thaw, burrowing animals and temperature changes in that order. Soil moisture generally appears to be very important, either directly or indirectly through its control on vegetation, for example in the formation of asymmetric valleys (Kennedy, 1969). The measured mean rate of soil creep was 2·1 cu cm/cm/year (approximately 0·35 cu ins/inch/yr) and that of soil wash 0·09 cu cm/cm/year (approximately 0·015 cu ins/inch/yr), i.e. about 5 per cent of the value for soil creep. Combined they could have lowered the basin of the Water of Deugh by an average of 8 mm (0·3 in) in the last 10 000 years. This strongly suggests that present processes, apart from any gullies or slips, are negligible so that there is every

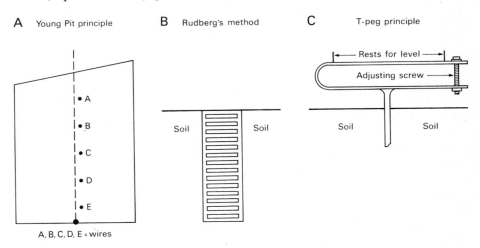

A   Young Pit principle   B   Rudberg's method   C   T-peg principle

Rests for level

Adjusting screw

Soil   Soil   Soil   Soil

A, B, C, D, E = wires

FIG. 4.13. Methods of measuring movement of soil profile (for explanation see text)

justification for considering that at least many aspects of the landscape may be relics of the past.

Kirkby emphasised that his results could only apply to the basin he studied or to closely comparable basins—and there would have to be comparability in slope, waste mantle, rock, climate and vegetation characteristics. One thus begins to appreciate how slow a process the extension of sound physical ideas to natural slopes is likely to be.

Erosion of slopes by moving waste

Before proceeding to the development of slope profiles, the possibility that these mass movements cause actual erosion of the solid rock below must be considered. Such a process of erosion was firmly believed by W. Penck (1924) to be an important factor in the erosion of slopes. It has also been cited as a possibility by Blackwelder (1942), who considered that movements of masses of boulders down steep dry stream channels in semi-arid regions may cause bruising and abrasion of the rocks beneath. The latter suggestion probably holds the key to the question, as the movements envisaged by Blackwelder are most likely to be slides in which the velocity is constant to the bottom. In all types of flowage the velocity is generally held to decrease to zero at the bottom of the layer in movement. It is obviously much more likely that there will be erosion by slides of material than by flow of material, as the actual particles in contact with bedrock will be moving much more rapidly. Generally speaking, few geomorphologists have thought moving debris to be an effective agent of sculpture on slopes, but the possibility cannot be entirely excluded, especially when sliding is the dominant movement.

The development of slope profiles

The development of slope profiles is one of the most difficult parts of the science of geomorphology. It may be said at the outset that we are far from possessing any conclusive ideas on the formation of slope profiles, in spite of the fact that the problem has been considered for at least half a century and different methods of approach have been tried. There are three fundamentally different methods. The first might be called the geometric treatment and consists of deducing, often mathematically, what can result from an initial slope under a set of assumed conditions. The second type of approach, which has been popular ever since the days of Gilbert and Davis, acknowledges the utter complexity of slope development and attempts to understand it in general terms by a process which might be termed inference based upon a wealth of observation. The third and more modern approach is that of statistical analysis of slope forms in relation to various possible controlling factors which can either be measured in the field or from a map. Essentially this method is looking for correlations between properties of the landscape and avoids assuming a pure cause–effect relationship.

The differences between the three approaches may be summarised as follows.

(*a*) If we could know and quantify all the factors, the prediction of slope forms could be reduced to a highly involved and perhaps impracticable form of mathematics. At the present time the geometric approach, because of our lack of knowledge, involves so many assumptions and simplifications that much of it seems divorced from reality.

(*b*) The method of inference based on field observations is, in a sense, crude. The complexity of problems can be considered and the past as well as the present included, but the answers are in such general terms and often involve such inspiration about processes that the method may seem unscientific to the protagonists of the other two approaches.

(*c*) The statistical method consists of measuring as many properties as possible and seeking correlations. It acknowledges the utter complexity of nature and the presence of feedback processes, pessimistically and probably rightly resigns hope of ever achieving a complete understanding in terms of cause and effect, and usually produces a series of statements concerning the probability of interaction between properties. It may throw some light on the past, but is usually concerned with landscapes thought to be at present in equilibrium.

Geometric studies of slopes

As a typical example of the geometric approach the work of Wood (1942) may be noted, not because it is the earliest example or the most mathematical, but because it is readily accessible. Ideas very similar to those used by Wood had been previously discussed by Fisher (1866) and Lehmann (1933) and, in so far as they were thought to apply to arid regions, by Lawson (1915).

In the simplest example Wood discussed the development of a vertical rock face under the influence of weathering. Weathering of the vertical face, called the free face, causes debris to accumulate as a scree or talus at the base of the slope. Gradually the talus grows until finally the free face is buried, and the whole slope consists of talus (Fig. 4.14). The latter is termed the constant slope. The form of the curved rock face behind the scree is explained by the rate of growth of the talus, and for the most idealised form of this discussion it is assumed that the volume of the scree is the same as the

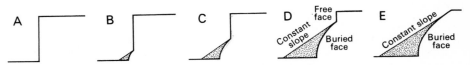

Fig. 4.14. Development of slopes (for explanation see text) (*after Wood*)

volume of rock weathered from the free face. At the early stage depicted in Fig. 4.14B the area of the free face available to weathering is much larger than that of the surface of the talus, and consequently the talus grows upwards rapidly. On the other hand, at the late stage shown in Fig. 4.14D the material from a very small area of free face is spread over a very large area of scree. Thus, progressively smaller increments of weathered debris have to be spread over progressively larger areas of talus. Consequently the rate of vertical growth of the talus is greatly retarded as the area of the free face is reduced. As a result, each portion of the free face is exposed to geometrically longer periods of weathering and suffers erosion corresponding to those periods. The final result is the convex buried face shown in the diagram.

If such a process occurs in the development of natural slopes, it will not be as simple as the ideal example outlined above, as several complicating factors occur and have been indicated by Wood. In the first place, the volume of the scree is not the same as that of the unweathered rock, as the latter consists of solid material and the former of broken pieces with spaces between. As a result, the rate of upgrowth of the scree will be somewhat faster than that visualised in the ideal case, and the buried face will be steeper throughout, although of the same convex form. If some of the scree is removed during its accumulation, as it may well be through washing out of the finer material and chemical weathering, which reduces the calibre of the material and consequently the angle of slope, there will be a tendency in the opposite direction. The rate of upward growth of the scree is slower than in the ideal case and the slope of the buried face becomes gentler. Again, the lower the angle of the original rock face, the gentler will be the slope of the buried rock face. It might be added also that rocks which produce coarse fragments will cause more rapid scree growth than those producing fine fragments, as the angle of the scree will be greater and the spaces in the scree greater in volume. The position may be summarised by saying that the coarser the debris produced by the parent rock, the more rapid the growth of the scree and the steeper the slope of the buried rock face beneath. The effects produced by the variations outlined above are illustrated qualitatively in Fig. 4.15.

FIG. 4.15. Variations in shape of buried face
A. Ideal case with scree volume equal to volume of rock before weathering
B. Scree volume greater than volume of original rock: coarse scree has steeper angle of rest
C. Some scree removed during formation
D. Original slope of rock less than 90°

Once the scree has been formed and so produced the constant slope, that slope tends to retreat parallel to itself, as any decrease below the normal angle of the scree causes the accumulation of more scree until the angle steepens again to the angle of rest, while any steepening causes sliding of the scree and the re-exposure and weathering of the free face with the production of more scree until the normal angle is once more attained. The slope, still retaining the angle of the scree, is pushed back into the solid rock behind the scree and continues to retreat parallel to itself.

So far, Wood's treatment of slopes is quantitative, although the ideas are expressed in words rather than in figures. The later development involves much more complex processes, the effects of which can only be stated in general terms. Fine material weathered and washed from the scree accumulates at the foot of the constant slope as a wash slope. The term seems to be better than Wood's term, waning slope, as this term smacks of the specialised ideas of W. Penck. The angle of the constant slope itself may decrease if the material composing it becomes more and more comminuted, as finer material has a lower angle of rest than coarser material. The comminution may be effected by a vegetation cover which reduces the effect of gravity on the fragments and encourages chemical weathering. As the retreat of the constant slope is retarded its upper parts are attacked by weathering on both sides and this results in a rounding with the formation of a convex upper part. Such late stage convexities must not be confused with the convexity sometimes found at the top of very young slopes, as this convexity is to be attributed to the stream cutting down faster than the slopes can develop, an idea very similar to the waxing development of slopes, visualised by Penck as due to continually accelerated uplift. The cycle of slope development, as visualised by Wood, is illustrated in Fig. 4.16.

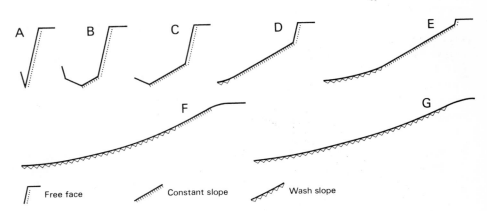

Free face   Constant slope   Wash slope

Fig. 4.16. Ideal course of slope development (*after Wood*)
Note. Increasing dominance of constant slope from stage B to stage D; increasing development of wash slope from stage D to stage G; formation of upper convexity at stage F

Ideas almost identical with those of Wood have been treated mathematically in a series of papers by Bakker and le Heux (1946, 1947, 1950 and 1952) and by van Dijk and le Heux (1952). Most of these papers are concerned with mathematical derivations of the form of the curves of the buried face caused by variations in the original angle of slope of the rock wall, in the angle of slope of the scree and in the ratio between the volume of the parent rock and that of the scree. These Dutch authors have treated both parallel retreat, which they call parallel rectilinear recession, and flattening, which they term central rectilinear recession. They conclude that there is a general tendency for the formation of slopes corresponding in angle to the scree angle of the debris of the rock concerned, provided that weathering and the simple removal of its products are the only processes operating, and provided that only a small amount accumulates as talus at the foot of the slope.

The applicability of these ideas to natural slope formation must now be considered. It is evident that many initial assumptions are made in these theories, especially in the mathematical treatment of them. Among the assumptions the following may be noted: homogeneous rock; weathering is the only process operating and the products thereof are readily removed; no climatic or base level changes intervene; the effect of the vegetation cover is negligible; there is a constant ratio between scree volume and volume of the parent rock; weathering is uniform over the whole slope; there is a terrace at the foot of the slope on which the scree can accumulate. Simplifying assumptions have to be made in the mathematical treatment to enable it to be applied. Thus the use of mathematics is probably limited to the working out of a few successions of ideal forms but 'whether the initial assumptions or the deduced profile development bears any resemblance to real landforms is dubious at best, because only one process out of several acting together is selected for analysis' (Strahler, 1950, p. 696).

The largest and most controversial assumption made, especially by the Dutch workers, is that weathering and the simple removal of its products is the process most important in the formation of slopes and that the mode of retreat of the slope is known. This is really equivalent to assuming what one wishes to find out, as the main questions in slope development are concerned with the nature of the processes, the forms which those processes produce and the nature of slope retreat. Again, although most of the mature slopes of our own landscape have passed the stage at which one would expect to find the constant slope still developed in scree, some of the youthful valleys of the highland parts of Britain ought to preserve them. Whether they do so is very doubtful, although a possible example is illustrated in Plate 4. Furthermore, the concave slope, often found at the foot of hillsides, appears usually to be a form cut into the solid rock and is not a constructional form composed of fine debris. Although some of the results of

PLATE 4. Carboniferous Limestone valley: the river Dove near Milldale, Derbyshire

these geometric methods are interesting, they are not wholly revealing, as they can only be usefully applied when full details of the processes operating are known.

Before turning to slope investigations depending on inference based upon field observations, some attention must be paid to the ideas of W. Penck. Penck's ideas are involved and not readily understandable even in English, probably because some of them are radically different from those of all other geomorphologists. Penck attempted to establish a relationship between the nature of a slope and the tectonic history of a region: convex slopes were formed in periods of accelerated uplift, straight slopes in periods of constant uplift, and concave slopes in periods of decreasing uplift or stability. The development of the latter type of slope affords a good example of his method.

Penck started with a straight rock slope and assumed equal weathering over the whole of it. The weathered fragments crumble and fall under the influence of gravity, all except the lowest (Fig. 4.17) adjacent to river level.

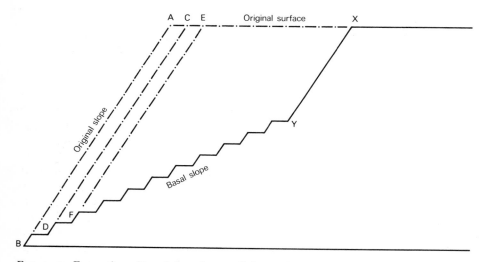

FIG. 4.17. Formation of basal slope by parallel recession (*after Penck*). The fragments at the base of each weathered layer (e.g. ABDC) would be infinitely small and would not give the irregularities shown above.

This lowest fragment cannot move as there is no gradient below it. Thus the slope profile moves back from AB to CD. In the next stage the profile moves from CD to EF, but again the lowest fragment is left in position as the gradient below is not sufficient to allow it to fall. The whole process extends until, finally, it may be assumed that the original slope has moved back to the position XY. The basal fragments left as successive slices of rock were weathered and removed form a new slope, which Penck termed the Haldenhang and which has been translated as basal slope. He is not at all explicit on what determines the gradient of the basal slope formed at the expense of the initial slope, but, presumably, it has something to do with the size of the particles to which the rock was reduced by weathering. On the new basal slope weathering takes place, but the material must be reduced to a finer calibre before it can move away than was necessary on the steeper initial slope. With the removal of the weathered layer from the basal slope the lowest particle again cannot move away, and there is the beginning of the development of a new slope facet of even gentler gradient below the basal slope (Fig. 4.18). The original slope thus breaks down into

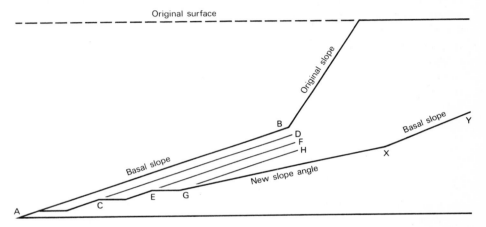

FIG. 4.18. Formation of flatter slope from basal slope (*after Penck*). As the slope develops the basal slope is weathered back to XY and the original slope retreats and disappears

a series of slope facets of gentler and gentler gradients. These facets all retreat parallel to themselves away from the river, each one of them acting as the local base level for the facet above. But, as each lower facet requires a successively longer period for the weathering and removal of a layer of debris, due to the fact that on the lower angles of slope the debris has to be more comminuted, the higher, steeper facets recede more rapidly than the lower, gentler facets. The net result is a broadly concave slope, with long, gentle lower slopes and short, steep upper slopes. Penck realised that debris accumulating on the lower, gentler slope facets would retard their development and at the same time increase the rate of development of the upper, steeper parts, which might become flattened, thus leading to an upper convexity. But, the upper convexity having formed, the supply of material would be reduced and, consequently, the lower parts of the slope would have time for the accumulated material to be removed from them and to recommence their normal development.

These conceptions would seem to be far too rigid to fit natural slopes. If the initial assumptions are studied carefully it is obvious that they include equal weathering over the whole face and simultaneous removal of the waste from the whole face, if parallel retreat is to be realised. It is very un-likely that these conditions could ever be realised naturally, as simultaneous removal would imply a rapid sliding of material from the whole of the slope as the only competent process. Again, it is not at all clear why one should get a series of successively gentler slopes each retreating parallel to itself. In fact, the slow migration of waste from the upper to the lower slopes would probably prevent any such action. Finally, as mentioned earlier, it is difficult to see what determines the gradients of the slope facets which are said to replace each other successively.

A full summary of the mathematical treatment of the geometrical app-roach will be found in Scheidegger (1961). On the whole it seems that it is premature to try to adopt too geometrical an approach, and the concentra-tion in the last decade has been on other methods of approaching slope problems.

Recently, however, an attempt has been made by Kirkby (in Brunsden, 1971) to provide a general mathematical treatment of slope development based primarily upon an analysis of the transport rates of debris, data which must be obtained primarily in the field by measurements of processes such as soil creep and wash. The changes of form will depend upon:

(*a*) the initial form of the slope;
(*b*) the variations in the rate of debris transport over the slope;
(*c*) the boundary conditions, i.e. what is happening to the divide at the top of the slope (this is usually considered fixed in the horizontal sense) and to the foot of the slope, which may undergo lowering and/ or horizontal movement;
(*d*) the curvature of the contours, because this will result in the receipt or removal of debris laterally (these convergence and divergence effects are discussed qualitatively in the next section);
(*e*) the rate of weathering over the slope and its rate relative to the transport rate.

This attempt differs from many earlier geometrical methods in that it depends on field estimates of transport rates rather than on a two-dimen-sional analysis of theoretical scree development. It differs from the statistical approach in that the basic model is deterministic (i.e. a cause-effect or process-response model) rather than probabilistic.

It seems that slopes tend towards a characteristic form, the time taken for this to be achieved depending on the initial form and the processes. Thus, if the characteristic form is concave, it will probably be achieved more quickly from an initial concave form than from a convex form.

The approach is essentially long-term and is not concerned with quickly achieved equilibrium conditions: hence, the concentration is on the effects of relief on transport patterns rather than the effect of hydraulic variables. Under ideal conditions it may be possible to reverse the argument and deduce process from form provided that we assume:

(*a*) that the characteristic form has been achieved;

(*b*) that we are correct in our guesses about boundary conditions.

And, one might add, that different process combinations cannot produce virtually identical forms.

It will be interesting to see how this type of long-term mathematical model is developed and whether it will open up new areas of information about the development of slopes with time.

Inductive studies of slopes

The method which geomorphologists usually call inductive is that in which causes are inferred from a detailed series of observations of individual examples. It would probably be better to call the method one of inference rather than one of induction. In many ways it resembles the art of detection.

There is a general consensus of opinion, though there have been those with different views, that slopes usually consist of an upper convex section, a middle straight section and a lower concave section. In areas of high relief and vigorous dissection the straight slope may be very strongly developed and the lower concavity reduced or absent. This was probably the reason why Strahler (1950), most of whose work was done in California, stated that a slope consists essentially of an upper convexity and a lower straight section, the lower concavity being either an illusion resulting from a bad choice of viewpoint, or a feature caused by accumulation of talus or structural control. If instead of accumulation of talus we substitute the removal of the erosive agency, usually a stream, from the foot of the slope, it might provide a better generalisation of the conditions under which lower concavities occur. Although it would need a large-scale statistical investigation to clinch the statement, there seems to be little doubt from general field observations that narrow valleys with streams in contact with the foot of slopes usually have straight slopes. Where there is no continuous erosion, that is in valleys with streams meandering from side to side and only undercutting slopes intermittently, in dry valleys, and in valleys where the delivery of debris down the valley sides exceeds either the stream's competence or capacity (see Chapter 5) to remove it, concavities usually occur. Thus the Californian valleys rarely have concavities; neither do the small shallow valleys in such areas as the till plateaus of East Anglia and the east Midlands: but most of the broad valleys of lowland Britain, the dry valleys of the Chalk and the Lower Greensand, the valleys of highland Britain which contain slumped masses of till which the stream is unable to remove, all tend to have concavities. There are probably exceptions to this, some of them due no doubt to the time-lag between a process starting or stopping and the development of the form in equilibrium with the new state of affairs.

The marked difference between the convex and concave sections has led to the assumption that different processes are at work shaping the slope in those sections. The concave profile of the lower part greatly resembles the long profile of a river and readily leads to the assumption that running water is probably the main process in operation on the lower part of the slope. But this assumption may not be true, and it is possible for slope variations to follow variations in one particular property of the soil mantle, such as the pore water pressure or the fineness of material which may have a progressive increase followed by a progressive decrease or vice versa from top to bottom of the valley side slope. Unfortunately, even if such is found, it is difficult to decide whether the progressive change causes the form or the form the progressive change.

The lower concave parts of slopes have usually been explained by some sort of hydraulic process by analogy with the concave long profiles of streams. Kirkby (1969) has pointed out that unchannelled sheet wash over a surface cannot produce a concave form and is likely to result in a convexity similar to that formed by creep. Only hydraulic processes operating in channels, in this case minute rills, have rapidly increasing erosive powers through downslope increases in volume and convergences. The junction of two streams of water results in the combined stream being able to transport the load of the two affluents over gentler gradients, as, for reasons stated in the next chapter, it is more efficient. It must not be imagined (Baulig, 1950) that one must expect to see such rills operating on the concave lower slopes of hillsides, for the rills may only be present after melting of snow or in periods of very heavy rainfall. The rills may be seasonally formed and destroyed, for example by summer rain and winter frost action in semi-arid claylands. They may have been more characteristic of past climate, in which case the slope form is a relic which present processes have not been able to alter very much. The concentrated rills do not form the whole of the concavity: they only cut shallow concave channels at intervals and other processes shape the interfluves between them, but those other processes are controlled in their effects by the profiles of the rills. The result of the rill erosion and its control over the other processes operating between the rills is a general tendency for the lower part of slopes to be concave. While the action of rills in this fashion can be understood in relation to semi-arid areas of bare soil or rock, it is difficult to imagine its application, at least under present conditions, to vegetated humid temperate regions.

The other possible explanations for the lower concavity concern the progressive change of some property of the soil mantle towards the foot of the slope. If the soil becomes finer towards the bottom through weathering, its angle of internal friction and hence the slope may decrease: if the soil becomes wetter down the slope, the pore water pressure may increase, the angle of internal friction and the slope both decrease. On the other hand, the cohesion may change and to some extent offset this. But, as Souchez (1966) pointed out, many concave slopes show no change in the grain size of the waste mantle downslope and the increase in water content is not commensurate with the observed change in slope. Furthermore, some slopes with very gravelly waste mantles show well-developed concavities.

Souchez seems to regard the slope as consisting of two elements separated by an angle: a lower, gentler-sloping element largely corresponding with the coefficient of sliding friction, i.e. adjusted just to maintain movement of the material concerned but not to initiate it, and an upper, steeper element corresponding with the coefficient of static friction, i.e. so that the slope is very close to the angle at which movement will be initiated. The differences between these two will be familiar to anyone who has shovelled into a gravel heap from the bottom, or filled the domestic hod from the coke or anthracite heap. The angle between the two elements would be blurred and rounded by the compressing flow, comparable with that in glaciers (see Chapter 13), which Souchez thinks occurs in very fine-grained soils through plastic deformation.

Even if we rule out such extraneous features as lower concavities merely stemming from the fact that many of our valleys were glacial troughs, the whole question remains complex because some lower concavities are cut in solid rocks and others are not. The development of a concavity in an accumulation of waste is perhaps more understandable than in solid rock, though most waste fan and scree slopes certainly start as straight slopes. In Britain it is beyond doubt that the lower concavity is often in solid rocks. Roads follow the lower concave part of valley sides just above flood level, and pits, for obvious economic reasons, are cut in the slopes immediately above roads—hence, exposures of concave bedrock slopes. Schumm (1964) maintains that the same is true for concave slopes on the Mancos Shale and other comparable shale formations of the arid American West. One cannot argue that the concave part represents protection from erosion by a creeping waste mantle supplied from above, for protection from the bottom upwards leads to the convex buried rock face of Wood's cycle of slope forms. All in all, it seems most likely, in order to cover both rock and waste forms and the great range of variation, to pin one's faith in some hydraulic process.

With that unspecific statement, let us turn to convex forms. A number of processes may contribute here: angles tend to be rounded by attack from two sides; pressure release jointing tends to parallel the landform in approximate curves. But the usual general explanations involve the action of soil creep or unconcentrated wash.

The latter hypothesis was put forward by Fenneman in 1908 and adopted by Lawson in 1932. According to Fenneman, who sought to explain the convex slopes above the heads of gullies, unconcentrated wash is so subdivided that it is a highly inefficient form of transport. It is therefore always overloaded, all its energy being used for transport so that none is available for downcutting. It would be better to say that the wash was loaded to capacity, as no stream can be overloaded, an idea that implies that a stream is carrying more load than it is able to carry. As the wash moves down the slope its volume increases, due to the influx of water from lower portions of the slope, but it remains, according to Fenneman, overloaded, as there is always an abundance of debris available for it to take up as its load. As the wash becomes greater in volume, there occurs a steepening of the slope, an assertion which must be considered unproven. The result of Fenneman's ideas on unconcentrated wash is an explanation of the convexity of the upper parts of slopes. The change to a concavity takes place, he considered, when the influence of base level, in this case the bed of the stream below, is felt. The idea that convexity of slope is due to the absence of the influence of base level has occasionally appeared in investigations on slopes and rivers, but it is a very difficult one to justify. On any convexo–concave slope, which is usually a graded system, movements of material on any part of the slope affect the other parts: thus, a slight downcutting in the valley affects the lower part of the slope and the repercussions are slowly transmitted to the upper convex parts of the slope. Thus, it is not true to say that the upper parts are not affected by the base level. The only example, where no effect of base level is felt, is probably the one in which the upper parts of the slope are separated from the lower by a vertical rock face. In this case factors affecting the lower slope may have no effect on the portion above the precipice, except perhaps after a very long period of time.

Fenneman's ideas have been criticised by Baulig (1940, 1950), who has pointed to difficulties other than the assumption that increasing wash involves an increasing slope. For any stream of water the condition of full load can only be realised if debris of the finest type is available, and the condition can only be maintained over any length of time if the production of the finest weathered particles at the surface equals the rate at which they are being removed by the unconcentrated wash. It seems very unlikely that such a situation could hold, as weathering into very fine particles is generally a slower process than transport. If the production of fine particles lagged behind their removal it would be impossible for the wash to be fully loaded with them and, hence, it might start downcutting.

A somewhat similar idea has been put forward by Horton (1945), who argued that, on the upper portions of slopes, there is usually a certain distance from the crest in which no erosion by wash takes place, as the wash is not sufficiently powerful to overcome the resistance of the soil to erosion. This section of overland or sheet flow corresponds to the upper flatter parts of slopes. The length of the section of no erosion will vary with a number of factors: heavy rainfalls will result in concentrations able to perform erosion before light rainfall; the permeability, or infiltration capacity, of the soil must be important, as impermeable soils result in less loss through percolation and hence concentrated, erosive flow is achieved higher up the slope than on permeable soils; vegetation will also have an important effect, as a fully developed soil with plenty of humus will be a much better absorber of water than one which has been stripped of its vegetation and the humus exhausted by over-cultivation. The significance of these factors in connection with soil erosion will be readily appreciated, but they also affect the form of natural slopes, the section of no erosion being virtually the convex upper portion of the slope.

If water action is not effective it is not logical to conclude that erosion of the upper parts of slopes does not occur once they have been formed and, therefore, that crests cannot be reduced beyond a certain point. The efficacy of other processes of mass movement must be considered, particularly the insidious, ubiquitous and slow movements caused by soil creep. Although Lawson (1932) thought that soil creep would be inhibited by friction, creep has been widely observed and its efficacy as a slow agent of transport believed by most investigators of slopes.

The fundamental paper on the upper convex parts of slopes and their relation to soil creep is that of Gilbert (1909). Gilbert based his hypothesis on the observations that the upper convexity was due to soil creep and that the thickness of the waste mantle does not usually increase appreciably downhill. If three points on a slope are considered, A, B and C (Fig. 4.19),

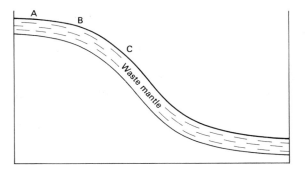

FIG. 4.19. Illustration of Gilbert's hypothesis of hilltop convexity. If weathering is equal everywhere on the slopes, twice as much waste passes C as passes B. As the waste does not thicken convexity results.

the amount of waste passing B is the thickness of a layer weathered in the section AB, but at C the passing waste consists of that weathered off from the section AB plus that weathered in the section BC. This may be put into more general terms by saying that the amount of waste passing any point on a hillside is proportional to the distance of that point from the crest. As the thickness of the waste mantle does not increase, it must be assumed that its velocity constantly increases downhill, for a constantly increasing amount of debris passes each point. As the only effective force is the component of gravity tangential to the surface of hillside, there must be a progressively increasing slope downhill, at least until other more effective agents begin to operate, usually the action of water.

Although Gilbert's hypothesis is the most satisfactory explanation of the convexity of hilltops, it has been criticised by Baulig (1940, 1950) in certain respects. It assumes equal weathering over the whole slope whether there is a protective soil mantle or not, whereas it is probably true that weathering is greatest at the top of the slope, as debris produced there moves downhill and affords some protection for the lower part of the slope. This constant re-exposure of fresh rock at the top of the slope is undoubtedly a factor in the rounding of hill tops (Birot, 1949), as the top will then weather back faster than the slope below. If weathering is not equal over the whole slope but decreases downwards, the supply of debris will not increase so rapidly with distance from the summit as in Gilbert's ideal case, and, therefore, the slope will be less convex. In the limiting case, where no debris is removed except from the very topmost parts, the slope should be straight right up to the zone in which active weathering is taking place.

Gilbert also assumed no change in the calibre of the waste, but its progressive comminution downhill must offset the need for a slope of continually increasing gradient. The reduction in calibre probably does not destroy the convexity, but modifies it to varying degrees.

The other major criticism is that Gilbert's hypothesis considers only two dimensions, and applies to a slope which is straight in plan, whereas many natural slopes are either convex, the end of a spur being a typical example, or concave, the head of a valley affording a good instance. Where the slope is convex in plan the area increases downhill and the supply of waste may spread out. Thus, each successive point down the slope will be passed by less debris than the corresponding point on a slope which is straight in plan (Fig. 4.20). The result should be that the convexity should not be so

Fig. 4.20. Tracks available for movement of waste
A. Slope straight in plan   B. Slope convex in plan   C. Slope concave in plan

pronounced on a slope convex in plan as on one straight in plan. On the other hand, at the head of a valley, i.e. on a slope concave in plan, the debris will converge and, therefore, the amount passing any point will be greater than that passing a similar point on a slope straight in plan. In this case velocities must be greater and the slope therefore steeper, so that slopes concave in plan should be characterised by the greatest convexity of profile. Natural slopes do not seem to show a variation in the amount and steepness of the convexity corresponding to the plan of the slope. It is likely, however, that the plan of the slope alters other factors at the same time as it modifies the process of soil creep. As Baulig rightly pointed out, wash will be favoured on slopes concave in plan, especially at valley heads,

and thus may help to reduce the gradient. Another factor which may operate is the effect of divergence and convergence of the waste mantle on the
weathering of the rock below. Where the debris diverges (i.e. on slopes
convex in plan), the waste mantle will probably be thinner than normal and
allow greater weathering of the rock below, thus steepening the slope. Where
the debris converges (i.e. on slopes concave in plan), the waste mantle will
be thicker and protect the rock on the slope from weathering. The decreased supply of weathered material downslope will tend to lead to a
diminishing of the convexity. It can be seen theoretically, therefore, that
conditions such as the plan of the slope, which affect the formation of the
convexity in one way, bring into operation other factors which tend to
cancel out those effects. In short, the formation of slopes, like most other
geomorphological processes, involves the complex interplay of a number of
interdependent variable factors.

Although it would be simple if it could be assumed that only soil creep
tends to cause the formation of the upper convex parts of slopes, this is
probably not true. Apart from the possible effects of unconcentrated wash,
which are probably of far less importance than soil creep, the effects of
cambering in suitable rocks must be considered. Cambering is a superficial
structure well displayed in the Jurassic regions of England and has been
specially investigated in Northamptonshire (Hollingworth, Taylor and
Kellaway, 1944). It consists essentially of the bending down into valleys of
resistant beds, usually sandstones or limestones, lying above weak incompetent beds, usually clays (Fig. 4.21). Obviously the operation of such a

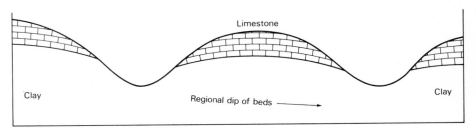

FIG. 4.21. Idealised cambering

process requires a special disposition of rocks, but the arrangement required is a very common one in many areas of Mesozoic rocks which have
not been strongly folded. Its effect is at least to aid the formation of the
upper convexity.

So far, the upper convexity and the lower concavity of slopes have been
considered, but no mention has been made of the straight slope which
sometimes intervenes between the two. To explain this, Baulig adopted an
explanation suggested by Lawson (1932) that the upper limit of concentrated rill action will vary with the season. In the wet season the middle
section of the slope may be under the influence of concentrated rill action,

while, during the dry season, it may be affected mainly by creep. The period of oscillation may be longer than seasonal: it may represent a Quaternary oscillation of climate as suggested by Baulig (1950) or a climatic periodicity of shorter duration but longer than a yearly variation. On the straight section neither process, the one tending towards concavity and the other towards convexity, is able to assert its full influence.

This does not seem to be altogether plausible as an explanation of the straight segments of slopes. The most general statement that can be made is that slopes will remain at constant angles, either in space or time, when all the factors remain constant, but, although this statement may sound impressive, it does not really take us much beyond the self-evident.

From time to time there has arisen the idea of a constant slope reflecting the angle of repose of the material. This is most easy to see in gravity-controlled forms such as scree. It is thought that different calibres have different angles of rest and the idea has been expanded in the case of rocks which have two very different debris sizes, for example the jointed blocks and crystal grains of granite, to suggest an explanation for abrupt breaks of slope, as between inselbergs and pediments in arid regions. In fact there are always two angles involved: the angle at which movement can start and the angle at which sliding movement ceases, corresponding with the coefficients of static and sliding friction. If these factors were fundamental in affecting slopes we might expect the slope to steepen until the static friction value was exceeded and then to collapse until the angle was adjusted to the sliding friction value. Thus slopes, depending on their particular state, might oscillate between two values. The two angles on granite in parts of Arizona studied by Melton (1965) were 28·5 and 36 degrees.

However, Melton was inclined to dismiss the idea even in the arid poorly vegetated Sonoran district of Arizona. Debris on granite slopes below 600 m (2 000 ft) remained fairly constant at 125–150 mm (5–6 in) diameter for slopes ranging between 16 and 28·5 degrees. It then increased in size to 200 mm (8 in) on slopes of 36 degrees, which is the limiting angle of a debris slope on granite. Above 600 m (2 000 ft) even this difference did not seem to occur. The facts that the debris size did not seem to correspond with the slope and that slopes much below the repose angle were met suggested that neither the boulder-controlled slope idea nor the repose angle idea held in this region.

It is also known that the binding effect of the root mats of vegetation on fine-grained materials will allow vegetated slopes to stand at angles exceeding those possible for the same material with no vegetation cover.

With our vegetated slopes it may be that straight slopes often correspond with waste slopes which have been grown over and are slowly changing in equilibrium. The binding effect of vegetation and the slowing down of movement (Schumm records that slopes on the Mancos Shale have creep rates fifty times as high as those recorded by Young on slopes of the same inclination in England) probably allow more weathering and change in water content and hence promote systematic changes in slope angles. Where movements are more rapid there is more likelihood of factors remaining constant for some distance on slopes, thus giving rise to straight slopes. This should mean that semi-arid regions, periglacial regions whether Arctic or Alpine, and tropical wet regions would show more straight slopes than humid temperate regions: they probably do.

With the hypothesis that the convexity is mainly due to creep and the concavity to concentrated rill action, a short discussion may be undertaken of the nature of the slope as it should vary with climate, lithology and age (Fig. 4.22). Different climates will affect the length of the segments of

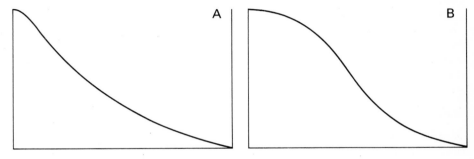

FIG. 4.22. Convex and concave slopes
A. Slope form favoured by heavy rainfall, impermeability or long development
B. Slope form favoured by light rainfall, permeability or youthful development

slopes affected by creep and by concentrated rill action respectively. Very wet climates should, therefore, produce slopes in which the area affected by water action is very large, thus leading to essentially concave forms. Dry climates should cause a much greater degree of convexity to occur, although these arguments cannot apply to some desert slopes, which often appear to be rock forms rather than rock covered with waste. Impermeability acts in the same direction as heavy rainfall, so that impermeable rocks should develop slopes of markedly concave character. Permeable rocks, on the other hand, should give rise to slopes in which the convexity is better developed, as it is on soft limestones such as chalk. The statement should be qualified by distinguishing between rocks producing permeable debris and those producing impermeable debris, as the resulting slope will probably reflect both the permeability of the debris and that of the rock. Thus, granite, which is not very permeable as a rock unless it is well jointed, produces coarse permeable debris, which is conducive to the formation of a well-developed convexity. The age of the slope probably affects its form in the following manner: as the slope weathers back the rate of removal of debris decreases so that the waste becomes more comminuted, vegetation being an important factor in this connection, and if the parent material is suitable, there will be a greater production of clay material due to chemical weathering. The increased proportion of clay material will affect the permeability of the slope, so causing concentrated rill action to extend higher up the slope with a corresponding increase in the concave portion and a decrease in the convex portion. In the early stages of slope development the rapid movement of debris prevents its full comminution and so increases permeability and the development of convexity in the upper part of the slope: later in the development of a slope the reverse is true.

The hypothesis of slope development outlined above is essentially that advocated by Baulig (1940, 1950). The normal slope in our climate consists of an upper convexity, due to soil creep, and a lower concavity, due to concentrated rill action, there being sometimes an intermediate straight section. The relative proportion of convexity to concavity will depend upon climate, lithology of rock and debris, and age of the slope. While the effects can be discussed in general terms, natural conditions are so complex that the interplay of all factors will be very rarely precisely the same.

Statistical studies of slopes

Within the last twenty years there has been an increase in the use of statistical studies of slopes to try to improve, through quantification, on the qualitative observations that had previously satisfied most field workers. Two main reasons have probably guided this development.

The first is the feeling that qualitative field observations are inevitably distorted. This is a valid criticism on a number of grounds. It is often very difficult to observe a total form in the field from an ideal point of view. Many field observations have been made to test a hypothesis already conceived. Indeed, this is inevitable: the alternative would be to go into the field, gaze around starry-eyed and hope that a problem might emerge like a bolt from the blue. However, it is possible for there to be an unconscious acceptance of favourable evidence and an unconscious rejection of unfavourable evidence. There may also be bias in the selection of areas: for example, if one wished to support an idea that all granites have gently rounded forms with tors, one would not go to Arran, or, if one did in ignorance, one would quickly leave. Again there may be differences between different observers, so that there can be no comparison between observations made by different people. One could go on multiplying possible spheres of bias. Hence, it was highly desirable that objective tests of unselectively selected data should be introduced. It is possible, of course, to overdo all this and to use a massive and elaborate technique to prove the self-evident, but this is the sort of abuse to which research is subject. The experience of palaeontologists is that the human eye is one of the best and fastest multivariate analysers known, in that a trained eye can almost instantaneously synthesise a variety of attributes and so name a species, that the identification is repeatable, and that palaeontologists of the same expertise usually reach the same conclusion. One must not become so paranoic about not believing the evidence of one's eyes that one needs a pair of calipers to tell the difference between a 10 penny and a 50 penny piece. Where the quantitative statistical approach really scores is in assessing whether small differences are significant or merely due to chance. The elimination of bias and the setting-up of objective slope analysis procedures has been thoroughly discussed by Strahler (1956).

The second and greater advantage of the quantitative statistical methods in slope discussions derives from the multitude of factors which may affect slopes. If one thinks of rocks, waste mantles, climate, hydrology, vegetation and man, and then of the different aspects of all of these that may possibly affect slopes, one despairs of getting very far into detail by means of inference from general field observations. There may be causes (independent variables), effects (dependent variables) and interactions (interdependent variables), so that on a slope it is possible that there could be ten, perhaps twenty-five, perhaps fifty, perhaps more factors, all playing their greater or smaller part controlling the form. Maybe we shall never be able to consider all of them, but we can decide by experience which are likely to be significant, formulate some means of expressing them quantitatively, and then by the use of statistical techniques assess the correlations between them and also the relative parts they play in slope development. Because of the virtual impossibility of prejudging every operational factor we shall probably never be able to acquire a complete explanation of slope forms, but the main factors should emerge from the analysis. Ultimately, such a complex model of slope development will develop that only a computer will be able to handle all the data involved in it.

The real beginning of modern attempts at statistical investigation of slopes was made by Strahler (1950). Strahler was concerned with maximum slope angles and obtained the orientation of these by rolling boulders downhill or by taking a series of slope readings in an arc straddling the direction of maximum slope, which would then be shown by the highest reading obtained. Maximum slope angles were averaged over lengths equivalent to a quarter or a fifth of the total slope length. A number of areas of different relief, climate, vegetation and soils were studied in this fashion. The maximum slope angles were found to have a normal distribution about a mean value and this led Strahler to the formation of a general law: 'Within an area of essentially uniform lithology, soils, vegetation, climate and stage of development, maximum slope angles tend to be normally distributed with low dispersion about a mean value determined by the combined factors of drainage density, relief and slope-profile curvature' (p. 685).

What this means is this. If all slope-forming processes are identical, mean maximum slope angles will be very much the same, the range of deviation from it will be small and as likely to be positive as negative; thus, the first part of the statement. The second part is best explained with a simple diagram. In Fig. 4.23A and B can be seen the effect on slope angles of varying only drainage density, i.e. stream spacing in the sectional view shown. Fig. 4.23C has the drainage density of Fig. 4.23A but greater and lesser relief, while Fig. 4.23D has the same drainage density and relative relief as Fig. 4.23A but different slope curvatures.

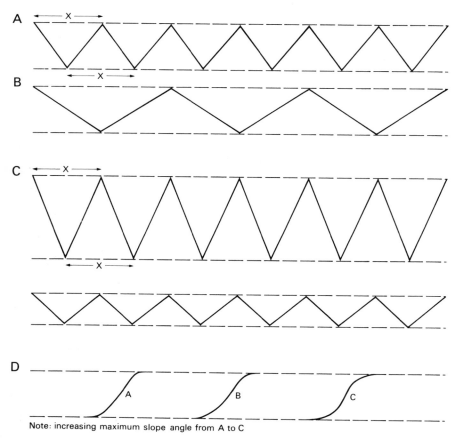

Note: increasing maximum slope angle from A to C

FIG. 4.23. Relations between drainage density, slope angle and relief (for explanation see text)

Strahler also showed that there was a correlation between maximum slope angle and channel gradient, which tended to confirm a long-held belief that channel slopes are adjusted to remove the debris supplied to them down valleyside slopes. In other words the system is in equilibrium. There are exceptions to this rule in that side slopes do not steepen with the increase of channel gradient very near the head of streams. It might also be possible that stream discharge and channel geometry, which together control stream efficiency (see Chapter 5), would somewhat obscure the simple correlation between maximum slopes and channel gradients. Again,

as Strahler pointed out, a vegetated slope at a given angle would contribute
less debris to a channel than a bare slope, so that it should be associated
with a lower channel gradient than the bare slope. These facts are probably
responsible for the correlation between maximum slopes and valley gradients being a rather broad regional one.

Strahler's work also led to some comments on the question of parallel
retreat or declining slope profiles, but these are more pertinent to the next
section of this discussion.

Much more complex analyses were made on badlands developed on a
silty sand spoil tip at Perth Amboy, New Jersey, by Schumm (1956a) and
on a series of drainage basins in Arizona, Colorado, New Mexico and Utah
by Melton (1957). These are both synthetic studies of whole relief and include drainage networks, channel and slope characteristics, so that it is
difficult to speak of slopes in isolation. Both studies aim at understanding
the complex interactions of the many factors involved.

Melton's work will illustrate this. The principle variables affecting slopes
which he measured were the following, though they control other features
of the landscape as well and so cannot be said to affect slopes only.

(*a*) The climatic effect was measured by Thornthwaite's precipitation-
effectiveness index, which had often to be interpolated from nearby stations
with meteorological records.
(*b*) Rainfall intensity was expressed by the amount of rain likely to be received by the worst storm of one hour duration occurring every five years.
(*c*) Infiltration capacity was obtained by means of a (relatively) lightweight
field apparatus that determined the amount of water absorbed per unit
area per unit time once the infiltration rate had reached a minimum. The
results may not have absolute significance but are comparable among
themselves.
(*d*) Soil strength was measured by the imprint left by a 12 lb (approximately 5 kg) iron shot dropped from a height to give a standard amount of
energy normal to the earth's surface. The height of dropping naturally
varied with the slope.
(*e*) Size distribution of soil was measured by standard soil mechanics
methods.
(*f*) Percentage of area bare of vegetation. This was assessed from the
readings along a 15 m (50 ft) tape orientated in accordance with the direction adopted by spinning a geological hammer and letting it fall.
(*g*) Roughness was a measure of all rock fragments exceeding 12 mm ($\frac{1}{2}$
in) in mid-diameter in an area with a radius of approximately 300 mm
(1 ft). The circles were located by random numbers along the direction
obtained from spinning a geological hammer.
(*h*) Maximum valley side slopes.
(*i*) Rock type.
(*j*) Local vegetation zone.

This illustrates both the randomisation of samples, at least in certain cases, and the attempt to appreciate the interaction of many factors. The effects of these variables on the basins and slopes were assessed by a complex series of statistical procedures. It was found meaningless to consider factors in isolation because rock and climate interact through vegetation, soil formation, runoff, infiltration, and soil creep on basin and slope forms. The slopes varied with rock type as might have been expected. Steep slopes were associated with a high degree of relative relief, high infiltration rates, low wet soil strength, low runoff intensity and high precipitation-effectiveness index.

Other studies have since been made of slope steepness often in connection with asymmetric valleys. They will be considered in the next chapter, but if any one factor has emerged as basically important in slope angles it is soil moisture: this controls weathering, hence grain size, pore water retention and angle of internal friction, which largely determines the shear strength; it also controls vegetation and hence infiltration, runoff, soil texture, erodibility through raindrop impact, and the degree of cohesion imparted by root systems. Many of these, of course, interact with each other.

One thing emerges clearly from these investigations, and from that of Carter and Chorley (1961) quoted below; slopes, channel characteristics and drainage network characteristics are all interrelated, often mutually interacting features of the landscape, linked through innumerable feedback processes. Hence they are difficult to discuss because there is no simple progression from an ultimate cause through a series of sequential effects. Further, our arbitrary divisions into chapters on slopes, rivers and drainage systems is an incoherent way of looking at nature, forced on us by the necessity of exposition.

The question of parallel retreat

Some authors, whose views have been outlined above, have included parallel retreat as an essential slope-forming mechanism, and the whole question was aired in a discussion in America many years ago (von Engeln and others, 1940). While the conditions in deserts and other tropical regions deserve special consideration and will be discussed in later chapters, it is appropriate to discuss here the possibility of parallel retreat in temperate humid climates.

Parallel retreat, in an ideal example, would involve equal depth of weathering all over a slope followed by instantaneous removal of the debris, with none of it accumulating at the foot of the slope to afford any protection of the lower part of a slope. Obviously these conditions can never be realised in nature unless all the debris is shaken off slopes periodically by earthquakes, an unlikely event, but one must consider the closest natural approximation to these conditions, i.e. when the debris migrates swiftly from slopes with no accumulation at the foot. Such conditions can only be realised when there is a powerful agent removing debris from the base, but such agents were not postulated by Penck, whose ideas imply only removal by gravity.

The simplest case of parallel retreat is an actively eroding sea cliff, from the base of which the debris is quickly removed by the waves. But the sea cliff only remains vertical as long as the waves are operating at its base: as soon as they are withdrawn the cliff starts to become degraded as can be seen by the examples behind raised beaches in various areas. Long abandoned cliffs, such as that behind the early Pleistocene beach deposits of the London Basin at an elevation of approximately 195 m (650 ft) OD, have been degraded to very gentle slopes indeed. Inland there is rarely any agent as effective in removing debris as wave action, but certain types of erosion approximate to the required type. Lateral erosion by rivers, by continually undercutting slopes, may effect an almost parallel retreat of the slopes above; springs sapping at the base of an escarpment, coupled perhaps with landslips, may also tend to give an approximation to parallel retreat, otherwise escarpments should be much more degraded than they are. But these possible examples of approximately parallel retreat involve basal sapping of the slope by one mechanism or another. A further process, which may have affected British slopes, is that of solifluxion over permanently frozen subsoil in the Pleistocene period. This may have involved a comparatively rapid stripping of a thin skin of waste from over the whole surface.

Confirmation from other areas of these general assumptions may be derived from studies such as that of the Colorado plateaus by Schumm and Chorley (1966). Here a series of particularly unresistant Mesozoic sandstones form scarps in a region of approximately 300–625 mm (12–25 in) annual precipitation and absolute temperature range of $-37°$ to $+45°C$ ($-35°$ to $+113°F$). These sandstones have low compressive strengths, high porosity and such poor cementation that one virtually leaves one's footprints in them. In many cases talus destruction exceeds talus production so that no scree slope and buried rock face can develop. Instead the scarps appear to retreat approximately parallel to themselves.

A study by Melton (1960) on valley asymmetry in Wyoming and Arizona bears on this problem. Asymmetry in homogeneous rocks can only develop because one slope flattens more than the other i.e. it implies that parallel retreat does not always prevail. Although many basic causes may produce asymmetry, Melton states categorically that there is one vital mechanism, the position of the stream in relation to the base of the slope. Where stream gradients are steep and downcutting is rapid, factors such as soil moisture and vegetation cover, which generally seem to produce asymmetry, are overruled. In fact there is a high degree of association between slope angle and the position of the stream in the valley. This is tantamount to stating that an active agent at the foot of the slope is necessary for parallel retreat.

89

Other authors would add that the relative relief must also remain constant, i.e. lowering of channels is equalled by lowering of divides.

Elsewhere it is difficult to see how parallel retreat can have operated. Any slowly moving waste mantle must give relative protection to the lower parts of slopes and expose the upper parts more, thus leading to unequal erosion and flattening of slopes with time. The comminution of debris which takes place with time also leads to the progressive flattening of slopes. If parallel retreat was the general rule, there should be a great similarity between all the slopes developed on one particular type of rock. Although there is some correspondence between rocks and slopes, in so far as general conditions are concerned, there is too much variation between the slopes developed on one rock for parallel retreat to be considered a serious possibility. Some interesting data on the problem have been given by Strahler (1950), who found a relation between the gradient of the long profiles of streams and the gradients of the steepest parts of the slopes leading down to those streams. As the gradients of streams decrease towards their mouths, so do the side slopes flatten correspondingly, a state of affairs visualised in Davis's cycle of erosion. Of course the lithology of the rocks, the climate and multicyclic development would probably complicate this picture. Observations were also made by Strahler of the slopes in a narrow valley at points where the stream was actively cutting into the base of the slope and at others where there was no basal erosion. The average slopes observed at points of these two types were 44·8 and 38·2 degrees respectively, thus indicating the degrading of slopes when the agent of erosion is removed from the base.

These observations have been confirmed by Savigear (1952) in a very interesting study of the evolution of slope profiles behind Laugharne Burrows in South Wales (Fig. 4.24). West of Laugharne at the mouth of the Taf estuary a period of cliffing was ended progressively along the coast by the development of the spit-like feature of Laugharne Burrows from west to east. Thus the cliffs have been longest removed from wave action in the west and for a much shorter time in the east. Although Savigear's paper contains much of interest to the advanced student of slope formation, the point that concerns the present discussion is the way in which slopes flatten from east to west on the abandoned cliff with age. The profile at A on Fig. 4.24 is very much steeper than the profile at B; it is also much more dominated by convexity than the profile at B which is concave through most of its length.

It might be possible to synthesise a general cycle of slope development in the following terms. The earliest stages of development see an increase of slope steepness with time, the middle stages see a period of parallel retreat with active streams and constant basal attack, the later stages see a decline in steepness with the removal of the erosive agent from continuous contact with the foot of the slope.

The first phase is suggested by the work of Carter and Chorley (1961) on

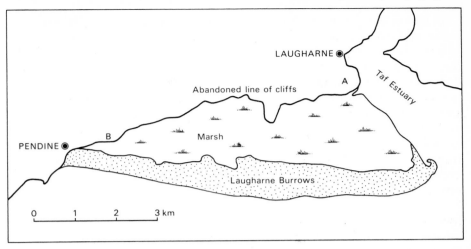

F<small>IG</small>. 4.24. Location of abandoned cliff behind Laugharne Burrows, South Wales

ephemeral stream development on a terrace in Connecticut. The stream
system was about 500 m long. They showed that there was a correlation
between stream order (see Chapter 6) and slope angle. If we regard the
development of streams of higher orders as time-dependent, which seems
reasonable, then the increase in slope angles associated with streams from
the first to the fourth orders seems to suggest increasing steepness with time
of valley slopes in the very early stages of development. There is a limit to
this process set by the angle of repose of the material and no steepening
seemed to occur between the development of fourth and fifth order streams;
there even seemed to be a decline between the fifth and sixth order streams
but it was not statistically significant.

Schumm's work on the silty sand waste tips of Perth Amboy suggests the
next stages. This area was mapped at a four-year interval, namely in 1948
and 1952, so that the time element could be considered in landform
development. There is evidence here that slopes remain constant while
relative relief (i.e. available relief) remains constant and no significant difference in slopes could be detected between the dates of the two surveys.
As this means in a broad sense that the area remained in the same stage of
development, it agrees with Strahler's law, and also with the results of the
work of both Strahler and Melton (1960). Later, when the active agent is
removed from the base, slopes may decline, as seems to be indicated by
the observations of Strahler, Savigear and Melton. With this Schumm
would also agree.

Of course, there are exceptions to such a generalised scheme as this.
One thinks immediately of the pediplain cycle and steep inselbergs, but it
is possible to accommodate this if the basal attacking agent is not a stream
but the line along which a process such as sheetwash is initiated.

It is true that different lithologies may seem to promote different processes and different forms of slope retreat at the same stage in the same climate. Schumm's work (1956b) on the Brule and Chadron formations in South Dakota points to such an example. In this case rainwash as the dominant process on the Brule seemed to promote parallel retreat, while creep seemed to cause a slope decline on the Chadron. However, this seems to have been because the actual rainfall lay between the threshold values for runoff on the two formations. In wetter or drier climates the actual values of rainfall could lie above or below the values required for runoff on both formations. In such cases slope evolution might well have been different. Thus rock type and climate, separately or combined, might upset the generalisations made above.

Approximate parallel retreat can be simulated in geomorphological laboratories, but there is always the question in such experiments of how far natural conditions are being reproduced. An experiment performed by Wurm in 1935 involved the subjection of a conical hill of unconsolidated material to a fine spray for a considerable period. The original angle of slope of the cone was 23 degrees and its slope steepened progressively throughout the experiment until, just before its final disappearance, its slope angle was 50–55 degrees. It must be remembered that an experiment such as this involves a mass of waste, not a rock form covered with waste; sprinkling with water droplets which are probably much coarser than scaled down rainfall; a continuous thorough saturation with water and a constant streaming off the whole surface; no vegetation; the operation of processes which probably approximate to very wet mudflows and the lack of slow and gentle processes such as soil creep and concentrated rill action. It is doubtful whether one can argue from such experiments to natural slopes.

Conclusion

If the reader is now confused and suspects that there are mutual contradictions in the various ideas that have been discussed above, he may rest assured that this is a reflection of the state of slope study. And yet we have only really considered very idealised cases. In fact one great simplification has been made very similar to that for which Davis has so often been criticised, namely the assumption that uplift is virtually complete before erosion starts. In this case it has been assumed at many points of the account that streams are fully incised before side-slope development begins, whereas slope development starts as soon as does valley incision, and the latter can be fast, slow, accelerated, retarded, intermittent or continuous.

The processes involved in slope formation are complex and interrelated, although they are probably understood in general terms. For instance, in discussing factors affecting the profile formed by soil creep, we mentioned some of the processes only in the simplest possible cases. Again, throughout this discussion homogeneous rock has been assumed to occur right to the top of the slope. Lithological variations within beds and between different beds must constitute a set of almost imponderable factors.

Because of the complexity of the problems, it is premature to apply a completely mathematical treatment at present, for, before a mathematical treatment can be successfully applied, the processes must be known in more detail. When they are more fully known, mathematics may be applied to work out the forms which would result. At that stage discrepancies between the natural profiles and the deduced profiles would indicate how accurate was the assumed knowledge of slopes. But, if slopes represent past as well as present processes, mathematics may be of little practical use. Yet another danger lies in the fact that because an assumed process could cause a profile approximating to natural slopes, it does not follow that the assumed process is the one operating in nature. There is no guarantee in geomorphological work that all similar forms are produced by one universal process, although it is probable that most of them are.

Finally a recommendation must be made that the serious student of slopes should read as much of the literature as possible to form his own opinions of the value of the various hypotheses.

5
The nature of a river valley

River development may be divided into two parts: the development of the features of an individual valley and the development of a system of drainage, composed of many individual valleys. The two are not wholly separable in nature, as they develop together, but, for convenience, the individual valley is treated in this chapter and the question of drainage development in the next.

The development of any valley involves such a complexity of factors that some of them must be assumed not to operate in order to simplify the discussion. Thus, the river valley in question is presumed not to be affected by movements of base level or changes of climate during its evolution. It is probable that the vast majority of valleys have been formed under the influence of different base levels and many of them must have experienced changes of climate, so that the discussion is to a considerable extent idealised, but the importance of the complicating factors will be treated in later chapters.

The treatment adopted is first to describe the features of rivers and their valleys and, afterwards, to attempt to explain these features in terms of their physical causes. There has been a considerable amount of research by hydraulic engineers, among others, concerning the nature of the flow of a stream and the ways in which its energy may be transformed. The application of this knowledge is of great assistance to the geomorphologist in understanding processes operating in a stream. But as yet it does not explain all the features of rivers, as these are due to factors other than the nature of the flow. Even without rejuvenation or changes in climate, such factors as lithological differences in the rocks and the nature of the vegetation of the adjacent banks and slopes have considerable effects on certain features of streams.

The initiation of streams

It is difficult to know what many initial surfaces were like, but one usually assumes them to have been irregularly hollowed so that streams developed leading into the hollows, which became ponds or lakes and then over-spilled into hollows lower down. The two segments of the stream so initiated would have had different origins: the first would require enough concentration of runoff for a rill to develop; the second is merely the overflow of a mass of water which is a stream from the start. It is with the first type that this section is concerned.

Horton's model of overland flow has been briefly mentioned in the last chapter in discussing convexity of upper slopes. It is an idea which probably applies best to semi-arid, poorly vegetated areas. Overland flow occurs when the precipitation is so great that not all of it can infiltrate into the soil. The infiltration capacity of the soil is governed by its permeability, which is a function of its depth, its distribution of particle sizes, any cracking that may be present in it, any prior saturation or wetting which still affects it. The distribution of particle sizes, including composite particles, is very much controlled by vegetation, which breaks up soil through its root systems, forms better crumb structure in soils by the addition of organic matter, thickens the soil mantle slowly, and increases water adsorption through the humus added to the soil. Vegetation also breaks raindrop impact. This is very important because open soil lashed by the rain has the fine particles stirred into suspension and redeposited as an impermeable surface skin which greatly reduces permeability—yet another geomorphological effect well known to gardeners. Thus it is very doubtful whether Horton overland flow occurs in many vegetated areas (Kirkby, 1969).

But it is likely that conditions of overland flow will occur on a bare initial land surface. The erosive power of this overland flow will depend on (1) the volume of the flow, which in turn depends on the nature of the rainfall, the infiltration and the length of overland flow, (2) the slope, (3) the roughness of the surface, and (4) the degree of turbulence involved. The resistance of the soil or waste mantle depends on grain size and cohesion. Owing to the irregularities of the surface there will be sudden surges of flow in particular areas and tendencies for flow to be concentrated along linear paths. Either can initiate erosion and so bring rills into existence, which, by concentrating flow, in turn accelerate erosion. As roughly parallel rills form a process of abstraction may develop, and for such rill abstraction the term micropiracy has been coined by Horton (1945). Horton imagined a set of parallel rills, one of which for some reason attained dominance and slowly abstracted the water from adjacent rills (Fig. 5.1A–D). By this process a slope originally followed by the rills is converted into a slope at right angles to the rills and directed towards the incipient stream: this is known as the process of cross-grading. So a stream may be born.

In areas of lower rainfall, greater permeability and more luxuriant vegetation throughflow may develop (Kirkby, 1969). Kirkby visualises some of the water recharging the main water table, a process which can operate only in those rocks in which such a feature is developed. Much of the remainder of the rainfall infiltrating into the soil is deflected laterally

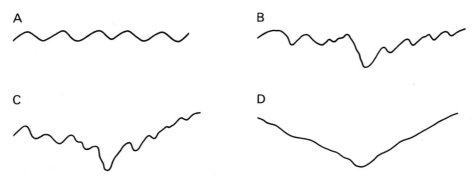

F<small>IG</small>. 5.1. Rill abstraction initiating a stream (*after Horton*)

downslope as throughflow, because of the decrease of permeability with depth. It is vastly slower than overland flow—of the order of 1 000 times as slow. There must be a limit to throughflow, and at this point overland flow starts, but the latter is far less common in vegetated areas and far less capable of initiating rills because its potential erosive powers are offset by the binding effects of vegetation. It is most likely to occur on the lower parts of slopes near the streams at their foot.

The dominance of throughflow helps us to understand a number of features. Water takes a long time to reach streams, with the result that throughflow acts as a strong regulating mechanism on peak flood discharges: overland flow occurring rapidly and almost simultaneously through small drainage basins will cause high peak discharges and high erosive potential in streams. The presence of throughflow means that the rill and gully network required for the drainage of humid temperate areas. is less dense than that required for the drainage of semi-arid areas in spite of their lower rainfall. The minutely dissected appearance of many arid hillside slopes has always posed a problem of explanation.

Again, if our original landscape has a coarse, permeable waste mantle, it is possible to imagine predominant throughflow emerging in the major depressions or valleys as a series of springs. These might, in conditions of strong flow, sap headwards, so initiating valleys in a different way and causing conditions in which rills may develop on the steep slopes leading into them.

Features of the long-profile of a river

Careful study of the long-profiles of rivers will reveal a variety of features in each profile and differences between the profiles of different streams. The general form of the profile is concave upwards, but the degree of concavity may vary considerably. The regularity of the curve may also vary between different rivers, some being concave only if the irregularities are ignored. Some may show regular concave sections separated by convex sections. Some may show steeper overall gradients than others.

If a motion picture of the stream in long-profile could be examined, it would be found that material is being eroded from some sections of the course, while in others it is being deposited. It is possible to infer that such processes are operating from the general nature of the bed: where the beds are very irregular and full of holes it is likely that erosion is taking place; where the beds are floored with alluvial material, whatever its calibre, deposition may well be occurring.

By a study of the beds of rivers, it may be observed that some rivers appear to be moving no solid material, others are transporting only very fine muddy material, while a few are rolling large boulders along their beds. At different times, when the discharge of rivers varies, the nature of the material transported, the load, may vary considerably. This is especially obvious in floods, when the amount of material transported far exceeds the amount transported in times of low water.

Transport, erosion and deposition

The energy possessed by any stream varies with the gradient of the stream surface and with the volume of the stream, but not all this energy is available for the erosion and transport of material from the bed and the banks. In fact the greater proportion of the energy of a stream is transformed in overcoming friction of two main types.

Part of the energy is transformed into heat in overcoming the friction between the river and the sides and bottom of the channel, or, in other words, friction between the stream and the wetted perimeter. The amount of energy lost in this way will depend partly on the roughness of the bed and banks, partly on the straightness of the stream, as bends increase the friction with the banks, and partly on the nature of the cross-sectional area of the channel. The efficiency of the cross-sectional form of the channel is often measured by a quantity known as the hydraulic radius, which is defined as the ratio between the cross-sectional area and the length of the wetted perimeter. The higher the ratio the more efficient the stream and the smaller the loss of energy due to external friction.

The ideal form of the channel for the efficient discharge of water is one which is semicircular in cross-section, but very few channels have this ideal shape. Although streams vary considerably, there is a tendency for the cross-section to be a rectangle. The most efficient form of rectangular cross-section is one in which the breadth is twice the depth. The least efficient is one in which the depth is very small and the breadth very great. Channels of these three types are illustrated in Fig. 5.2, and, as they are all of the same cross-sectional area, a comparison of their hydraulic radii will indicate their relative efficiencies. The semicircular cross-section has a hydraulic radius of 1·00 cm, the rectangle twice as wide as deep a hydraulic radius of 0·89

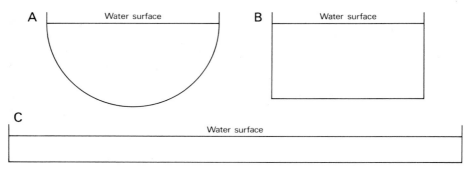

FIG. 5.2. Different forms of stream channels (for explanation see text)

cm, and the shallow rectangular section one of 0·54 cm. It is immediately apparent that the first form offers less resistance to the flow of water than the others, so that more energy is available for other purposes. If the depth of the shallow rectangular channel was reduced to one-sixth of the value shown in Fig. 5.2C, the hydraulic radius would be reduced to a little less than 0·10 cm. A thin film of water of enormous breadth must therefore be most inefficient, and the concept may help in the understanding of the inefficiency of unconcentrated wash, which has been mentioned in the previous chapter in connection with slopes, although the comparison should not perhaps be extended too far as the nature of flow may be different.

Not only does the efficiency of the channel change with the nature of its cross-section, but also with the size, even though the type of cross-section remains the same. This can be most clearly illustrated with a series of actual rectangular cross-sections, all twice as wide as they are deep (Fig. 5.3). The middle one, C, is the same as that shown in Fig. 5.2B. The hydraulic radii for the various rectangles are: A, 0·25 cm; B 0·50 cm; C, 0·89 cm; D,

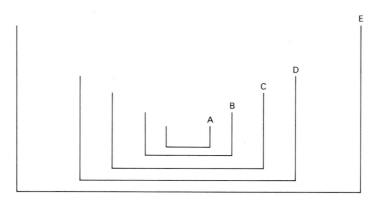

FIG. 5.3. Effect of size of channel on efficiency (for explanation see text)

1·25 cm; and E, 2·00 cm. The rapid increase in efficiency helps one to understand why large streams are better transporting agents than small streams, and why each individual stream is a better eroding and transporting agent in flood than at low water, a point which will be reconsidered later.

The second cause of transformation of energy by the stream is an internal one and is primarily due to shearing between turbulent currents. Every fluid has a velocity threshold at which flow changes from laminar to turbulent. Laminar flow may be visualised as the sliding of near-horizontal sheets over each other. Geomorphologically it is insignificant in streams being confined to the boundary layer at the bed and banks. It is not possible for this type of flow to support suspended load. Turbulent flow is much less regular and consists of a series of chaotic, secondary eddies superimposed on the main flow. The threshold at which the change from laminar to turbulent flow takes place varies with the viscosity: it increases with increasing viscosity, but, as temperature rises reduce viscosity, it is lowered by increasing temperature. Two stages of turbulent flow may be recognised. The first is streaming flow, which is the normal occurrence in streams: the second is shooting flow such as is found in rapids. It involves very high velocities and high erosive potential.

The losses through internal friction are hardly separable from those caused by external friction, as an increase in the roughness of the bed will cause a greater retardation of the bottom layers, a greater difference in velocity between the bottom layers and those above, and hence more powerful turbulence with an accompanying transformation of energy due to internal friction.

As the turbulence itself is responsible for setting particles into motion, the energy required to overcome friction cannot strictly be separated from energy used in transporting material, but in a rather artificial way it can be said that the energy not transformed in overcoming friction is available for the transport of material. The amount is very small, as it has been estimated that 96–97·5 per cent is used in overcoming friction (Mackin, 1948). Thus, even small differences in the amount of energy transformed by friction greatly affect the transporting power of streams.

The load in a stream can be carried in three main ways:

(*a*) as dissolved load;
(*b*) as suspended load;
(*c*) as bed load. This last is also called the traction load and may be moved by rolling, sliding and saltation (= the movement of particles in a series of jumps).

The percentage contributions to total load made by these various classes vary widely with the nature of the stream, the climate, the lithology of the rocks and the season at which observations are taken. Some figures quoted by Morisawa show that in North America the suspended loads and dissolved loads are roughly comparable on the average, but that, omitting one or two exceptional cases such as the Colorado and the Laurentian drainage, there is a general range from about 75 per cent dissolved and 25 per cent suspended in the North Atlantic drainage to about 25 per cent dissolved and 75 per cent suspended in the Mississippi. These figures apply to a continent considerably affected by human interference. In a natural state it is likely that the dissolved load would be relatively higher. The high quantity of dissolved load is not surprising if considered in conjunction with the assumed importance of throughflow mentioned above: the slow movement of water through the soil, especially in vegetated regions, virtually assures its saturation with dissolved salts.

Bed load is very difficult to measure and must vary widely with different streams and different times of observation. On the whole suspended load probably accounts for well over three-quarters of the total of bed load plus suspended load in most streams.

The entrainment of particles is brought about by drag effects caused by local differences in velocity or by turbulence. If there is a difference in the water velocity affecting a particle either horizontally or vertically, there is a difference in the force exerted and a drag effect exists. Alternatively, upward components in turbulent flow can literally lift a particle off the bottom.

The relations between particle size and velocity which govern the picking up and deposition of load are shown in Fig. 5.4. The velocity for the

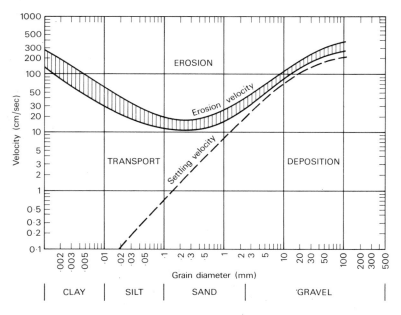

FIG. 5.4. Erosion and deposition in relation to stream velocity (*after Hjulström*)

erosion of loose particles is shown as a band because much depends on the shape, density and attitude of the particle and on the flow characteristics of the stream. The curve of settling velocity should also really be a band because it is very much affected by grain characteristics. It will be noted that the lowest erosion velocity is that required to move fine sand. It is not surprising that higher velocities are required to move larger particles, but it may seem strange that the erosion velocity also rises sharply with particles finer than sand. This is because such particles tend to form smooth beds and are usually held together by strong cohesive forces. The other important information to be derived from the graph is the great difference between the erosion and deposition velocities of fine particles, whereas the two are much more equal in the case of gravel. This effect is responsible for the suspended load.

Gilbert showed that the relationship of velocity and transport was complex. For example, there is a distribution of large particles which ensures maximum turbulence and hence maximum erosion. If the large particles are too close, the turbulent eddies interfere with each other and lose effectiveness: if too far apart they do not produce a maximum of turbulence. Similarly, in mixtures of finer and coarser particles, a bed of fine particles can make the movement of coarse particles easier, provided there is not so much fine material that the coarse becomes embedded in it. Again, in wide shallow channels there may be a lot of turbulence created by the strong velocity gradient near the bed and this will increase erosive power.

The effect of the different types of load in absorbing the energy of the stream varies. Dissolved load hardly affects the issue and suspended load, once set in motion, requires little energy for its continued movement. In fact the increase in viscosity it produces may decrease turbulence and so decrease the amount of energy converted to heat in overcoming internal friction.

Two qualities of a stream must be carefully distinguished in this connection: its competence and its capacity. Competence is measured by the weight of the largest fragment which the stream can transport and varies with the velocity in any given stream. It is usually said to vary with the sixth power of the bed velocity. The capacity of the stream is measured by the total weight of material transported and varies with the calibre of the load. According to Mackin, it varies approximately with the third power of the velocity if a fair proportion of all grain sizes is available, with a higher power if the material is all fine-grained and with a lower power if the material is coarse. The principle probably does not apply to the finest material, as, when a stream exceeds a certain specific gravity through the introduction of very fine material, it is subject to totally different laws of flow (Birot, 1952). A stream may be called fully loaded with respect to a given calibre of load, when it is carrying all the material of that calibre that it is able to move. Obviously, the term 'fully loaded' will have different values depending on the grain size of the material. The stream trans-

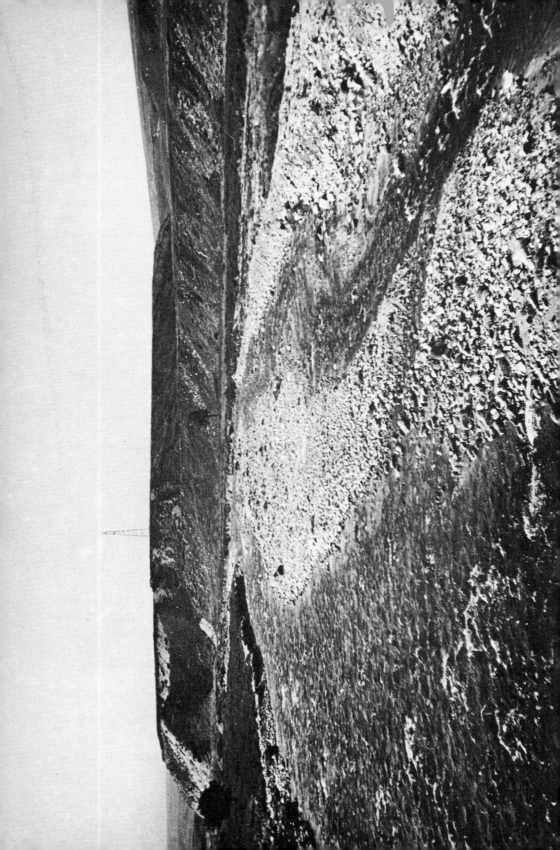

porting the very finest material, because of the change in the nature of its flow mentioned above, cannot have the term applied to it, as there seems to be no real limit to its load. It changes from a stream to a mudflow and finally to a pasty mass of mud.

These basic physical principles assist in an understanding of conditions observable in actual streams. The variations in the nature of the load transported in different streams must obviously depend not only on the energy of the streams but also on the nature of the material available for transport. Thus, two streams with identical gradients and identical discharges may not be transporting the same load. One stream may be flowing through an area of weak sandstones, which supply plenty of fine debris down the valley slopes and from the bed itself: this stream may be transporting a considerable load. The second stream may be flowing in an area in which the debris is composed entirely of coarse boulders: it may be moving no visible load at all (Plate 5). The first stream is competent to move the debris available: the second, even though identical in other respects, is not competent to move the boulders. If a stream is not competent to move load of the calibre available, the velocity increases until either it moves the material present or the excess energy becomes dissipated through increased friction. The effect of competence can be readily observed in many mountain streams of such areas as Wales. At low water they flow over beds of large stones and boulders which they cannot move, but any fine material introduced by such factors as cattle trampling the banks is immediately transported downstream.

Erosion and deposition introduce somewhat different problems. Erosion takes place where the stream has an excess of energy, but excess of energy does not always result in erosion, just as it does not always result in transport. If the country rock is unconsolidated or weakly cemented, erosion may not differ greatly from transport, as individual particles may be lifted up by the turbulence near the bed of a stream. If the rock is susceptible to chemical erosion, the process goes on whatever the velocity of the stream. But hard resistant rocks represent a difficult problem. If the stream is carrying a large bed load of coarse gravel, it is easily appreciated that such a load, by impact with the rock, may loosen fragments from it, but many streams carrying mainly load in suspension and solution do not appear capable of eroding the underlying rock except by solution. Even with a powerful stream transporting a large bed load, a lot of preliminary work by other factors is probably necessary before the stream succeeds in eroding rock fragments. Most rocks have planes of weakness, joints, bedding, cleavage or schistosity, and the action of chemical weathering may well effect a general loosening of the fragments before the stream is able to dislodge them through impacts between the bed load and the rock. If they are

PLATE 5. Terraced valley train, Strath Rory, Ross and Cromarty

sufficiently loosened, a stream of clear water may be able to dislodge fragments through the action of turbulence.

It is also thought that two processes involving water alone might be able to erode rock without any preliminary weakening being caused by other mechanisms. The first of these, which is alleged to be capable of forming potholes, is called evorsion and is a direct effect of the force of the water on the bedrock. Not all potholes are formed wholly in this way because the well-known grinding action of the stones contained in them must also be taken into account. The second is cavitation (Scheidegger, 1961). If the forces acting in the stream reduce the local pressure in the water below the vapour pressure of water, bubbles may form. When pressure increases these implode and the resulting shock waves may hammer the bed and banks. This process, which is known to affect machinery such as turbine blades, may operate in areas of very high velocities in streams, for example waterfalls and rapids.

When, because of a decrease in energy usually caused by a decrease in gradient, the stream is no longer competent to transport its debris, it must start to deposit it. Factors other than gradient also affect deposition, as any diminution in volume, whether due to percolation through the banks or due to a seasonal decrease in the discharge, may have the same effect. The first debris to be deposited will be that of largest calibre, succeeded at lower points in the stream by progressively finer material, while the very finest material may still be transported even though the energy of the stream is greatly reduced. It can be seen from this that coarse material is not only transported with more difficulty, but is transported for a shorter distance than finer material. It therefore takes longer for such material to reach the mouth of the stream. In practice, the coarse debris is made finer by being ground together in the bed load of a stream and is usually not transported to the mouth until it has been comminuted. The Shoshone river in Wyoming (Mackin, 1948) affords a good example of the process: it rolls andesite boulders, 20–30 cm (8–12 in) in diameter, along its bed and in a few tens of kilometres they are reduced to half their size by the constant grinding action. Thus, material which would otherwise have been quickly deposited, becomes capable of being transported in the lower gentler reaches of a stream by being comminuted.

In any inspection of a natural river the nature of the bed and the land adjacent to the river may indicate erosion or deposition, while the stream itself may appear to be incapable of performing either. But if the same stream is looked at in flood it will present a very different picture: the large and apparently immovable boulders are being bounced along the bed and in the lower reaches material is being deposited.

In floods the enormous increase in discharge over the same gradient results in greatly increased velocities and hence in greatly increased competence. The ratio between low water discharge and flood discharge may be very great. The Seine, for example, has a maximum flood discharge

34 times greater than low water discharge at Paris, while the Loire at Briare, some distance upstream from Orléans, has a maximum flood discharge 261 times greater than the low water discharge (examples quoted in Baulig, 1950). To comprehend the amount of transport which may be effected by a stream in flood, it is really necessary to know something of the climate and hydrology of the upper part of its basin. Such factors as the amount of rainfall, the relation between rainfall and runoff, which depends to a considerable extent on the permeability of rocks and soil and the type of vegetation, and the spring melting of winter snows are all significant in the study of the discharge which may be achieved in floods.

The tremendous increase in the erosive power of mountain streams in maximum flood can be illustrated well by the catastrophic floods on Exmoor in August 1952, which have already been mentioned in connection with the landslips provoked by the accompanying rain. On the southern side of Exmoor the upper Exe rose 4–6 m (12–20 ft) in a short period and caused enormous damage, but the most startling figures are those for the East and West Lyn rivers, which drain a small area of northern Exmoor via a common outlet at Lynmouth. The total area drained by the two streams is a mere 100 sq km (38 sq miles), but the calculated peak discharge of the two small streams was fourteen times as great as the average discharge of the Thames and almost as great as the largest discharge ever recorded on the Thames (Kidson, 1953). The enormous discharge was due not only to the heavy rain, but also to the fact that prolonged rain before had saturated the soil and made it incapable of absorbing further moisture. It is estimated that two-thirds of the rain falling in the one night ran off the surface. This enormous but short-lived flood performed tremendous erosion in a short space of time: 40 000 tons of boulders were moved into the Lynmouth area in one night and a deltaic deposit in the town was estimated to contain 150 000 cu m (200 000 cu yds) of boulders, debris and soil. Not only was the total amount of material moved enormous, but individual boulders of more than 10 tons weight were transported by stream. The size of these boulders clearly indicates the great increase in the competence of the stream during flood.

In a flood such as that of the two Lyn rivers, the transport of debris is not achieved by a straightforward flow of water. The debris torn from the banks, including boulders and trees, appears to form temporary dams as it accumulates in narrow and tortuous sections of the river. Water is ponded back behind such dams until there is sufficient head to cause a breach. The same breakdown of dams may occur in reservoirs, if the flooding rivers are in areas of settled country. With the break of each dam an enormous surge of water occurs and, in such surges, the largest boulders are set in motion down the stream.

The effects of large increases in discharge were also to be seen in the flood of the river Whangaehu, which caused the terrible railway disaster of 24 December 1953 in North Island, New Zealand. The cause was not simply a very heavy rainfall, but the bursting of a lake in the volcanic crater of Mount Ruapehu. A cascade of water and debris tore a channel 45 m (150 ft) deep down the mountain side, and the waters of the Whangaehu rose 6 m (20 ft), unfortunately dislocating the railway bridge at the time a train was crossing. A broken pier of the bridge weighing 150 tons was moved 40 m and one of the railway coaches 8 km (5 miles) downstream.

Examples such as these illustrate the great importance of the occasional catastrophic happening in river erosion. They are as important in this connection as they are in the formation of slopes.

The flood, the main effect of which may be erosion upstream, may be largely occupied with deposition downstream. There are probably two main reasons for this. The first is to be found in the normal decrease in gradient downstream, which results in a decrease of energy and hence loss of competence. The second is the result of the form of the valley. Upstream, the channel of a river may well be bounded by comparatively steep hillside slopes, so that even a large flood discharge can be accommodated in a channel of reasonably efficient characteristics (Fig. 5.5A). In the lower reaches of a river, however, a floodplain usually occurs, in the middle of which is the actual river channel (Fig. 5.5B). The channels are usually not

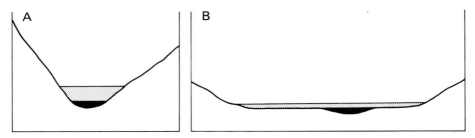

F IG. 5.5. The effect of floods on channel form: A, upstream; B, on the floodplain (flood water is shown dotted)

large enough to accommodate flood discharges and water spreads on to the adjacent floodplain. As a wide sheet of shallow water, usually with a low velocity, it has a most inefficient form of cross-section and consequently there is a considerable increase in the amount of energy transformed in overcoming friction. The result of the decreased gradient and the inefficient channel form downstream is a reduction in both competence and capacity, which leads to deposition. The coarsest material is deposited on the bed of the channel due to the decrease in velocity and volume: sandy material is deposited at the sides of the channel, where the velocity is much reduced, and goes to form the levées or natural embankments which are characteristic of streams in floodplains; on the floodplain itself the water may be so slack that even fine material is deposited; spreads of coarse material may also be left on the floodplain where the stream in flood actually breaches its banks rather than merely overflows them.

The twofold effect of erosion upstream and deposition downstream is very well illustrated by the type of change in streams caused by soil erosion. Poor agricultural techniques result in a greater proportion of runoff and more extreme variations in river discharge, which cause gully erosion upstream and the spreading out of the eroded material in the lower reaches. Some very good examples from the Mississippi basin have been described by Happ, Rittenhouse and Dobson (1940).

Although it is undoubtedly true that catastrophes such as floods perform a very large amount of geomorphological work, magnitude has to be set against frequency in any assessment of which flood is most important in general denudation (Leopold, Wolman and Miller, 1964). This involves virtually multiplying magnitude by frequency to discover at which level most work is done. For example, if the maximum flood occurring every fifteen years performs ten times the amount of work of the maximum flood occurring every year, then it is obvious that annual floods in the long run perform more work than the fifteen-year floods. Whether this is true depends to a great extent on the variability of the stream under consideration. The less variable the discharge the greater the amount done under near-normal conditions. In practice, because of the bare surfaces and high peak rainfalls of semi-arid regions, the exceptional will play a larger part under such conditions than it will in humid temperate conditions, where less variable rainfall and a complete vegetation and soil cover will damp down the variability. Such conclusions relate mainly to the suspended load in streams. With dissolved load the position is somewhat different, for dissolved load depends on the availability of soluble material and the length of time it is in contact with the runoff, so that the dissolved load may increase even less than the suspended load with the magnitude of the flow. The dissolved load depends much more on the thorough percolation of rocks and waste mantles by the rainfall, and is hence favoured by normal rather than extreme conditions.

Thus there is a general tendency to regard the annual or biennial peak discharge as a geomorphological factor of the greatest importance. This is reinforced by the fact that bankfull discharge (i.e. the state when a stream is full to the brim) occurs every year or every two years, and it seems that this is the state most likely to be a major control on channel form for water that spills over on to the floodplain is virtually wasted. Yet, we must not be completely obsessed with the total amount of work. The very occasional catastrophe may initiate processes which are beyond the powers of the more frequent but lesser peak conditions. This is obvious in cases like the Lynmouth flood, where mass slope movements are initiated, and also in the effects of storm surges on coastal landforms (see Chapter 8).

Grade, equilibrium or the steady state

Engineers concerned with channels and geomorphologists concerned with streams have from time to time discussed a balanced state in these systems.

In earlier discussions the factors generally considered have been gradient, load and discharge: in later discussions more complex interactions between many factors have been considered. The general questions are whether there is some sort of ideal form of the long profile of a river and whether a useful geomorphological concept can be framed concerning it (Dury, 1966).

In the introduction it was observed that some streams showed a smoothness of profile, usually a concavity of varying degree, while others still possess irregularities in their profiles. Such irregularities may be caused either by initial changes of slope in the depression followed by the stream or by the presence of outcrops of more resistant rocks across the stream (Plate 6). It has been seen that erosion will occur on the steeper sections and deposition on the gentler so that there will be a tendency towards the reduction of differences in gradient and the formation of a smooth profile. Tributaries transporting heavy loads over steep slopes may bring such an amount of coarse material into the streams that an increase in the gradient below the point of confluence may well be necessary for the evacuation of the debris, and may persist for a long time.

Irregularities in the initial surface will probably be smoothed out before those caused by hard rock outcrops, so that a stream draining an area of homogeneous rocks may attain a smooth profile before one draining an area of diversified rocks. In the latter instance, although the hard rock bands may hinder the attainment of a smooth profile over the whole course, such a profile may be attained by the reaches of the stream on the softer rocks (Fig. 5.6). The smooth profile sections will really represent a local

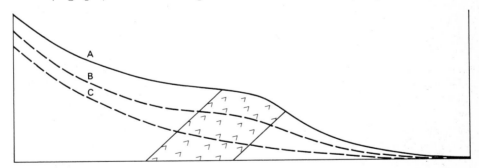

Fig. 5.6. Grading in reaches due to a resistant rock outcrop. A, B and C represent successive phases in the formation of the graded profile

balance between erosion and deposition: they will be so graded that the stream can transport its load over them without significant erosion or deposition. There are two necessary qualifications to this statement: the

Plate 6. The Orrin Falls, Ross and Cromarty. Developed on resistant Old Red Sandstone conglomerates

profile is probably related to flood conditions, as it is then that the major activity of streams takes place; the balance between erosion and deposition is only true for a short-term view, as over a long period slow changes take place for reasons which will be discussed below. The hard rock bands may defy erosion for very long periods of time. In the example shown (Fig. 5.6) there may be little coarse debris from the area of weaker rocks above, so that the stream will have no tools for any very active erosion. But very slowly the hard outcrop will be lowered and, as it is lowered, the section above will be continually regraded to suit the falling local base level formed by the hard band. Ultimately the hard band will be so lowered that the whole profile of the stream is smooth from source to mouth. Although at any given moment it is so adjusted as to transport the load of the stream, over a long period it will still change slowly. When a river reaches this condition it is said to be at grade and the profile is known as a graded profile. A comprehensive definition has been given by Mackin (1948, p. 471):

> A graded stream is one in which, over a period of years, slope is delicately adjusted to provide, with available discharge and with prevailing channel characteristics, just the velocity required for the transportation of the load supplied from the drainage basin. The graded stream is a system in equilibrium: its diagnostic characteristic is that any change in one of the controlling factors will cause a displacement of the equilibrium in a direction that will tend to absorb the effect of the change.

Mackin's concept of grade probably represents the ultimate development of the ideas that were largely formulated by Davis. It is a broad-scale view and may be criticised as a theoretical concept, because it does not contain criteria by which the state of grade can be easily recognised in the field. There are really three ideas implicit in the definition: the profile of the stream is organised to transport debris; there is no mention of erosion ceasing although it may well proceed very slowly; any change in any factor in any part of the stream transmits an effect through the whole profile. Before discussing alternative definitions, the factors which control the general form of the stream profile must be examined.

Fundamentally the lowest point of any stream is fixed by sea level, or, as it is usually called, base level. Similarly, in a stream with graded reaches, the lowest point is fixed by the local base levels formed by such things as hard rock outcrops. In the regrading of the upper reaches of the example described above, it was seen that the form of the upper part of the profile hinged on the slow degradation of the hard outcrop.

In any natural stream under the influence of base level there are a number of factors which, in the broad and ideal view, change systematically downstream and so tend to the formation of a concave long-profile. They are:

(*a*) An increase in discharge downstream, which is usually greater than the increase in load, except in certain rare cases such as a stream flowing from mountains with precipitation to deserts virtually without precipitation. Increasing discharge means that a given load, and in most cases a larger load, can be moved with the same velocity over a lower angle of slope.

(*b*) A decrease generally in load calibre downstream, so that a given mass of load can be transported over gentler slopes.

(*c*) The increasing discharge implies an increasing cross-sectional area. We have already seen that, provided this does not change shape, an increase in size means an increase in efficiency.

(*d*) As the depth is greater and the load finer, the bed will become smoother downstream so that energy transformation due to internal friction caused by turbulence will decrease. Hence flow over gentler slopes is possible.

In an ideal example, absolutely progressive increases in discharge and comminution of load are possible, but they are rarely if ever achieved in nature. Tributary streams ensure that the discharge increases intermittently, and the nature of the load may also change rather suddenly from the same cause. Departures from the ideal concavity (Fig. 5.7A) may thus be formed. The examples of the junctions of the Saône and the Isère with the Rhône (Baulig, 1950) show two of these effects. Below the Saône junction the gradient of the Rhône is less steep than either the Saône or the Rhône above the junction, as, presumably, the increase in efficiency allows the combined load to be transported over a gentler slope. Below the Isère junction the Rhône gradient is less steep than that of the Isère but steeper than that of the Rhône above the junction (Fig. 5.7B), as the Isère brings in a heavy load of debris from the Alps for the removal of which an increase in gradient of the Rhône is required.

It is rare to find a natural stream in which the calibre of the debris and the discharge remain approximately constant from source to mouth, but examples of rather a special nature have been reported from Iceland by Lewis (1936). They are meltwater streams flowing from glaciers over outwash plains of their own making. Except in the immediate vicinity of the snouts of the glaciers, the streams have straight profiles with a gradient of over 8 m per km, a lack of tributaries, and appear to be at grade. They are somewhat different from normal streams, as they have no valleys and more than a normal tendency to braiding. Braiding, the splitting of a simple stream into distributaries which unite and divide continually, results in a very inefficient form of river. When a stream divides the effect is the opposite from what happens when a tributary joins the main stream: the two parts are together less efficient than the original stream, so that where division takes place there may be a tendency for a steepening of gradient in the distributaries.

Apart from the general tendency towards concavity, it was observed that

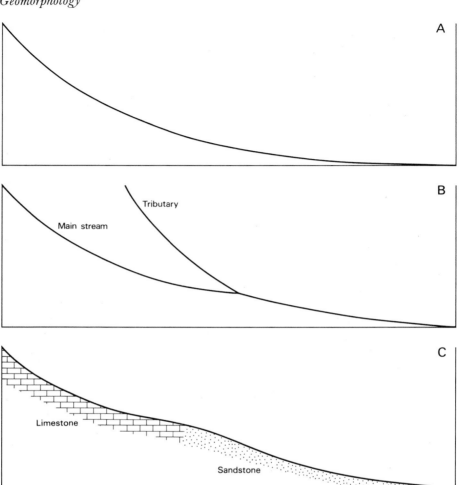

FIG. 5.7. Types of graded profiles (for explanation see text)

some streams preserve a steeper overall gradient than others. This is largely a reflection of the nature of the load delivered to them and transported by them. The coarser and more abundant the load, the steeper will be the gradient at which grade is reached. The Shoshone river of Wyoming, already noted as a river with a load of very coarse material, appears to be graded, although it has a gradient of about 5·5 m per km (30 ft per mile). The calibre of the load will depend on the nature of the rocks and the amount of weathering and erosion suffered by the fragments before they reach the stream. The amount of the load will depend to a considerable extent, at least after the river is graded and engaged in transporting material delivered to it down the valley slopes more than in downcutting, on a complexity of factors. For any given rock the steepness of the side slopes will govern the rate of delivery of waste to the stream, but the same slope on different rocks will not ensure an equal supply of waste, as the resistance of the rock to weathering and the size of the weathered fragments may be different. Probably of even greater importance will be the nature of the vegetation of the valley slopes, as this will affect the rate of movement of material downhill. Some interesting figures for the rate of erosion from various types of land in a two-year period have been given by Happ, Rittenhouse and Dobson (1940): the amount eroded per sq. km for land covered with oak forest was about 9 tons, with Bermuda grass pasture about 35 tons, from land cultivated across the slope with cotton about 12 500 tons, from land cultivated down the slope with cotton about 35 000 tons, and from barren abandoned land about 29 000 tons. Although these figures represent the operation of man as an agent of erosion, marked but probably smaller differences must occur with different natural vegetation, so that the rate of supply of debris to streams and hence their gradients may vary considerably.

Lithological variations of major character in a drainage basin may also have an effect on the nature of the graded profile. Even after the major irregularities caused by differences in rock resistance have been smoothed out, differences in the rate of supply and calibre of the debris from different rocks may affect the nature of the graded profile. The effect may be very slight as the load is carried continuously down the stream, so that additions of a new type of debris will cause slow modifications rather than abrupt changes. One may imagine a stream flowing from a clay tract to a limestone tract: in the limestone area there may be virtually no addition of load except in solution. If there is a progressive increase in discharge, the gradient may well become gentler than it would if debris were being continually added to it. On the other hand, a stream flowing in its upper reaches on limestone and in its lower reaches on sandstone should, in the graded state, possess an increase in gradient at the junction as the nature of the load added will change from load in solution to a considerable load of sand particles (Fig. 5.7c). Under natural conditions other factors related to the lithology would have to be considered: at the limestone–sandstone junction there may be such an addition to the discharge due to spring flow that the change in gradient required for the transport of the sandstone debris may be very small.

The magnitude of the complicating factors discussed should not be exaggerated, as, in spite of all, the average river has a concave long profile. According to Mackin, however, the possible variations are sufficiently large and non-systematic to make it impossible to fit a simple mathematical curve precisely to the long profile of the river. The subject will be discussed in a later chapter in connection with rejuvenation, as it is in the restoration of former long profiles that mathematical curves have been mainly used.

There has been a general tendency in recent years to abandon the term 'grade', if not the idea lying behind the concept. Partly this is a semantic and not a geomorphological question. Some have suggested the replacement of 'grade' by 'equilibrium' or, because equilibrium may not be everywhere complete, by 'quasi-equilibrium', or, because equilibrium is changing, by 'dynamic equilibrium'. Some of these suggestions seem to derive from the work of Kesseli (1941), who pointed out both time and space difficulties confronting the equilibrium concept.

If, as has been suggested above, channel characteristics are related to bankfull discharge, there cannot be equilibrium except on the one or two days a year when such discharge is achieved. Secondly, as flood conditions progress downstream, it is not strictly true to say that a whole profile can be in equilibrium when the conditions moulding it are not contemporaneous. However, as Mackin maintains, this is like looking at the 'weather' of a river, whereas the term 'grade' refers to its 'climate', its mean condition. There cannot be equilibrium on a very short term view and, as most people since Davis have pointed out, it changes in a long-term view: hence it must be a term adapted to some intermediate viewpoint.

More pertinent is the question of how one recognises a graded stream and, if one cannot do so easily, of whether the concept has any use (Dury, 1966). Various ideas have been put forward, such as smoothness of the concavity of the long-profile, but they are difficult to apply on an objective basis. Take the simple example postulated earlier in this chapter of a small stream flowing on to a bed of large boulders with a steep gradient, a relic perhaps of the Ice Age. If that stream accelerates until its energy is transformed into heat through turbulence, even though it is unable to remove its bed load, it still reaches equilibrium by means of the increases in velocity and turbulence. Hence, it is at grade in a general sense, even though it may have a pronounced convexity in the long profile. One could extend this argument to suggest that any stream is at grade, even one with a series of rejuvenation heads. In fact, the general tendency today is to regard the time out of equilibrium as very short, unlike Davis who regarded it as extending through the youth of the cycle. Perhaps this change has come about because Davis considered that the change to equilibrium must affect the passive resistant member of the system, i.e. the bed, whereas it is much more likely that it will affect principally the fluid, mobile member, i.e. the stream, in one way or another and, hence, equilibrium will be soon established.

Strahler and others have suggested replacing the concept of grade or equilibrium by the concept of a steady state in an open system. Such systems regulate themselves, so that over a period of time the discharge and load entering the system, i.e. those provided by rainfall and waste sheet movements, are balanced by discharge and load leaving the system, i.e. the stream and its entrained debris. Within the system the form of the channel in all its aspects, roughness, cross-section, gradient and pattern, changes to absorb any variation that may occur, such as the amount and calibre of the sediment locally yielded to the stream. In fact, all the causes of irregularity noted above, such as rock variations, tributaries bringing in sudden and heavy accessions of load, the constant profile outwash streams, the increase of turbulence to restore equilibrium when there is no competence to move the load, are contained within the steady state idea of grade. Even this concept, however, is at best medium-term, because in the long run denudation and lowering occur and are irreversible at least by exogenetic processes.

These ideas all really add up to the following points of view. Grade as an idea involving the slow modification of the long profile with a long lag in the ungraded state is not acceptable as a concept. Grade is achieved almost instantaneously. As a corollary, it is pointless trying to recognise grade, because it is universal. One might almost say that this makes it useless as an idea. River profiles are broadly concave, although all sorts of irregularities may exist and be maintained in them.

All this is a far cry from the early ideas of a balance between erosion and deposition in a stream which were propounded in the eighteenth and nineteenth centuries by continental engineers, notably Surell (for a summary of these earlier ideas see Baulig, 1950). At any moment the balance is almost perfect, so that it is not surprising that the state of balance or grade was held by some writers, notably Philippson, to denote the end of erosion by the stream.

Such a concept seems to have been derived from a short-term view of the activity of a river, as even a period of several hundred years might see little or no erosion. But, when one considers the thousands and millions of years through which geomorphological processes operate, the concept is probably not true. It has been seen how the gradient of a graded river is to a considerable extent determined by the nature of the waste delivered to it down the hillside slopes, and the variations possible with differences in rocks and vegetation have been discussed. Variations in the rate of delivery of waste are also caused by the general process of the flattening of slopes as the landscape becomes older. A river may become graded to transport the material available at an early stage in the degradation of a land mass and have little energy left for the purposes of downcutting. But later, as the slopes become less steep, the debris on them will move more slowly and, consequently, be delivered to the stream as finer material. At the same time the slower movement will result in less frequent exposures of rock for weathering with the result that the overall supply will be decreased. The smaller amounts of debris and its finer state will result, as the course of erosion proceeds, in the rivers having a surplus of energy, which can be used for lowering their beds. As the changes in the waste supply are slow, the rivers are able to maintain grade during the whole process. Corrasion of the bed by chemical weathering can also go on after grade is reached. It appears best, then, to regard grade as a state changing slowly with the course of erosion.

The whole discussion on grade illustrates the difficulties arising from differences between the short-term, dynamic point of view and the long-term geological point of view, from looking at sections of streams and from looking at whole drainage systems. It also illustrates how labels, such as grade, equilibrium and steady state, although sounding erudite, confuse the issue by concealing the ground common to all the ideas.

Features of the cross-profile of a river

It is impossible to divorce the study of the cross-profile from the study of the long-profile, and, in considering the supply of waste, mention has already been made of the effect of hillside slopes, which really form the cross-profile of a valley. Similarly, floodplains might best be described as features of the cross-profile, although they have been mentioned in connection with flood effects.

Different rivers will show three main sorts of differences in cross-section: in the shape of the channel, the shape of the floor of the valley, and the shape of the valley as a whole. Channels may vary considerably in cross-section, though on the whole they appear to approach a shallow rectangle. Valley floors differ widely, as in some there is virtually no flat floor between the channel and the valley sides, while in others there may be varying widths of floodplain. Valley profiles will vary from narrow steep-sided trenches to gentle open forms and, in addition, the symmetry of the cross-section may not always be the same.

The changes in the cross-sections of the valley as a whole are really a function of the nature of the rocks and the age of the valley, if it is assumed that there has been no drastic change in the process operating on them. The problem is really one of slope formation, which has been discussed at some length in the previous chapter. The slopes will be steep if the stream is actively downcutting, if the rocks are very resistant to erosion, or if the stream is engaged in eroding laterally at the base of the slope. Slopes become more gentle in older valleys, in ones where the rocks are less resistant to erosion, or where, for some reason, the stream is not actively cutting the foot of the slope. Just as a river may become graded in sections so may a side slope preserve a complex profile due to the outcrops of less and more resistant beds. In time, but it may take a very long time if the rocks are excessively resistant, the valley side slopes will probably become smoothed out just as the long profiles of rivers are smoothed out.

Studies of valley asymmetry have already been mentioned in connection with the question of valley side slopes flattening with age. Melton (1960) attaches prime importance to what he terms the erosional environment, that is the presence or absence of a stream at the foot of the slope, as the immediate cause of valley asymmetry. Behind this lies the search for all the mechanisms that may cause streams to erode at the foot of the slope. The cases quoted by Melton from Arizona, where steep-gradient streams

(6–23 degree channel gradients) cutting down rapidly and removing talus from the foot of the slopes inhibit any asymmetric development in spite of vegetation differences, seem to support the idea that differences in basal undercutting are the prime causes of asymmetry.

Asymmetry in the cross-profiles of valleys is often due to structural causes, being very common in areas of gently dipping beds. Such profiles denote a tendency for the stream to migrate down the dip, as it is easier to erode sideways and downwards in the softer beds than to erode vertically downwards into the harder beds (Fig. 5.8). Sapping by springs at the foot

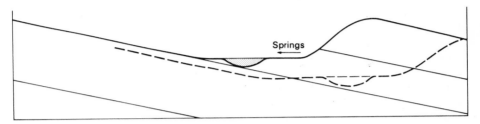

Fig. 5.8. Uniclinal shifting. The pecked line indicates a later position of the valley

of the escarpment down dip from the stream may also assist in the process.

There is a limit to this process because the stream cannot erode indefinitely downwards. Once its overall gradient becomes slack, it will probably start meandering back into the dip slope, in which process it may be assisted, especially where dips are low, by the push from the spring-fed streams deriving from the foot of the escarpment. There is an excellent example of these effects in the Chalk escarpment of East Anglia. In this region, especially in Cambridgeshire, the spring lines, the main one on the Totternhoe Stone and a lesser one on the Melbourn Rock, are a long way in front of the main Chalk escarpment and are not sapping its base at all. As a result it is very degraded compared with, for example, the Chiltern or South Downs escarpments.

Apart from obvious structural examples, such as those described above and valleys along strike faults in areas of uniclinal dip, there are many areas of asymmetric valleys in rocks which seem to be homogeneous and devoid of significant directional structural influences. Various general explanations of these have been put forward. One was based on the fact that the earth's rotation deflects moving bodies to the right in the northern hemisphere: this was held to be responsible for the undercut right banks of some streams, for example those of the Lannemezan plateau at the foot of the Pyrenees in south France. On the whole, the efficacy of this Coriolis force effect is not believed today.

Locally valleys may become asymmetric through secular downwarp assisting lateral shifting during the development history of the streams, for example in east Kent as suggested by Coleman (1952).

There are many examples left where no structural or lithological controls operate. Such cases have been investigated by Melton (1960) and Kennedy (1969), who summarise much of the earlier writing on the subject. This often pointed to supposed asymmetry differences between temperate and periglacial climatic zones.

In periglacial areas snow and ice may protect north-facing slopes and allow greater erosion on south-facing slopes. This difference is very similar to the explanation proposed by Taillefer (1950) for the Lannemezan valleys, where it was alleged that snow driven by westerly winds in the Pleistocene accumulated on the east-facing lee slopes leaving the unprotected west-facing slopes to be steepened by strong periglacial erosion, principally solifluxion. Again, the stream may find it easier to undercut the south-facing slope in a periglacial region, because the latter is frozen, and hence immune from erosion, for a shorter period of the year than the north-facing slope.

On the other hand, in temperate and especially in semi-arid areas the better conservation of soil moisture on the north-facing slopes would lead to a more complete vegetation cover there (this was established quantitatively by Melton) and hence to less runoff than on the more open south-facing slopes. This could lead to a greater flattening of the latter slopes. Material from the slope would accumulate as talus at its foot, thus pushing the stream against the north-facing slope so accentuating its steepness.

However, the climatic division into two types of asymmetry has been shown by the observations of a number of workers to break down in practice (Kennedy, 1969). The asymmetry characterised by steeper north- and east-facing slopes is the more common of the two in all climatic zones. Kennedy accepts three basic ideas: Melton's idea that the position of the stream is vital; Schumm's (1956) contention that creep tends to flatten slopes and that rainwash, sheetwash and gullying tend to steepen them or keep them constant; Kirkby's idea (1963) that the number of moisture changes and freeze-thaw cycles tends to accentuate creep on the drier of two well-vegetated slopes. In the northern hemisphere this will be the south-facing slope. These tendencies may, of course, clash and in such a situation one can dominate the others.

The problem has become complex. In a cold climate, provided that creep remains dominant, the north- and east-facing slopes will be steeper (Kirkby); if the slope is bared by thawing and so exposed to periglacial wash and gullying, the asymmetry may be reversed (Schumm). If slopes are undercut by a stream in humid temperate climate (Melton) the relaxation time before they are again flattened by creep will be less on the south-facing than on the north-facing slopes (Kirkby); hence the usual north-facing steep slope form of asymmetry of these regions. In arid areas steepening of the south-facing slope by basal undercutting may initiate rill and gully erosion which will tend to maintain the steepness, whereas creep may continue to operate on the opposite slope (Schumm). On the other hand, if such a process difference is not initiated, greater runoff on the south-

facing slope may flatten it more than the lesser runoff on the better vegetated, north-facing slope (Melton). Thus, in arid and semi-arid areas
asymmetry may be more variable.

Add to this the suggestion that the deeper the valley the greater the
asymmetry, because of the greater potential energy possessed by slope-
forming processes, and we are back in a fluid state of probable truth rather
than in an ossified condition of unsupported generalisation.

Cross-sections of the channels of streams show a marked absence of the
ideal semicircular form and, instead, a preponderance of a rectangular
form. Why this should be so is not altogether clear, but the erodibility of
banks may have a considerable effect. Any tendency to form a semicircular
section would cause a considerable increase in velocity and turbulence,
resulting in undercutting and collapse of the banks and thus leading to a
more rectangular form. The resistance to erosion will vary with the type of
material forming the banks: generally resistance decreases with decreasing
grain size up to a certain point, a grain size of 0·2–0·6 mm (Hjulström,
1935), while below that size a considerable increase in resistance takes
place. Virtually this means that sands are the least resistant rocks and
clays very much more resistant, owing to their coherence. Thus, in unvegetated land, the widest and flattest channels would be expected to occur
in sands, as these are easily undermined and collapse readily into the stream.
This expectation is largely borne out by observation. Vegetation acts as a
binding agent, especially if a good root system occurs, and tends to decrease
the undermining and collapse of banks. In their study of accelerated erosion,
Happ, Rittenhouse and Dobson (1940) quoted an example where the
clearance of vegetation from banks resulted in the increase in average
width of a stream from 8 to 13 m (27 to 43 ft). If bank undermining and
collapse are important causes of the rectangular cross-section of streams,
one would expect to find the deepest and narrowest stream sections in areas
of very resistant rocks and probably in clay lands.

There is a tendency in stream studies to say that something happens
because it has to happen, as we shall see in talking of meanders below. This
attitude may be involved in the shape of the channel. Gilbert showed that
the wide, shallow, rectangular section was the most efficient transmitter of
bed load if not of total load, because of the sheer area of bed and because of
the high turbulence generated by a rough bed and a shallow stream on a
steep gradient; hence, the statement that, if a heavy bed load has to be
carried, the stream will accommodate itself to the conditions by broadening and shallowing its channel. The broad, shallow channels of outwash
streams, in which most load is usually bed load, are conformable with this
statement.

Lateral erosion

Floodplains and meanders denote that the stream is tending to erode
laterally rather than directly downwards: in some examples it can prob-

ably be said that all the erosion is lateral and none vertical. The conditions under which lateral will exceed vertical erosion are far from clear.

It is sometimes stated that lateral erosion begins when a stream is no longer able to cut downwards. The idea appears to have been derived from the writing of Gilbert (1880), who stated that 'whenever the load reduces downward corrasion to little or nothing, lateral corrasion becomes relatively and actually of importance' (p. 120).

Stated thus, the argument appears to contain a flaw, because, if a stream has insufficient energy to cut downwards, it should also have insufficient energy to erode laterally. The defect in the argument has been elaborated by Kesseli (1941): if a stream completely fills its channel (Fig. 5.9A), a lateral shift from the position AB to CD involves the erosion of a wedge of material, BOD, and the deposition of a similar wedge, AOC. If the stream has no energy for erosion, it is difficult to see how it can shift laterally at all, as the amount of energy gained by depositing the wedge, AOC, will only provide enough energy to maintain the wedge, BOD, in motion, whereas considerably more energy is needed to set it in motion. In the ideal case, then, the problem is difficult, but it becomes still more difficult if the stream does not fill its channel (Fig. 5.9B). In this case, the energy acquired from the deposition of the wedge, AOC, will be insufficient to cause an equal displacement of the stream, which, not filling its channel, would have to set in motion and remove a wedge, XOY, which is considerably larger than the wedge, AOC.

A possible flaw in the argument is the adoption of a form of channel approximating to semicircular in cross-section. If a more natural form is assumed (Fig. 5.9c), a shift of one bank from BD to WY would involve a deposition in the area, ACXV, with the result that the stream would become narrower in proportion to its depth (ABDC to VWYX) and, therefore, more efficient. In practice, of course, the position would be a little more complicated, as the stream would deepen as well as narrow, but this would help the general effect.

In addition, in many floodplains the stream bed may consist of coarser or more coherent material than the banks, and the stream may be not competent to erode the bottom but competent to erode the banks.

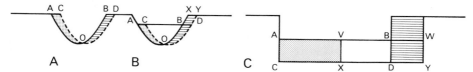

F<small>IG</small>. 5.9. Erosion and deposition involved in the lateral displacement of a river (for explanation see text) (A and B *after Kesseli*)

The meandering habit of streams

From the foregoing discussion of lateral erosion it is difficult to believe in the idea sometimes expressed, that meanders develop when a stream ceases downcutting, as they involve a great lengthening of the stream, a reduction of its gradient, and, hence, a decrease in its energy. It seems more likely that meanders develop early, that both lateral and vertical erosion take place together, that even the floodplain may be formed before the stream ceases downcutting and may itself suffer regrading to lower gradients.

Many difficulties confront those seeking to explain meanders, and by no means the least is the explanation of the origin of meanders. In the outline of the cycle of erosion (Chapter 2) it was stated that chance irregularities initiated meanders, but the explanation is not altogether satisfactory. If chance irregularities were the cause there should be streams with odd meanders here and there, whereas meanders are usually present as a system. There have also been explanations depending on the propagation downstream of an oscillatory tendency, initiated perhaps by a stream, for some reason, starting out of centre in its channel. If this were the true explanation, there should be streams with well-developed meanders upstream and poorly developed ones downstream, a regular progression existing between the two.

More is known of the factors affecting the nature of meanders, as experiments on a model of part of the Mississippi have been performed (Friedkin, 1945, quoted in Baulig, 1948). These experiments showed that increases in discharge resulted in a broadening of the meander belt, defined as the distance between lines tangential to the outer curves of the meanders, and an increase in the distance between the crests of individual meanders, or their wave length. An increase in slope tends to have the same effect, though a great increase reduces the meandering habit. An increase in bed load also tends to cause the stream to straighten. Thus, meandering is promoted by large discharges, gentle gradients and small bed loads.

The probable importance of discharge in affecting the width of the meander belt had been noted before in statistics relating the width of the stream to the width of the meander belt. Factors other than discharge affect the width of streams, but, generally speaking, the wider the stream the greater the discharge. The observations of Bates (1939) are probably the most comprehensive in this respect. For free meanders, which are those characteristic of floodplains, the ratio between the width of the meander belt and the width of the stream decreases as the width of the stream increases: streams 30 m (100 ft) wide have a meander belt sixteen times the width of the stream; those 300 m (1 000 ft) wide have meander belts twelve times the width of the stream; those 900 m (3 000 ft) wide have meander belts eleven times as wide as the stream on the average. For incised meanders, where the stream is bounded by valley side slopes throughout the ratios are respectively 41, 18 and 17·5. The greater widths of the meander belts with incised meanders probably reflect the greater amount of erosion necessary before cutoffs can take place. Even in the same stream the width of the meander belt sometimes increases when the river passes from non-resistant rocks on to resistant rocks (Blache, 1940), where cutoffs are only achieved after a greater amount of erosion.

The tendency towards a lateral and down-valley migration of meanders is attested by the number of slipoff slopes and undercut cliffs seen on natural meanders. The explanations suggested involve the nature of flow in a meandering stream. The thread of deepest water in a meandering stream does not follow the median line of the stream but is deflected towards the concave banks and away from the convex banks (Fig. 5.10A). The tendency

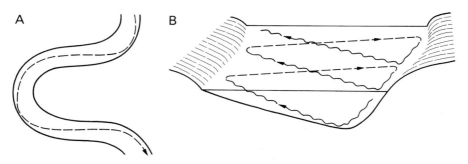

FIG. 5.10. Water movement in meanders
A. Direction of current of greatest velocity in a meander
B. Helicoidal flow in a meander

for a stream to continue to flow in a straight line results in a slight increase in the level of the surface on the outsides of the bends. The head of water so formed should result in a cross-channel movement being imparted to the stream: it appears to be imparted to the bottom layers, as these are flowing less rapidly than the upper layers and are, therefore, more easily displaced. The flow across the bottom from the concave bank to the convex one is supplemented by a return flow in the other direction at the surface (Fig. 5.10B). The motion is not absolutely at right angles to the current but must possess a downstream component varying with the velocity of the stream. The result is a helicoidal, or corkscrew, motion, which should tend to shift material from the concave bank to the convex bank, thus assisting in erosion on the outside and deposition on the inside of bends.

In small streams in the laboratory material can be observed moving across the river from one bank to another, but whether this is due to helicoidal motion is not certain. According to Matthes (1941), however, such flow takes place in distorted models and in channels which are very deep in relation to their width, but has not been observed in any part of the Mississippi system which he has studied. Neither was it observed in the experiments with a model of part of the Mississippi. Matthes believed that the material eroded from one side of a river on the outside of one bend was deposited on the same side of the river on the inside of the next bend. The difference in scale may be fundamental in accounting for the differences between the cross-channel movement of material in laboratory streams and the apparent lack of such movement in large natural rivers in the Mississippi system.

In the Mississippi experiments it was observed that the thread of maximum current tends to leave the concave banks during floods, so that the point of maximum erosion moves downstream (Fig. 5.11). This characteris-

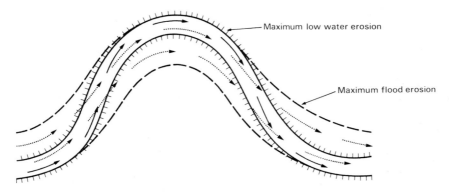

Fig. 5.11. Nature of a meander at low water (solid arrows indicate flow) and in flood (banks and arrows indicating flow are pecked)

tic is used by boatmen, who, apparently, sometimes shelter in the quiet water adjacent to the concave bank during floods. The concentration of flood erosion on the downstream side of the bends, and the greater power of such erosion through increased turbulence, makes it probable that much of the downstream migration of meanders is effected during floods.

Lateral erosion also appears to be achieved mainly during floods, though as a result of a different process. Although the highest velocities in flood are not found against the concave bank, the general rise in the level of the stream causes a rise in the adjacent water table and thus a thorough saturation of the banks. As the flood subsides, the banks, no longer held up by the pressure of water against them, collapse into the stream. At the same time the thread of maximum velocity moves back towards its low water position against the concave banks and possibly assists in the removal of the material dumped into the channel by the collapse of the banks.

Such processes as these could explain the lateral movement of a meandering stream and the downstream migration of the meanders themselves. Once again, the importance of floods appears to be very great. The floodplain resulting from such processes is essentially an erosional feature, although it may be covered with some thickness of alluvial deposits. It is different genetically from floodplains formed by a youthful stream aggrading parts of its course during the formation of a graded profile and from the purely depositional floodplain, which results from aggradation in the lower reaches of a valley caused by a rise in base level. Floodplains of the latter type will be discussed in a later chapter.

There has been a tendency in recent years in the study of meandering to abandon as impossible at present the search for the ultimate explanation of meanders. Instead the features and associations of meanders have been looked at. It has been observed that the jet stream meanders in the atmosphere, that the Gulf Stream meanders in the Atlantic, that meanders can form in streams carrying no debris over ice as well as in streams on waste and rocks. Thus, meandering is seen to be something necessarily associated with flow, but why this should be so is not known.

Even in a stream between straight banks the thread of maximum velocity may not be a straight line but pass from deeper pool to deeper pool. If these are on opposite sides of the channel then it is apparent that a sinuosity can develop in a straight channel. Such deeper pools are usually separated by shallower areas known as riffles. All this is analogous to the state of affairs in meanders, where the pools are on the outsides of bends and the riffles are between the bends. Pool and riffle development, like many channel features seems to be related to bankfull discharge. Furthermore, the spacing between pools is as a rule five to seven times the width of the stream both in straight and meandering streams. The riffles are areas of shallow water, steeper gradients and inefficient channel section: the pools are deeper, gentler and more efficient. Thus energy is dissipated more rapidly over riffles than it is in the pools.

Leopold, Wolman and Miller (1964) use these facts to suggest a partial reason why meandering is the natural state. They assume that a river is a steady state in an open system and that energy dissipation should be constant through the system: this constant dissipation rate is called the energy grade line. It has been stated above that energy dissipation rises through the riffles and falls in the pools. To smooth this out two courses are open to the stream. It can either iron out the pools and the riffles, or it can arrange to dissipate more energy in the pools. One way of doing the latter is by increasing frictional losses through increasing the curvature of the channel and hence the turbulence. Thus, provided a stream has pools and riffles to start with, it can approximate to the energy grade line by meandering, which must be regarded as the stable condition because any tendency to revert from it would make the rate of energy dissipation irregular again.

This is not a complete explanation because it does not answer the questions: (*a*) why the energy grade line is not achieved by the stream clearing away its pool and riffle system, and (*b*) why pools and riffles occur in the first place. In fact, it redefines the problem of meanders largely as the problem of pools and riffles.

At this point we should mention braiding, which is the state of dividing and reuniting, or anastomosing, channels (Plate 5). It is seen at its best on glacial outwash, pediments and in miniature form on sandy beaches at low tide. It is a form of flow associated with steeper gradients for given discharges than is meandering—Leopold, Wolman and Miller show that this gradient division is clearcut. It is also associated with very variable discharges and banks of readily erodible material, which contribute greatly to bed load. Under these conditions, which are conducive to broad shallow channels, the necessary increase in efficiency for moving the load at low water can only be achieved by dumping material in the bed and so making the channels narrower and deeper. Bank erodibility must be associated with this, because, in a stream with resistant banks, the channel bars would be swept out during high discharge, especially if they were more erodible than the banks. In the typical braided stream they are not. Once again, in the case of braiding the explanation is virtually that the stream has to adopt this state as a result of the conditions present in it.

A further aspect of meandering is the underfit nature of many modern meanders, a phenomenon which has been stressed in a series of papers by Dury culminating in 1965. Dury called attention to the fact that many stream meanders are misfits in a pattern of much larger valley meanders as shown in Fig. 5.12, where the stream staggers over the alluvial fill of a valley meander system cut in solid rock. Dury has interpreted these features as due to great alterations in discharge consequent on climatic change. In

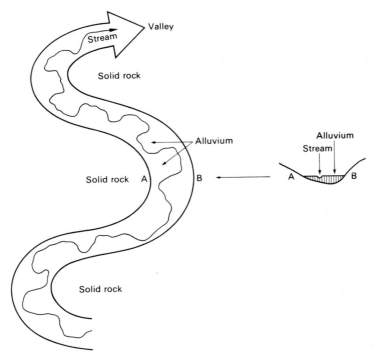

Fig. 5.12. Valley meanders and stream meanders

his later estimates he postulated discharges twenty to sixty times as great as at present to explain varying degrees of underfitness, and to account for the dimensions of the valley meanders. Various climatic elements may have contributed, but the most likely is the decrease of rainfalls of long duration and high intensity. It is thought that many of the valley meanders date in a general sense from the last glacial episode.

Underfit meanders had been noted before. Davis (1909) in his discussion of the Seine, the Meuse and the Moselle spoke of the Meuse as 'staggering with most uncertain step around the valley curves' (pp. 595–6). Davis also attributed this to a great decrease of discharge, but thought that, if this had been climatically caused, it should have affected the Seine and the Moselle as well. Instead, he attributed the loss of discharge to the capture of parts of the Meuse system by the other two rivers, when the Meuse was with difficulty keeping pace with an alleged uplift in the Ardennes, a hypothesis which has probably few adherents today.

In an interesting study of the Ontanagon area of Michigan, which has only emerged from Lake Nipissing in the last 10 000 years Hack (1965) distinguished between intrenched meanders (the valley meanders of Dury) and alluvial meanders (the stream meanders of Dury), and spoke of compound meanders consisting of one set of meanders superimposed on another set of greater wavelength. In part of the area studied Hack was able to show that large wavelength meanders had formed in an area which had emerged little more than 4 000 years ago, and hence could not be due to any great climatic change. Hack was inclined to relate meander wavelength to bed and bank materials, for observations suggested changes with calibre, the wavelengths being much shorter in finer materials. Thus, if a set of valley meanders are aggraded with fine alluvium, a new set of smaller-scale meanders will develop on that alluvium without changes of discharge and climate. In fact Hack rejected Dury's hypothesis for the area he studied.

No doubt the position is complex and, as it has so often been shown in the past that similar features may be produced by different causes, it would probably repay one to keep an open mind on this question. If several causes can produce underfitness, we are not likely to learn them by assuming that there is one universal cause.

Conclusion

The last fifteen years have been a period of information collection and the study of correlation between various aspects of valley and stream form. As a result we know a lot more about these features and are less prone to make over-simple generalisations than we were in the past. At the same time we remain far from an understanding of many of the basic causes of the phenomena described.

6
The development of drainage systems

Most earlier discussions of drainage patterns considered their development in relation to the structure and the lithology of the underlying rocks. In later years, possibly because this approach was too qualitative, drainage patterns have been treated more as geometric patterns and attempts have been made to derive relationships of a more or less universal nature between various elements of the patterns. This approach was started by Horton (1945).

Drainage systems in relation to geology

In the discussion of the development of drainage, geomorphologists use two sets of terms to characterise streams, one a set relating to the history of the streams, and the other a series of descriptive words usually specifying relationships between streams and structures.

Genetic stream terminology

The genetic terms, consequent, obsequent, subsequent and insequent, have been used and defined in Chapter 2, but, as they have been sometimes misapplied, some further discussion is necessary. Their correct use has been the subject of papers by Johnson (1932) and Baulig (1938, 1950).

Consequent streams are those which have their courses determined by the initial slope of the land. Obviously, any depression, the floor of which is sloping, will have two types of consequent stream: the main stream following the axis of the depression might be termed a longitudinal consequent, if a fuller description is needed, while the streams flowing down the sides of the depression into the main consequent are probably best termed lateral consequents. For the second class Davis proposed the terms lateral or secondary consequents, but, as the latter term has been used in a different sense by some writers, it is better to adopt the term lateral consequent. It will be noted that the concept of a consequent stream, as initiated by Davis and followed by most geomorphologists, does not include any reference to geological structure. It is true that consequents often flow in the same direction as the dip of the beds, especially in the well-known example of scarpland drainage, but rarely at the same angle as the dip. But they need not flow with the dip. Johnson quoted the drainage of a gently tilted fault block, where the streams draining both the gentle back slope and the fault face are equally consequents, as they depend for their direction on the nature of the initial surface.

Subsequent streams have been consistently interpreted as streams 'developed by headward erosion under the guidance of weak substructures' (Davis, 1909, p. 260). Davis added a clause to the effect that the weaknesses would be exposed in the valley sides of the consequent streams, but this limitation is not necessary, as such streams may develop by headward erosion from any other sort of stream and even from the coast. While the majority of subsequent streams follow the outcrops of weak strata, and hence correspond in direction with the strike of the beds, those which follow shatter belts and similar lines of geological weakness will have no set relationship with the strike.

Resequent streams, which were not mentioned in Chapter 2, are those which appear to be consequent in direction, but are hardly likely to be the direct descendants of the original consequent streams. In the simple example of the dissection of a broad anticline (Fig. 6.1) the original con-

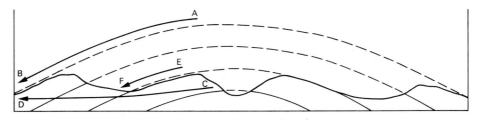

FIG. 6.1. Consequent and resequent streams (for explanation see text)

sequent stream, AB, slowly becomes entrenched and later occupies the position CD. The dipslope exhumed by the general course of erosion gives rise to streams, such as EF, which have the same direction as the consequents, but are of later generation. Such streams have sometimes been called secondary consequents, but, in view of the alternative and original meaning of this term noted above, it is better not to apply it.

It was noted in Chapter 2 that obsequents were originally defined as streams having a direction opposite to that of the consequent streams in their vicinity, but that the term has been traditionally interpreted to mean merely a stream flowing against the direction of dip. It will be used in the latter sense in this chapter, as the former meaning involves considerable difficulties in application at times. Whichever meaning is used, the term is not really genetic but descriptive and, in spite of the form of the word, it should really be one of the descriptive terms noted below.

By virtue of their definitions the terms, consequent, subsequent and resequent, carry considerable implications as to the history of erosion of any region. Thus consequent, if applied, means that the form of the initial surface is known, while resequent carries the same implication. Subsequent means merely that the stream was generated by a line of geological weakness, usually observable in the field or on a geological map, so that it can be used more readily than the other two. Even so, there may be streams flowing in the same direction as true subsequents, but along no apparent geological weaknesses. In this case the term should ideally not be applied until the weaknesses are demonstrated.

Really, then, there are two occasions on which the genetic terms can legitimately be used: if by a thorough study the erosional history of a region is known with reasonable certainty, their use is justifiable; or if a theoretical example is being discussed, they may be used as the discussion of such examples involves development from a given surface. The usage can be illustrated by two theoretical examples.

The original surface of the land is assumed to be the upper surface of a series of beds arranged in a system of pitching folds. In the initial stage (Fig. 6.2A) the main consequent stream, C, follows the pitch of the syncline, while lateral consequents, LC, develop on the flanks of the syncline. If the hard bed capping the structure is easily breached near the crest of the anticline, subsequent streams, S, may develop if weak beds are thus revealed, while obsequents, O, will develop down the inward-facing scarps.

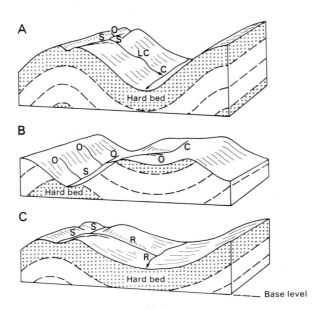

FIG. 6.2. Development of inverted relief and later resequent drainage on an area of folded rocks (for explanation see text)

If the disposition of hard and soft beds is suitable, erosion by the consequent, C, may be so inhibited that slowly the subsequent drainage achieves mastery (Fig. 6.2B). The subsequent drainage may in turn be checked by a hard bed—it has just reached it in Fig. 6.2B—while the courses of the original longitudinal consequents may be relatively easily eroded. Ultimately by a complex and little understood process the drainage may revert to a pattern very much like the original one (Fig. 6.2C), but, as the streams are hardly likely to be the direct descendants of the original consequents, they are best described as resequents. In the example illustrated, the resequent pattern will probably prove to be stable, as it is developed so near to base level.

A somewhat more complicated structure is shown in Fig. 6.3, in which a broad initial depression exists in an area of faulted rocks. The main con-

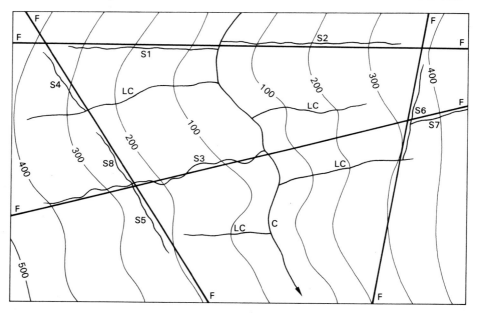

FIG. 6.3. Development of drainage on faulted rocks (for explanation see text)

sequent stream will follow the axis of the depression, C, and lateral consequents, LC, will develop in chance hollows on the flanks of the main depression. The obvious lines for the subsequents, S 1 to S 8, will be the fault lines: some of these will be parallel to the lateral consequents, e.g. S 1 and S 3, some will be parallel to the main consequent, e.g. S 4 and S 8, some will be opposite in direction to the main consequent, e.g. S 5, some will be tributary to the main consequent, e.g. S 3, some will be tributary to lateral consequents, e.g. S 4, while others will be tributary to other subsequents, e.g. S 5 and S 8.

In the well-known example of the development of scarpland drainage the streams of each class are all at right angles to those of another class and generally conform to the pattern of dip and strike, but, even in the slightly more complicated examples discussed, neither of these relationships is seen to be true.

Descriptive stream terminology

When real drainage, the history of which is unknown, is being considered, it is convenient to have a set of descriptive non-committal terms to describe the streams, as the genetic terms can obviously not be applied. Terms originally proposed by Powell, such as cataclinal and monoclinal, have found no wide usage, as they are unnecessarily obscure. Much simpler expressions, usually relating the streams to some structural feature, may be used. Longitudinal and transverse can specify relationships with fold trends; anticlinal, synclinal, dip, strike and antidip are all adjectives, the application of which makes the position of the stream in regard to structure immediately obvious (Wooldridge and Morgan, 1937). Such terms are to be preferred to the genetic terms where geomorphological ignorance exists.

The adaptation of streams to structures

It is not immediately apparent why streams flowing down dip should breach through the hard bed into the beds below. The key to the understanding of the matter is to be found in a consideration of the ratio between angles of dip and the gradients of streams which are approximately graded. Even rapid outwash streams have gradients of the order of only 8–12 m per km (40–60 ft per mile), while well developed lowland rivers have average gradients very much less than this. The river Cam above Cambridge, for instance, has an average gradient of less than 1 m per km (5 ft per mile) for most of its course. From the geological point of view, such features as chalk downs are gently dipping, their angles of dip being frequently of the order of 2 degrees. Translated into gradient, this means a dip of approximately 35 m per km (180 ft per mile). Thus, even a bed 300 m (1 000 ft) thick may be traversed by a lowland river in a distance of less than 10 km (about 6 miles) (Fig. 6.4A).

The same sort of relationship holds with streams flowing down the flanks of folds, but the contrast is even more marked, as angles of dip of 10–30 degrees are common even on gentle folds (Fig. 6.4B).

The ratio between stream gradient and dip is one of the main causes of the feature known as inverted relief, in which structural vales, or synclines, form hills, and structural hills, or anticlines, form vales. Breached and eroded anticlines are very common features, and their main explanation is probably to be found in the disposition of beds. It has sometimes been said that stretching and cracking on the axis of the anticline will cause it to be

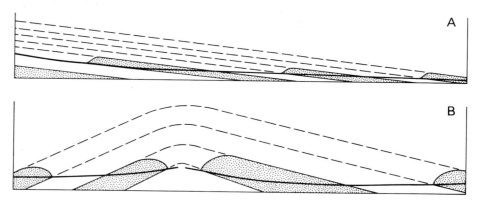

FIG. 6.4. Relation between angles of dip and stream gradients on gently dipping beds, A, and fold, B. Resistant beds (stippled) form cuestas as shown

a ready prey to erosion, but the same stretching must occur in a syncline, though it will be at the bottom of the bed instead of on its upper surface. Although this may contribute to the dissection of anticlines, the fact that anticlines often lift weak beds to considerable heights above base level, where erosion is stronger, is probably the main factor involved. In some folds the order of the beds may be such that inversion is unlikely to occur.

As a spectacular example of a synclinal mountain Snowdon is well known, though equally instructive, but less spectacular examples are common in southern England. The most comprehensive example occurs in the vicinity of Lewes, Sussex (Fig. 6.5), where the river Ouse crosses two minor folds, the Mount Caburn syncline and the Kingston anticline.

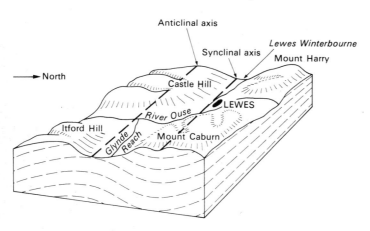

FIG. 6.5. Relations between relief and structure near Lewes, Sussex

PLATE 7. The Vale of Fernhurst, Sussex. An eroded anticline exposing Weald Clay between Hythe Beds escarpments, from the southern one of which the view was taken

On the western side of the river the relief is normal, the syncline being followed by the valley of the Lewes Winterbourne and the anticline forming the high ground to the south; to the east of the river the relief on both folds is inverted, the upland mass of Mount Caburn occupying the synclinal line and the valley of the Glynde Reach the anticline. Thus, all four cases are present in close juxtaposition: normal and inverted relief on both an anticline and a syncline.

Another good example is to be found in west Sussex and western Surrey, where a section drawn northwards from Midhurst shows the breached Fernhurst anticline (Plate 7), one of the most attractive anticlinal vales in southern England, the valley of the river Wey in the Haslemere syncline and the unbreached anticline forming the mass of high ground at Hindhead (Fig. 6.6).

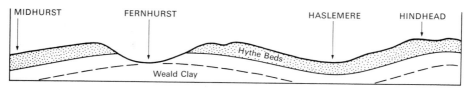

Fig. 6.6. Relief and structure north of Midhurst, Sussex

Splendid examples occur in the area to the north of Weymouth in Dorset, where a series of periclines occur on the northern flank of the main Weymouth anticline. The Sutton Poyntz pericline (Fig. 6.7) affords almost an

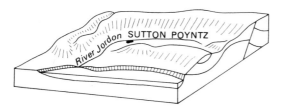

Fig. 6.7. The anticlinal vale of Sutton Poyntz, Dorset

ideal example, as the little river Jordon has cut out an almost enclosed lowland on the Kimeridge Clay, flanked by Portland Stone escarpments, except on the north where faulting causes the Chalk to appear as the scarp former.

The development of subsequent streams

In Davis's view the subsequent streams, once started by breaching of the structures, slowly extended headwards until they gained dominance over the whole drainage pattern. By virtue of their more readily eroded courses, they disrupt and integrate the original consequent drainage by a series of river captures.

Various points of evidence may be left to indicate that river capture has taken place. In the diagram (Fig. 6.8) a simple example, in which a tribu-

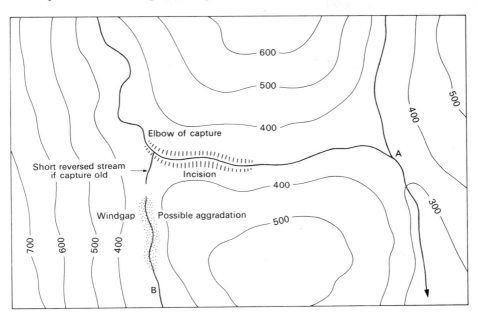

FIG. 6.8. The signs of river capture

tary of stream A has captured the headwaters of stream B, has been used. The signs that capture has occurred are as follows:

(*a*) At the point of capture there is a marked elbow in the capturing stream.
(*b*) The old course of the captured stream is indicated by a wind gap.
(*c*) The capturing stream may be incised immediately below the point of capture, as, through the acquisition of the headwaters of the captured stream, it gained in erosive power.
(*d*) The beheaded portion of the stream may be insufficiently powerful to move the material delivered down the valley slopes into it, and aggradation may result.
(*e*) A short reversed stream may lead from the wind gap to the point of capture, especially if the capture is of some age.

These signs of capture, which should ideally be present, are rarely found, apart from the elbow of capture and the wind gap. Even the wind gap may not show conclusive evidence of having been once occupied by a stream, as after a considerable period of time it may become only a high level col in a ridge, the exact significance of which is not easily decided. There may be left only a marked bend on the capturing stream to indicate that capture has occurred.

But a bend on a stream is not sufficient evidence for a hypothesis of river capture to be demonstrated, as more concrete evidence is required. If gravels derived from rocks only to be found in the basin of the headwaters of the original stream are found in the wind gap, the probability of capture is very great, but such evidence is rarely present.

In the absence of such conclusive evidence, the geomorphologist should attempt to analyse the probability of capture on geological grounds. In effect this means that he must advance reasons why one stream should have cut down faster than the other. There may have been a greater discharge in one, or the distance to the sea may have been shorter, but usually the greater erosion in one will be found to depend on the fact that it crossed areas of less resistant rock.

It is not sufficient, however, to consider the present outcrops of the beds, as the outcrops at the level at which capture is presumed to have occurred may have been different. In Fig. 6.9B a stream is superimposed over the

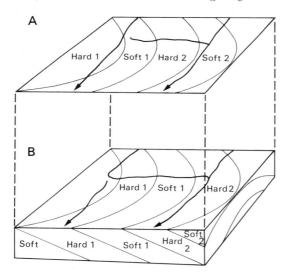

FIG. 6.9. Reconstruction of outcrops at the level of a river capture (for explanation see text)

end of a pitching anticline and the remnants of another stream of similar type denote that it has been captured by a tributary of the former. From the present outcrops it is difficult to see why this capture should have occurred, as the capturing stream is flowing over a resistant bed and the captured stream over a non-resistant bed, the beds in the figure having been labelled hard and soft for the sake of simplicity. But, if the outcrops at the level at which capture took place are reconstructed (Fig. 6.9A), it can be seen that the capturing stream flowed across a softer bed, while the captured stream was able to cut down less rapidly into the hard bed over which it flowed. The capture is thus made plausible by reconstructing the geology at the level at which capture took place, usually indicated by the height of the wind gap left by the capture.

An actual example of very similar conditions is found in the Weymouth district, where the drainage appears to have been superimposed from a marine terrace at about 70 m (240 ft) O.D. across the pitching end of the Weymouth anticline (Sparks, 1953). At present the drainage is dominated by the river Wey, but there is no obvious reason why this stream should have been favoured, as its course on the whole lies across the Cornbrash outcrop. But, if the outcrops at 70 m (240 ft) are reconstructed, it can be seen that the Wey flowed at that level mainly across the Oxford Clay, a non-resistant bed in which it was able to entrench itself easily.

The importance of reconstructed outcrops may be illustrated by reference to the capture of the Candover by the Itchen, east of Winchester (Fig. 6.14), which will be referred to later in another connection. The Itchen was superimposed across the Winchester anticline at the point where the fold reaches its maximum amplitude, while the original Candover was superimposed across the flank of the anticline. At Winchester, almost the whole of the Upper Chalk had been planed off the fold by a marine transgression in early Pleistocene times, so that the Itchen quickly cut through the remainder of the Upper Chalk and was then able to cut down rapidly in the less-resistant Middle Chalk, while the Candover had still to contend with a considerable thickness of resistant Upper Chalk on the flank of the anticline (Wooldridge and Linton, 1955). This type of hypothesis, which is often applied, leaves out of consideration the fact that, although the capturing stream may reach less resistant rocks in the anticline first, it still has to cut down through the more resistant rocks on the flanks.

The type of river capture described is largely explained by differences in rock resistance, but another factor becomes of importance in areas of permeable rocks: the possibility of underground diversion preceding and aiding surface capture. The conditions envisaged are illustrated in Fig. 6.10: active erosion by the spring at the head of stream A is causing it to approach closely to the higher level valley occupied by stream B. While the

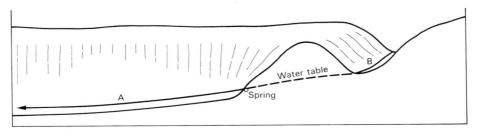

FIG. 6.10. Underground diversion of drainage

spring is still far away from stream B the water table between the two valleys will have a roughly symmetrical shape, allowing approximately equal flow into the two valleys. But as the spring head of stream A saps back, it will slowly appropriate the lion's share of the water table until very little water reaches stream B. Finally, if the streams are close enough and if the differences in elevation are sufficiently great, the water table may slope down from stream B into the spring head of stream A, causing the desiccation of the former. Thus in permeable rocks the closer the spring of one stream approaches another stream the greater are the chances of diversion: the more permeable the beds the better for the capturing stream, as the water table will be flatter. Such captures should, in fact, be facilitated in very well jointed limestones, in which there is a free underground flow of water.

A most remarkable example of underground diversion of drainage has been described by Howard (1938) from the Yellowstone National Park (Fig. 6.11). Tower Creek follows a fault zone separating breccias from

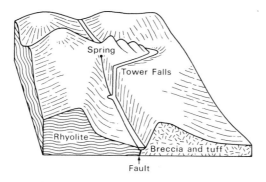

FIG. 6.11. Tower Creek, Yellowstone National Park (*after Howard*)

rhyolite, except where it takes a loop on to the breccias in one place. This loop on the breccias hinders the downcutting of the stream and where the stream rejoins the fault zone there is a waterfall 40 m (132 ft) in height. Immediately above this point a powerful spring issues along the line of the fault and is evidently fed by seepage from the upper, higher part of Tower Creek. This spring is actively eroding headwards along the fault and will presumably in time short-circuit the loop that takes the stream on to the breccias. All the active preliminary part of the diversion will have been performed by underground abstraction.

An example where the capture is almost complete can be cited from the Sutton Poyntz pericline, which has already been described as a good example of inverted relief. Skirting the western rim of the pericline and cut in Chalk to the north of the faults and in Portland Beds to the south, is a peculiar feature known as Coombe valley (Fig. 6.12). For most of its

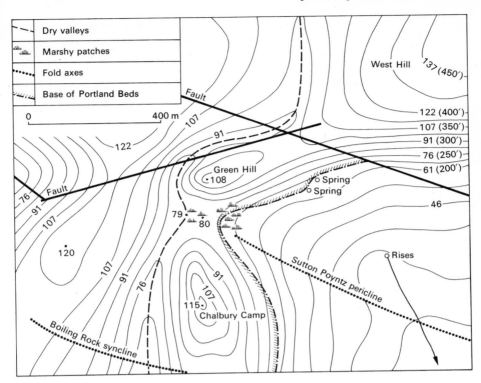

Legend:
- ――― Dry valleys
- Marshy patches
- ••••• Fold axes
- Base of Portland Beds

0   400 m

West Hill 137 (450')

122 (400')
107 (350')
91 (300')
76 (250')
61 (200')
46

Fault

Green Hill
·108

Spring
Spring

79
80

120

Rises

91

107
115·)
Chalbury Camp

Sutton Poyntz pericline

Boiling Rock syncline

Fig. 6.12. Coombe valley, near Weymouth (*after Sparks*)

length it is a normal steepsided limestone valley, but between Green Hill and Chalbury Camp the eastern side has practically disappeared and is only a metre (3 ft) higher than the level of the valley floor. The stream originally flowing down the valley has been diverted underground, as the Sutton Poyntz vale has been extended in the non-resistant Kimeridge Clay, through springs occurring at the junction of the Portland Beds and Kimeridge Clay in the northern part of the vale. The surface capture has almost taken place at the axis of the Sutton Poyntz fold where the Kimeridge Clay reaches its highest elevation and where, therefore, there was the least thickness of resistant beds above to be eroded.

Although the two examples of underground diversions quoted are both on a small scale, they illustrate well two of the intermediate stages of capture. Such processes must have operated commonly in river captures, especially in limestone areas but also in other types of permeable rock.

Although the disposition of beds can usually afford reasons why any given capture should have occurred, any subsequent stream to effect capture must erode headwards for a considerable distance along a belt of weakness. There is no direct way of assessing the actual rate at which such headward erosion takes place. Davis thought it to be a slow process, so that a full dominance of subsequent drainage might only be achieved in a second cycle of erosion. Obviously much must depend on the nature of the rock and the amount of rock to be removed, so that there will be enormous variations in the rate of headward erosion.

Some indication of the relation between the rate of downcutting by the consequent stream and the rate of headward erosion by the subsequent may be got in general terms from certain areas in the following way. Let it be assumed that a consequent drainage was initiated from a surface, the contours of which can be reconstructed (Fig. 6.13), and that consequent A was

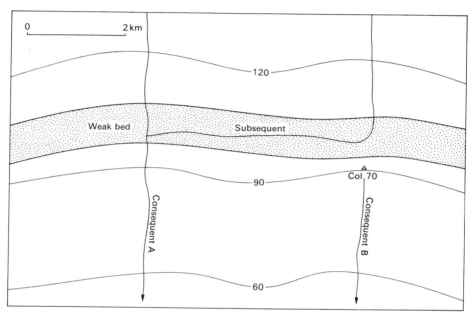

Fig. 6.13. Method of assessing relative rate of development of subsequent streams (for explanation see text)

able to cut back and capture the upper part of consequent B by means of a subsequent tributary working along a non-resistant bed. If the height of the wind gap below the elbow of capture is known, the following statement can be made: while a subsequent stream was working back a distance of 4 km (2¾ miles), the consequent was able to cut down from 90–70 m (300–230 ft), an incision of 20 m (70 ft). The relative rate of incision will depend on height above base level, climate and position in the course of the consequent stream where capture takes place. Although such a measure is only in relative terms and subject to great variation, it appears to be a first step towards assessing rates of subsequent development. As yet there are few areas where enough is known about the height of the surface from which the drainage was initiated to enable such calculations to be made, as insufficient research has been done. A notable exception is the Hampshire Basin, the geomorphology of which has been closely studied (Wooldridge and Linton, 1955).

In the Hampshire Basin the capture of the Candover stream by the upper subsequent Itchen may be used for a calculation (Fig. 6.14). The part of

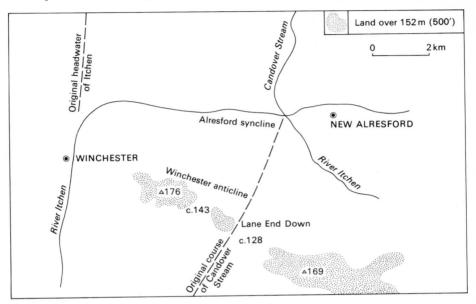

FIG. 6.14. The capture of the Candover by the Itchen

the Hampshire Basin concerned was planed off in early Pleistocene times, and a south-flowing superimposed drainage was initiated on the emerged floor of the early Pleistocene sea. The drainage cut straight across the folds represented in this district by the Alresford syncline and the Winchester anticline. Two main consequent streams drained the area, the Itchen and the Candover, but the former, for structural reasons already discussed, cut down faster and a subsequent tributary worked back eastwards along the Tertiary deposits in the Alresford syncline to capture the upper part of the Candover, which had not succeeded in cutting down so fast through the anticline at Lane End Down. The height of the early Pleistocene plain, from which the drainage was superimposed, seems unlikely to have been above 180 m (600 ft) OD at the point of capture and the floor of the wind gap east of Lane End Down is about 125 m (420 ft) OD. Assuming this to be the original level at the time of capture, it can be seen that the Itchen cut back 7 km (4½ miles), while the Candover cut down 55 m (180 ft), i.e. there was one km of headward erosion by the subsequent while the consequent cut down 8 m (i.e. approx. 1 mile for 40 ft downcutting). If the Candover originally occupied the higher gap west of Lane End Down, the relative rate would be somewhat higher. The process went on early in the cycle when downcutting is at a maximum: towards the end of the cycle the amount of headward erosion in proportion to downcutting would probably be higher. Somewhat similar rates of headward erosion appear to be indicated by hypotheses of capture in some adjacent areas of similar rocks. The rate appears to be high, but it must be remembered that the lithological contrasts are great and that permeability may have had something to do with the captures.

Although in outlining the development of drainage on folded rocks, it has been said that a resequent pattern may form ultimately to the exclusion of the pattern dominated by the subsequent streams, the evolution of the normal cycle of erosion appears to involve the increasing dominance of subsequent drainage as a general rule. Such a state can be inferred by the number of important rivers flowing along the outcrops of non-resistant beds. There appear to be certain limited areas, however, where the dominance of the subsequent streams is shortlived.

As an example some small streams and valleys in the northern part of the Weymouth lowland (Fig. 6.15) may be taken. The area is one of superimposed drainage, comparable with the Hampshire Basin (see below), and there appear to have been two south-flowing consequents, one through

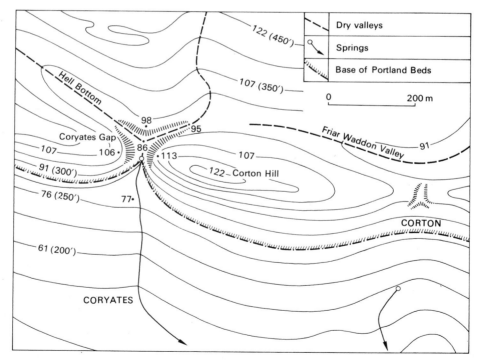

FIG. 6.15. Coryates gap near Weymouth (*after Sparks*)

Coryates gap and the other through Corton gap (Sparks, 1951). But, because the rocks are strongly diversified, these were soon disrupted by the development of subsequent streams, the northern one being along the line of the Friar Waddon valley and Hell Bottom and cut into Wealden and Purbeck Beds and the southern along the Kimeridge Clay outcrop through Coryates. Between them the resistant Portland Beds form a marked ridge. Like most of the southern coastal districts of England, the area has suffered repeated rejuvenations, the effects of which were transmitted to the area along the river Wey. Because the rock is much less resistant the subsequent stream occupying the Kimeridge Clay vale cut down more rapidly than its northern neighbour. The Portland Beds ridge was pushed northwards and the springs at Coryates and Corton abstracted more and more drainage from the Friar Waddon valley underground through the permeable Portland Beds until the northern valley dried up. An actual surface link-up has taken place at Coryates gap, and any water flowing down Hell Bottom, which is now dry, would pass through the gap as, indeed, it did in the flood of 1955 (Arkell, 1956). Thus, the original drainage pattern has been partly restored.

The process requires rather special conditions. Chief among them are rejuvenations, for in the period immediately following a rejuvenation the balance between streams is disturbed and the possibility of recapture appears to exist. Permeable rocks are also probably of great significance as capture takes place effectively underground while the actual surface streams are still some distance apart.

The hypothesis seems likely to apply to certain limited areas near the coasts. Recaptures such as those described provide only a minor exception to the general rule that the subsequent streams become dominant, as both the streams and the areas involved are very small.

Drainage patterns

The adaptation of streams to structure results finally in the drainage network acquiring a pattern, which is largely a reflection of the structure. Various names have been proposed for particular patterns. In areas of homogeneous rock, especially clays, the pattern will have a ramifying pattern usually known as dendritic (Fig. 6.16A); in areas of gently dipping beds, such as occur in scarplands, the pattern might be described as trellised (Fig. 6.16B); on dissected domes a radial and concentric pattern may develop (Fig. 6.16C). A full discussion of the commonly occurring patterns has been given by Zernitz (1932). While such terms for pattern are descriptively useful and various types of pattern tend to predominate, a too rigid classification should not be attempted, as there are as many variations as there are variations of structure.

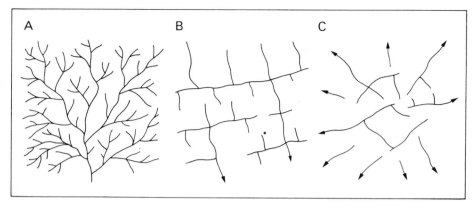

FIG. 6.16. Drainage patterns: A, dendritic; B, trellised; C, radial

Drainage patterns discordant to structure

The drainage evolution discussed so far has all been in conformity with the structure. But some areas show a discordant pattern, which is usually betrayed by streams cutting across folds. The discordant pattern has been called an inconsequent pattern, and the accordant pattern a consequent pattern, but, in view of the possible confusion with consequent streams, the terms accordant and discordant or conformable and non-conformable appear to be better. The condition of streams cutting indifferently across anticlines and synclines can arise in four main ways:

(*a*) Through the capture of a stream flowing down the flank of an anticline by a stream in line with it on the other side of the fold. Such head-on capture may occur occasionally, but it cannot be used to explain a regional discordance of drainage pattern.

(*b*) Through glacial diversion, a subject which will be discussed more fully in Chapter 13.

(*c*) Through antecedence, whereby a stream manages to keep on cutting down through a fold rising across its path.

(*d*) Through superimposition of the drainage from an unconformable cover of rocks of different structure, which have usually disappeared through erosion.

The last two causes deserve to be discussed in some detail, as the hypotheses have been widely used by geomorphologists.

It might just be possible to suggest that, under certain conditions of asymmetric folds and unconsolidated rocks, streams may come to flow across the axes of folds without any abnormal development. From the example shown in Fig. 6.17 it could be argued that the original consequent streams, A and B, would cut down until they reached the positions A′ and B′, at which stage stream A would cross the fold. It seems unlikely, however,

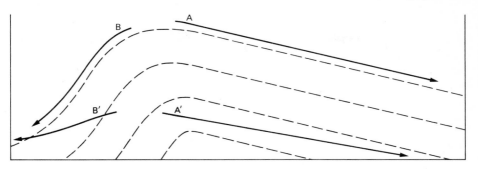

FIG. 6.17. Development of a stream across a fold by normal erosion (for explanation see text)

that the streams would develop in the way suggested, because stream B with its steeper gradient would tend to push the watershed back towards stream A, thus keeping it approximately on the line of the axis of the anticline.

A more probable cause of local discordance may arise in the following manner. If at any stage of the dissection of an area of folded rocks a massive clay bed is encountered, the streams, which formerly occupied the synclines or anticlines, may commence to wander freely, regardless of structure, as structure only affects stream courses when it introduces varying lithology. Thus large meanders may develop across folds (Fig. 6.18A), and as the stream continues to incise itself, it may cut down into the harder rocks revealed at greater depth in the anticline (Fig. 6.18B). In this way streams may

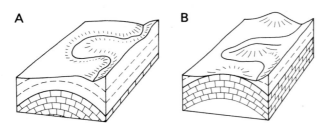

FIG. 6.18. Local incision of a stream across a fold

become locally discordant to structure. Examples of this type have been reported from the Central Weald, notably the crossing by the Eden-Medway of the Penshurst anticline (Wooldridge and Linton, 1955).

Local peculiarities of this type, which do not require hypotheses of antecedence or superimposition, should be readily soluble by field work. But the majority of discordant drainage in unglaciated regions has been explained usually by superimposition, but occasionally by antecedence.

ANTECEDENT DRAINAGE

Antecedence, as a hypothesis, should be the last resort of a geomorphologist seeking to explain an inconsequent pattern of drainage, as it is, except in ideal cases, undemonstrable. It involves guesses as to the rate of uprising of folds and the rate of downcutting by rivers, about neither of which is very much known. It will be more plausible as an explanation in examples where the rivers are large and the amount of upwarping small, but becomes highly improbable when applied to small brooks crossing comparatively large folds, as for example the drainage of the Weymouth district and the Isle of Wight.

If a fold rises across the course of a river, the reactions will depend upon the rate of upwarping compared with the rate of downcutting. If the upwarping is extremely slow (Fig. 6.19A), there may be little effect upon the river and it may maintain its smooth, graded profile as before. If the upwarping is more rapid, but not sufficiently rapid to divert the upper part of the stream completely (Fig. 6.19B), a convexity may develop in the profile of the river where it crosses the fold. At the same time there should be aggradation on the upstream side of the warped area, as the gradient of

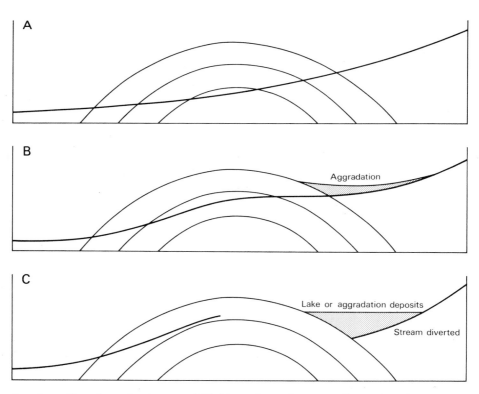

FIG. 6.19. Reactions of a river to a fold rising athwart its course (for explanation see text)

the river will be decreased there. As the fold continues to rise, the aggradation should be tilted up towards the fold in an ideal case. Finally, if the
river is insufficiently powerful to maintain its course across the fold, a lake
should be formed above the fold and this will overflow at the lowest available outlet (Fig. 6.19c). The remnants of a lake or of lacustrine deposits,
depending on the age of the diversion, will be all that is left to indicate the
inadequacy of the stream's power to cut down through the fold, apart,
perhaps, from a notch in the crest of the fold representing the former course
of the river. It is really examples of the second type, where the drainage has
been affected but not diverted, that afford the most likely conditions to
demonstrate antecedence in the field.

Some of the most convincing evidence for antecedence has been presented
by Wager (1937) to explain the nature of rivers crossing the Himalayas,
notably the rivers Arun and Tista. The Arun may be taken as the typical
example. In its upper parts it flows with the structural grain of the country
from west to east but later turns south to flow across the Himalayas through
a series of deep and almost impassable gorges (Fig. 6.20). There are really

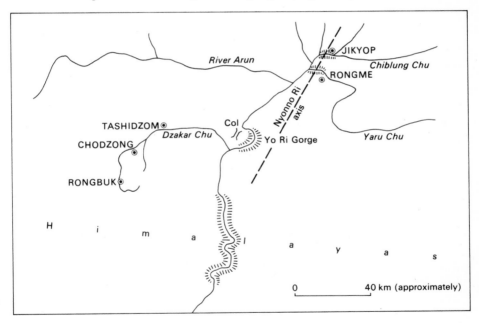

FIG. 6.20. The upper parts of the course of the river Arun (*after Wager*)

only two possible explanations, antecedence and capture, as glacial diversion
is not supported by the field evidence cited by Wager, and superimposition would imply that the streams were initiated on a surface touching
the highest mountain peaks of the present landscape, i.e. at an elevation
of almost 9 000 m (30 000 ft).

147

If the long profile of the Arun (Fig. 6.21) is examined, it can be seen that the course is convex where the river crosses the Himalayas, a feature which would be expected if the river had only maintained its downcutting with difficulty. Further, the Arun is flanked in its upper reaches by a series of broad and readily recognisable terraces. At the junction with the Yaru Chu

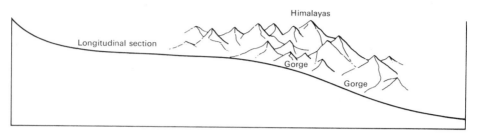

FIG. 6.21. The long profile of the river Arun (*after Wager*)

the Arun is flowing at the surface of these spreads of gravel, but it becomes progressively entrenched until, just before the Yo Ri gorge, it is 200–300 m (700–1 000 ft) below the surface of the gravels. Unfortunately, the area has not been accurately surveyed and the vital piece of evidence is lacking: it is not known whether the surface of the gravels rises towards the Himalayas or remains horizontal. If these gravels are the aggradation resulting from the Himalayas rising across the Arun, they should rise towards the mountains.

A tributary of the Arun, the Dzakar Chu, shows other interesting features in the relations between its terraces and the profile of the present stream. Again there are no measurements available, but the nature of the relations is known. Upstream from Tashidzom the terraces decrease in height above the river, the normal relationship between terraces and the stream after rejuvenation, but a little above Chodzong they begin to increase in height again and at Rongbuk they are well above 30 m (100 ft) above the river. This behaviour is abnormal, as the terraces should merge with the present profile upstream. It clearly suggests that the area drained by the upper part of the course has been differentially uplifted, thus enabling the river to cut down more vigorously. The obvious mechanism is upheaval of the Himalayas.

There are also indications in the area of uplift along the line of the Nyonno Ri axis to the east of the Arun. Both the Chiblung Chu and the Yaru Chu cross the northern part of the ridge in marked gorges, and there is evidence of a lake 90 m (300 ft) deep in the upper part of the area drained by the Yaru Chu. Furthermore, fine sediments, which appear not to be the foreset beds of a delta, are found on the eastern side of the Nyonno Ri ridge and dip away from it at 20 degrees. All these facts are indications of upheaval along the Nyonno Ri line.

There is further possible evidence of the upheaval in the Yo Ri gorge of the main stream, which is found where a loop of the Arun has become entrenched into the hard gneisses of the Nyonno Ri ridge. A little to the west of the Yo Ri gorge a broad col has been excavated in soft schists and forms the main route north, as the gorge itself is impassable. If the upper longitudinal Arun had been captured by the southward-flowing part, one would have expected the latter stream to erode headwards along the easier course, whereas, in fact, it appears to follow the most difficult course.

Apart from this fact, a good case could be made out for capture, as the lower Arun has strong advantages over the longitudinal stream north of the Himalayas. It has a direct course to the sea over a very steep slope, the southern flank of the Himalayas, which should give it great erosive power. The southern flanks of the Himalayas also receive a very heavy monsoonal rainfall, which would help the lower Arun to cut down rapidly. A capture of this type would explain the convexity of the profile, the way in which the river is entrenched in its own gravels north of the mountains, unless they prove to be tilted towards the Himalayas, and the gorges, but it would leave unexplained the avoidance by the Arun of the col floored with soft rocks to the west of the Yo Ri gorge and the relations between stream profile and terraces on the Dzakar Chu. It is, of course, also possible that capture has been the main process concerned and that the streams have survived only a slight and fairly recent continuation of the Himalayan uplift.

If a block of country athwart a terraced river is uplifted and the river is easily able to maintain its course, there may be none of the signs shown by the river Arun. There may be, however, a distinct upwarping on the terraces as they pass over the area being uplifted. Idealised conditions of this sort are illustrated in Fig. 6.22, in which the higher terrace is distinctly

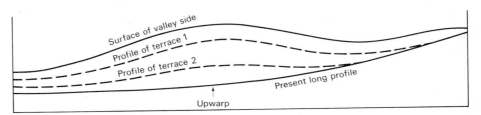

FIG. 6.22. Effect of upwarping on terrace profiles

more arched than the lower. Evidence of this type has been used to suggest that the Rhine is antecedent to a certain amount of warping of the Rhine block, through which it flows between Bingen and Bonn.

One needs to have fairly well-preserved terraces to be able to substantiate a hypothesis of upwarping, as the reconstruction of the profiles is a difficult task. Few rivers, at least in north-west Europe, have their terraces preserved as continuous physical features or as continuous spreads of gravel. Usually the valley side spurs preserve traces of flattenings, which represent the terraces. Even if the river possesses rejuvenation terraces, the fragments thereof may have been considerably lowered, especially in the case of the older, higher terraces. In addition, there is the possibility of non-significant fragments of terrace, formed by the river swinging laterally during a phase of downcutting, being found. The reconstruction which is placed on the terraces depends largely upon the way in which the lines joining the terrace fragments are drawn. Thus, in Fig. 6.23 the fragments

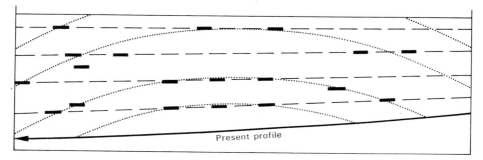

FIG. 6.23. Reconstruction of terrace profiles from terrace fragments (for explanation see text)

could be interpreted either as the remnants of a series of unwarped terraces (pecked lines) or as a series of warped terraces (dotted lines). Very often the geomorphologist in the field has even less evidence than that displayed in the figure.

SUPERIMPOSED DRAINAGE

Superimposed or, as it is sometimes called, epigenetic drainage is a very common phenomenon. Indeed, superimposition appears to be almost as common as normally developing drainage. The mechanism is simple (Fig. 6.24), all that is required being an unconformable cover of younger strata of different structure on which the drainage is initiated. The unconformable cover may be a series of consolidated rocks; it may be a

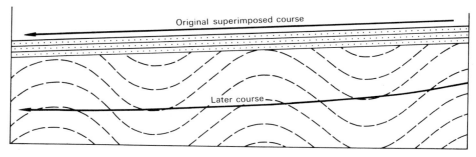

FIG. 6.24. Ideal development of superimposed stream

thin veneer of marine deposits covering an abraded marine platform; or it may even be an extensive deposit of river gravels. Superimposition, at least of a local type, is almost bound to occur in many cases where rivers cut down through a considerable thickness of rocks, and examples of local type have already been indicated.

As a hypothesis, superimposition is eminently plausible in the explanation of regional discordance of drainage. If there are many transections of folds by streams, the hypothesis of head-on capture involves numerous assumptions of the arrangement of streams in line on different sides of the folds. Antecedence would imply that all streams were capable of cutting down through all the folds, which usually display as great a variety of amplitude as the streams do of energy. Superimposition merely requires the demonstration of a former cover of unconformable rocks, either in the form of outliers or in the preservation of the eroded surface on which they originally rested. In Great Britain, superimposition has been used to explain the drainage of large areas, especially the coastal regions. In the north and west, the possibility of a former cover of Chalk extending over the highlands has proved to be one of the most useful geological assumptions ever adopted by the geomorphologist. Chalk outliers are preserved beneath lava flows in Antrim and in some of the volcanic areas of western Scotland: it is thus quite probable that the Chalk originally extended over much of the country. It is the ideal surface to suggest as the one on which the anomalous drainage of such areas as Wales and the Lake District was initiated. In the south and east of England, the hypotheses of superimposition have usually rested on the existence of erosional platforms below 200 m (700 ft), especially the early Pleistocene bench which surrounds the London and Hampshire Basins.

As a small example of a drainage pattern which is still consequent, but which could be superimposed if sea level fell 3 or 6 m, (10 or 20 ft), the Sussex coastal plain between Worthing and Littlehampton may be cited. The structure includes two folds (Fig. 6.25), the Littlehampton anticline, which has been planed down except for Highdown Hill on its northern flank, and, to the north, the Chichester syncline in which the lower beds of

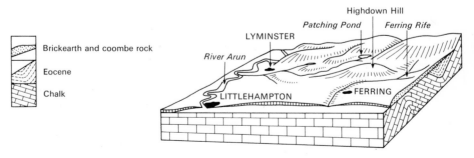

Fig. 6.25. Drainage on the Sussex coastal plain near Littlehampton

the Eocene are preserved. A small stream, the Ferring Rife, rises on the syncline and flows generally southwards to the sea near Ferring. At present it flows over a surface composed mainly of Pleistocene deposits, coombe rock and brickearth, but its course is not far above the planed-down chalk of the Littlehampton anticline, as can be seen in the lower part of the beach where the chalk is exposed at low tide. Given a slight rejuvenation, the Rife will become superimposed across the Littlehampton fold, and, after a much longer period of time, all traces of the Pleistocene deposits on which it was initiated will disappear.

In areas of superimposed drainage the deposits of the surface of initiation have usually disappeared. Such is the case over the whole of Wessex, including the adjacent regions of the Weymouth lowland and the South Downs. Wessex represents a very clear example of superimposed drainage (Wooldridge and Linton, 1955).

The folds shown on Fig. 6.26 are typical of those common in southern England. They are generally of amplitudes less than 300 m (1 000 ft); they are short and somewhat arcuate, so that the ends are slightly offset; they have steeper dips on their northern limbs than on their southern. A consequent drainage developed on this structure would have flowed west–east following the major troughs between the different groups of anticlines,

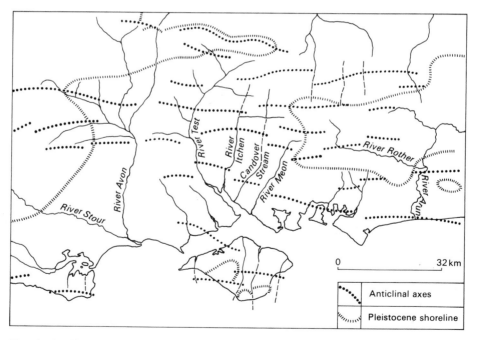

FIG. 6.26. The superimposed drainage of Wessex: the pecked lines represent reconstructed former river courses (*after Wooldridge and Linton*)

whereas the actual drainage flows towards the Solent, from the north in the Hampshire Basin and from the south in the Isle of Wight.

Although no trace of the deposits from which this drainage was superimposed are preserved, the level of superimposition can be identified from conditions in the London Basin. In that area, early Pleistocene deposits have been discovered on the dip-slopes of both the North Downs and the Chilterns, where they occupy a bench at 195–165 m (650–550 ft). In the Hampshire Basin, although no deposits have been traced, the bench, at the same elevation as that in the London Basin, is identifiable, and traces of it have also been observed in the South Downs, where, however, it is much more dissected, due to proximity to the sea. The position of the old shoreline behind the bench in the Hampshire Basin is indicated in Fig. 6.26. The area outlined is that of a gulf with its axis along the line of the Frome and the Solent, its northern shoreline in the latitude of Basingstoke and parts of its southern shore, or perhaps islands, in the Isle of Wight.

When the early Pleistocene sea finally left this gulf, the drainage of the area was effected by a main longitudinal consequent, the Frome–Solent, and by a series of lateral consequents draining Hampshire and the Isle of Wight. The transverse consequents have become entrenched into the folded Cretaceous and Lower Tertiary rocks underlying the area and form a typical example of superimposed drainage. They are in marked contrast to the rivers of the area not planed off by the early Pleistocene sea, where the drainage is essentially longitudinal, as it has become slowly adapted to the structure over a very long period.

Since the superimposition of the drainage in Wessex, there has been insufficient time for it to adapt itself again to the structures encountered in the rocks. A certain amount of new subsequent drainage has developed, especially where beds of contrasting lithology were exposed, a good example being the headward erosion by the upper Itchen to capture the Candover (Fig. 6.14). But, in spite of limited adaptation to structure, the superimposed pattern is very clear indeed.

The structure of the beds from which the drainage has been superimposed is not always as simple as that of the early Pleistocene cover in the Hampshire Basin. In some of the western parts of the country, where the drainage may have been let down from a Cretaceous cover on to the underlying Palaeozoic rocks, a complex pattern may have developed in the cover rocks. One may imagine, for example, a well-adjusted pattern of scarpland drainage forming on a thick cover of tilted sediments overlying a massive igneous intrusion (Fig. 6.27). If the development is sufficiently long, a rectangular pattern of drainage, with subsequents, obsequents and consequents, will evolve and, after the passage of a long period of time, may be superimposed on to the underlying igneous rocks. If the igneous rocks are jointed in the normal rectangular manner, any new streams developing in the igneous rocks will be difficult to distinguish from the rectangular superimposed pattern inherited from the cover rocks.

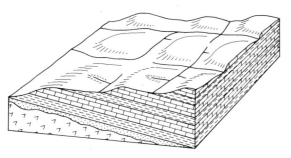

FIG. 6.27. The form of a drainage pattern superimposed from gently dipping strata (for explanation
see text)

A superimposition of comparable type appears to have occurred on
Dartmoor (Green, 1949). The rectangular pattern developed on a cover
of Cretaceous and Eocene rocks lapping on to the eastern side of the granite,
and has since been superimposed on to the granite. The main consequent
streams flowed eastwards, the chief of them being represented by the Dart
from Two Bridges to near Ashburton and the old abandoned course of the
Dart from Ashburton to the sea at Teignmouth. The subsequent streams
were north- and south-flowing, but only the south-flowing ones are at
present important in the landscape, as a general tilt of the area to the south
has brought about the suppression of the north-flowing streams. The effects
of the tilt to the south were to steepen the gradients of the south-flowing
streams and, thus, to give them more erosive power, and to decrease the
gradients of the north-flowing streams, thus depriving them of energy.
Since the superimposition, streams have developed in the granite mainly
in conformity with the joint pattern, which is here north-west to south-east
and north-east to south-west. Although the superimposed drainage and the
drainage developed later on the granite are both rectangular in pattern,
they may be distinct because of the different orientation of the straight
stretches.

In some areas where the superimposition appears to have been from a
simple veneer of sediments covering a marine abrasion platform, detailed
studies of the geomorphology have shown that the process of superimposition has been repeated many times over. Generally, most areas of southern
England show evidence of a slow, spasmodic emergence, at least from the
time when the early Pleistocene terrace at 195 m (650 ft) was cut, and probably from even earlier times. Such a course of events is represented in the
relief by the remnants of a number of marine-cut terraces arranged in a
staircase form in coastal districts. A diagrammatic representation is shown
in Fig. 6.28. In this example the drainage was first initiated at the level of the
terrace A, when the shoreline was in position 1. A fall in sea level, during
which the shoreline moved to position 2, exposed a new section of the sea
floor, B, over which the streams were extended. Successive lowerings of

the sea caused the shoreline to migrate to positions 3 and 4, leaving further sections of the sea floor exposed as lower benches, C and D. Thus the streams draining the original area are slowly extended over successively lower marine terraces. When the thin marine deposits have been removed from

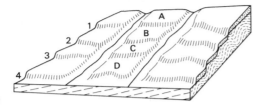

FIG. 6.28. Extension of consequent streams over emerged sea floors

these terraces by erosion, the streams will be superimposed on to the rocks below. In this way the streams are extended seawards by repeated super-impositions. The effect of this type of origin, which may be exemplified in the streams draining the dipslope of the South Downs (Sparks, 1949), is to give fewer chances for the development of subsequent streams. In this theoretical example, if the streams had been superimposed through their whole length at stage A, and if weak beds had occurred anywhere along the course, there would obviously have been more time for the integration of the drainage by subsequent streams than in the example in which the sea floor is uncovered section by section. As such a process of extended super-imposition has probably taken place in many areas, for successions of marine terraces are common on many of our coasts, the result has been a better preservation of the superimposed pattern than would otherwise have been possible.

In the examples quoted it has usually been easy to see the likelihood of superimposition, because the stream trend and the structural trend are almost at right angles. Such relations are normal in south-eastern England, and also in such other well-known areas of superimposed drainage as South Wales and the Appalachian region of the United States. When, however, there is only a slight angle between the stream and structural directions, or when the two directions are identical, the problem of recognising that super-imposition has occurred is much more difficult.

An example of this type occurs in the Chalk district of north France, especially round Dieppe (Sparks, 1953), the conditions found being illus-trated schematically in Fig. 6.29. The folds concerned are of the same type and the same age as those affecting Wessex, but the trend changes from west–east to north-west–south-east in northern France, i.e. at right angles to the present coast. A general consequent pattern of drainage developed on such a structure would have a north-west–south-east alignment, but so would a pattern composed of a series of extended consequents superimposed

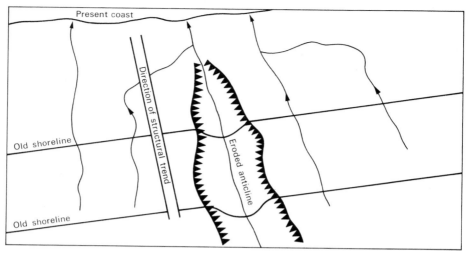

Present coast

Direction of structural trend

Old shoreline

Eroded anticline

Old shoreline

FIG. 6.29. Superimposed streams and structures parallel to each other

from marine deposits, if the shorelines had been approximately parallel to the present one.

Where these conditions exist, it is necessary to look closely at the drainage and to concentrate on single anomalies, as regional discordances of drainage and structure do not occur. In the example, one stream follows the axis of a pitching eroded anticline: in France the anticline is the Bray anticline and the stream the river Béthune. At first sight it could be a normal subsequent stream, but the hypothesis does not appear suitable upon closer inspection. With normal drainage development the anticline should be first breached, by a lateral consequent, at the point of its maximum amplitude, because at this point the softer beds below the Chalk capping the fold are at the greatest elevation above base level. From this point, somewhere towards the southern margin of the area shown in the diagram, short subsequent streams should extend headwards in both directions, one towards the south and the other towards the coast. But the stream shown draining the anticlinal vale does nothing of the sort: it flows northwards along the axis of the pitching fold and, even when the fold disappears, it continues its course to the sea. Such behaviour is difficult to explain by a normal process of drainage development. When, however, the old shorelines are reconstructed, they appear to have embayments, where the weaker rocks exposed in the anticline occurred in the cliffs. When sea level fell, the original stream, which may in the south have been the remnant of an old subsequent, flowed over the sea floor along the axis of the embayment, and it kept this course because similar embayments appeared in later and lower shorelines. In this process the direction of the streams may have been helped by strike faulting on the steeper north-eastern limb of the fold, so that this may not be the perfect case. Nevertheless, so few examples of this type have been reported, that this example will serve to illustrate the difficulties in establishing a hypothesis of superimposed drainage where the superimposed streams are parallel to the structures.

The areas cited demonstrate the way in which superimposed drainage may vary, but they far from exhaust the examples of superimposition. It is advisable to study the drainage of other regions, such as Wales, the Lake District and the Appalachians, references to which are given later.

Drainage systems in relation to geometry

The quantitative analysis of drainage networks really stems from Horton (1945). It has been developed to enable comparisons to be made between different drainage basins, to enable relationships between different aspects of the drainage pattern of the same basin to be formulated as general laws, and to define certain useful properties of drainage basins in numerical terms. Many of the indices derived are in the form of ratios, or dimensionless numbers, so that comparisons can be made irrespective of scale, for example between normal-scale drainage and the miniature systems of some badlands.

For these purposes a ranked hierarchy of streams has to be used. Three of these are illustrated. Fig. 6.30 shows Strahler's modification of Horton's original system. It is taken first as being probably the simplest and most used system. The fingertip or headwater streams which receive no tributaries

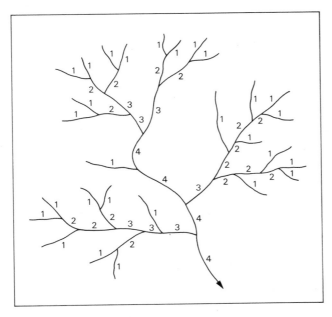

Fig. 6.30. Strahler's system of stream ranking

are called first order streams. Two first order streams unite to form a second order stream. Two second order streams unite to form a third order stream and so on. Where two streams of different order join, for example a first and a third order, the combined stream retains the order of the higher order stream contributing to it. The result of this system of ordering is that it does not reflect any increments except approximate doubling of the discharge, assuming that streams of the same order in the same drainage basin are approximately equal in discharge.

Horton's original system is somewhat more complex than this (Fig. 6.31) in that the stream of maximum order in the drainage basin is determined and is then extended back to its farthest source. In Fig. 6.31 it is

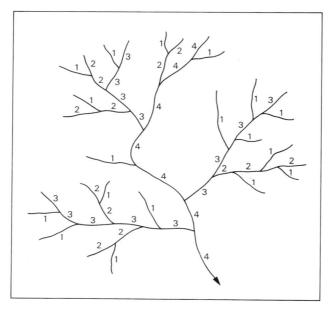

FIG. 6.31. Horton's system of stream ranking

shown as a fourth order stream right back to its source. Similarly the third order streams tributary to the fourth order stream are also extended back to their farthest sources as third order streams, and so on. The system of ranking should be clear from the figure. This means that Horton ranks whole streams, however artificially they may have to be defined, whereas Strahler ranks segments of the drainage system. With Horton's system there is an element of subjectivity in that a choice has to be made at some bifurcations between streams that are virtually equal. Horton himself suggested that at the last bifurcation the straight line continuation should normally be taken as the higher order stream.

A third method of ranking has been described by Shreve (1967) and differs from the previous two (Fig. 6.32). This consists of adding the rank numbers of the two streams contributing to a junction to arrive at the rank number of the stream below the junction. Thus, at any point on a drainage network the order of a stream is given by a number which represents the

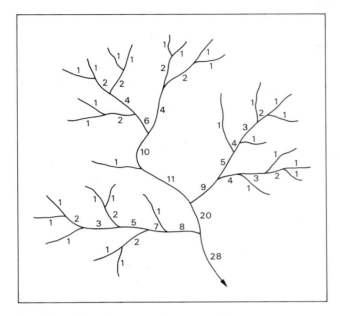

FIG. 6.32. Shreve's system of stream ranking

total number of the first order streams which have contributed to it. If we assume that first order streams are of approximately the same magnitude and that discharge is neither lost nor gained from any source other than the tributaries (which is not true), then the Shreve number is roughly proportional to the discharge in the segment of a stream to which it refers.

From the study of streams ordered by either the Horton or Strahler systems certain general tendencies or 'laws' may be derived.

If the logarithm of the number of streams in each order is plotted against stream order the points lie approximately on a straight line. In practice one plots the data on semilogarithmic paper to avoid looking up logarithms. This property is known as the law of stream numbers, which states that the number of stream segments of each order form an inverse geometric sequence with order number. If, in a given drainage basin, such a straight line relationship holds, it follows that the ratio between the number of streams of one order and that of the next higher order is also a constant. It is known as the bifurcation ratio and, in drainage basins which are not greatly distorted by geological factors, it usually varies between values of 3·0 and 5·0.

159

Similarly, if the logarithms of the mean lengths of the stream segments of different orders are plotted against stream order, the result is usually an approximately straight line. This is the law of stream lengths. It states that the mean lengths of stream segments of the successive stream orders approximate to a direct geometric sequence in which the first term is the average length of a first order stream. (It must be remembered that, as these analyses are almost invariably made from maps, one is really using the horizontal equivalents of lengths and areas.)

The law of basin areas follows the same general pattern. In a drainage network the mean basin areas of the orders approximate to a direct geometric sequence in which the first term is the average area of a first order basin. It should be noted that, just as with stream length, individual basin areas can deviate greatly from the mean for the order in question.

A comparable relationship was found between the average slopes of the stream segments of different orders and is known as the law of stream slopes. These form an inverse geometric series with order number, comparable with the law of stream numbers. It goes without saying that great lithological diversity, which affects gradients through load calibre and yield, can upset this relationship. So too can a basin which has had a multicyclic history, or one with a markedly horizontal structure of hard and soft beds, especially if the low order streams flow for the most part on a hard bed.

A similar relationship has also been observed between the average relief of the basins of different order streams and the stream order. Relief is defined merely as the difference in height between the highest and lowest points of the basins concerned.

In addition to these so-called laws, various useful indices can be derived from measurements made in drainage basin analysis. One cannot list all of them because ratios between all sorts of things can be produced, although not all have much if any use. Among the more useful measures are:

(a) *Drainage density.* This is simply the total channel length divided by the total area of the basin in whatever units are most suitable: km per/sq km is probably most useful for normal drainage and possibly m per sq m for miniature badlands. Wide ranges in drainage density values occur in the United States: the lowest, 2 to 2·5 km per sq km (3·0–4·0 miles per sq mile) are found on resistant sandstones in the Appalachians; values of 5 to 10 (8–16 miles per sq mile) occur in deciduous forest areas on rocks of moderate resistance in the central and eastern parts; 30 to 60 (50–100 miles per sq mile) is typical of the drier parts of the Rockies; 120 to 240 (200–400 miles per sq mile) has been reported from badlands and 660 to 780 (1 100–1 300 miles per sq mile) from miniature badlands (Strahler 1968).
(b) *Stream frequency* is the number of stream segments per unit area. Like drainage density it is also a measure of the texture of the drainage net and hence of the dissection of the land surface.

(*c*) *The length of overland flow*, i.e. the distance down the steepest slope from a point on a divide to a point on a channel, is related to drainage density. For first order basins it is approximately one half of the reciprocal of drainage density. It is a measure of erodibility.

(*d*) *The constant of channel maintenance* is a restatement the other way round of drainage density. It defines the area required to sustain one unit length of channel. By means of it one might be able to predict whether there was any likelihood of a drainage system extending headwards.

(*e*) *Basin shape measures.* Various indices can be used: the ratio between basin area and the square of basin length; the ratio between basin area and the area of a circle of a circumference equal to the perimeter length of the basin; the ratio between the diameter of a circle with the same area as the basin and the length of the basin.

These dimensions, ratios and laws are useful in various ways. At the lowest level they eliminate subjectivity. 'Finely dissected' is not likely to mean precisely the same thing to different people, but a drainage density expressed numerically, e.g. 5 km per sq km, does have precise meaning. Simple comparisons can also be made, for example, between drainage densities on rocks of the same type under different climatic and vegetation conditions.

At a higher level the quantification makes the study of drainage networks susceptible to statistical analysis, which cannot be applied to verbal descriptions, unless these are ranked in some crude way. Thus the way is opened for complex analyses of the interactions between many factors within drainage basins, and also between different drainage basins. With the mechanisation of these analyses by computers the most elaborate analyses can often be made in a few minutes of machine time.

Some of the measures have important bearings in applied geomorphology. For example, the degree of peaking in flood discharge, assuming that rainfall has occurred equally all over a basin, will be influenced by the proportion of throughflow to overland flow and by the structure of the basin network. The first of these is reflected in the drainage density and its associated measures, such as the length of overland flow, and the second in the general pattern of stream orders in the basin. This can be illustrated if one compares the sort of network shown in Fig. 6.30 with one in which a long second order stream receives a series of closely-spaced first order tributaries arranged in a herringbone pattern. In the first case, assuming a rather low drainage density, the waste mantle will first regulate the discharge as throughflow and afterwards the flood discharge will be transmitted at varying times to the main stream through the complex network. The flood will have a broad, flat peak. In the other basin the high drainage density implies rapid runoff, which will reach the second order stream almost simultaneously from all the tributaries. The quick rise and fall of peak discharge following rain or melting snow in low order basins can be readily observed in clay areas in Britain, especially if one compares it with the slower rise and fall in the higher order basin to which the low order basin is tributary.

There are, however, certain pitfalls and difficulties in the geometric analysis of drainage basins.

The measurements have to be comparable. It may seem that everyone must measure the same things in the same way, but the first point that comes to mind is that the number of channels shown on maps of different scales is not the same. Therefore, basins mapped on different scales may have apparently different orders, which merely reflect map information. Comparison of data from maps on different scales must be avoided, unless some arbitrary method of correction is employed. Even when the maps are on the same scale, one worker may measure the blue lines of the streams, another may project the channels back to the point at which the contours cease to be inflected. The work must be standardised. There may be differences between maps on the same scale because of different surveying authorities: there may be differences between maps on the same scale made by the same authority, because of greater or lesser care on the part of the original field surveyors, and because the contours which are surveyed and those which are interpolated may vary with the range of elevation in question. This is true in Britain. Again, one part of a map may be mapped with different intensity from another, especially when a boundary is involved. This can readily be observed in the case of road densities on the 1:1 000 000 Michelin Grandes Routes maps. It would be instructive to calculate route densities in those parts of France and Germany which appear on the map devoted primarily to France and on that devoted primarily to Germany. Probably none of these snags invalidates the system, though we must expect a certain haze of imprecision about the figures. A map scale of about 1:25 000 is considered adequate for most drainage basin analysis.

Care must be exercised if, in studying the smallest gullies, these are measured in the field and added to an analysis made otherwise from maps. The gullies are likely to occur on steep slopes and theoretically their lengths should be multiplied by the cosine of the angle of average slope to make them comparable with the map data.

The advantages of statistical analyses lie in their ability to reveal significant differences which are too subtle and too complex to be discernible by eye. If misused, they can become a lengthy and exhausting way of demonstrating the obvious. There is no point, for example, in going to great lengths to show that a number of very elongate basins can be correlated with the strike and the relative resistance (measured in some way or another) of alternating hard and soft beds when a glance at a geological map and memoir would have satisfied any normal person that this was indeed so.

Care must be taken that the central object of study remains the drainage, and does not become the geometric properties of networks as exemplified by drainage systems. The latter seems to me to be a branch of mathematics. Similarly, as with all mechanical procedures, especially when machines are available, there is the risk that work will expand to fill the available time and that complex analyses will be done without regard to their potential usefulness in adding to geomorphological knowledge about drainage systems.

Certain inferences that have been drawn from 'random walk' models must be considered. These are really games of chance. Their application to drainage systems can be illustrated as follows. Imagine a row of equally spaced points: these can represent the sources of first order streams on a watershed. One can imagine each point having a turn, as in ludo, and progressing a fixed distance at an angle to be determined by chance, provided that it is not less than 90 degrees, i.e. it must be downhill. Obviously a series of staggering paths will be generated (Fig. 6.33) and collisions will occur.

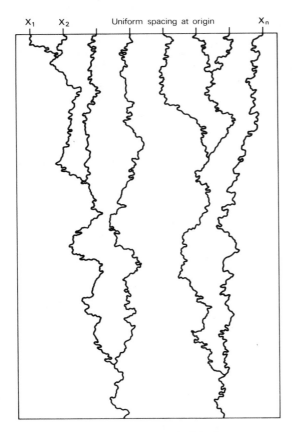

FIG. 6.33. Random walk model of drainage system
(*after Leopold, Wolman and Miller*)

163

After a collision the joint path is extended as a single unit. The collisions and elimination of paths are analogous to stream abstraction. The calculation of many such branching patterns allows one to determine the most probable system under the conditions postulated. Such artificial networks (Shreve, 1966, 1967) have properties and relationships closely approximating to the Horton 'laws' described above. It has consequently been suggested that some of these relationships are attributes common to all systems of randomly developed networks and not really laws of orderly stream development. Of course, it does not follow that stream networks are generated at random, even though random walk models approximate to them in many properties. Many would argue that streams do not develop from the head downwards but from the mouth upwards.

Most of the generalities mentioned in this section have been derived in the United States from small basins in weak rocks that rapidly adjust themselves to changing conditions. The distorting effects of structure and lithology have occasionally been mentioned. Provided that the assumption is made that some present property or properties of the drainage basin are controlling the stream and slope geometry, these can be measured and the quantitative description of the landforms completed by a statistical investigation to suggest the most likely explanation. If, on the other hand, the lithological diversity is great and the erosion history complex as in many parts of Europe, the quantitative description defines more clearly the problem which still has to be solved in terms of inference from field observations.

7
The effects of rocks on relief

The correspondence between the relief and the underlying rocks is often so close, as a comparison of geological and topographical maps will show, that the nature and the arrangement of the rocks are fundamental in the development of landforms. The structure of the rocks, in the narrower modern sense of the word, affects the general pattern of the relief, while the lithology of individual beds influences the relief in detail. But there are areas where the relief does not seem to be affected by the rocks, so that a geological map is not the only requirement for a study of relief.

The effects of rocks on relief can only be studied in outline here. Even an introduction to the broader field aspects of the subject would involve a whole book (Sparks, 1971), while it is quite obvious that detailed laboratory and field testing of rock properties is going to be an important and increasing future field of research.

The effects of structure

Where beds of differing resistance are involved in folding or faulting, the processes of erosion may cause the pattern of the outcrops to be reflected in the relief. Thus, alternate resistant and non-resistant beds in a gently dipping series may be etched out by erosion to form a typical scarp and vale landscape, like that of southern and eastern England. Symmetrically folded beds may lead to a pattern of long parallel ridges and valleys, relief of this type being found in the Jura. Pitching folds may be reflected in a pattern of converging and diverging ridges like that of certain sections of the Appalachians. The arrangement of the relief is merely the pattern of the interfluves, so that in an area of well-adjusted drainage, the study of the drainage pattern is automatically the study of the relief pattern, both being basically dependent on structure.

There are, however, certain conditions in which the structural pattern will not be reflected in the relief. If the tilted or folded rocks contain no significant lithological contrasts, the structure may not influence the relief appreciably. In the East Midlands, for instance, the Upper Jurassic consists of three clay formations, the Oxford, Ampthill and Kimeridge Clays, which, being of approximately uniform resistance, generally form a wide belt of low-lying ground, part of which is occupied by the basin of the Fens. Similarly, the Silurian and Ordovician rocks of parts of Wales and the Southern Uplands of Scotland are comparatively uniform in lithology, so that even the complicated folds of the Southern Uplands are not always reflected in the relief.

Much also depends on the history of erosion of any particular region. In the Davisian cycle of erosion the maximum adjustment of drainage to structure occurs at a late stage and one would expect the structural pattern to be most evident in the relief at the same time. In youthful landscapes erosion may have been acting for an insufficiently long period for the structure to be developed into relief. In the extreme example of an uplifted surface of marine denudation, on which subaerial erosion is only just starting to work, there will be virtually no structural influence on relief. A clear example of this effect can be taken from the coastal plain of west Sussex and eastern Hampshire (Fig. 7.1). The plain was eroded by marine

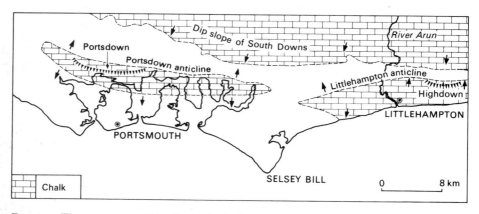

FIG. 7.1. The structures hidden beneath the Sussex and Hampshire coastal plain. Arrows indicate direction of dip

action at two main levels, about 40 m (130 ft) and about 7·5 m (25 ft) OD, the Chalk exposed in the Littlehampton and Portsdown anticlines being completely planed off except for fragments of the northern limb of each fold, which form Portsdown (120 m (409 ft) OD) and Highdown (80 m (269 ft) OD) respectively. There is little relationship between structure and relief, as the former is hidden beneath a cover of Pleistocene deposits. If the area were to be uplifted, the structure should appear in the relief as the streams removed the Pleistocene deposits and re-exposed the Chalk.

PLATE 8. The Malvern Hills from Herefordshire Beacon. The hills are developed on resistant Pre-Cambrian gneiss between Triassic sediments to the right and Silurian and Devonian sediments to the left

The effects of lithology

The resistance of an individual rock to erosion depends on a number of factors such as hardness, permeability and jointing. Whether the resistance of a rock is effective in forming relief depends largely on the lithology of adjacent rocks and on the time during which erosion has acted. Two sets of factors may, therefore, be separated: those governing the resistance of rocks and those governing the effect of resistance on relief.

Factors governing resistance to erosion

(a) HARDNESS

The effect of rock hardness on resistance to erosion is impossible to measure. Provided that all other things are equal, including resistance to chemical erosion, the harder rocks will obviously form the higher ground. The Pre-Cambrian rocks of the Malvern Hills stand up abruptly from the Devonian and Triassic sediments of the surrounding plains (Plate 8); Pre-Cambrian rocks in the Charnwood Forest project through the Trias of the Leicester district; the volcanic rocks of the Ochils form a great escarpment where they are faulted against Carboniferous sediments (Plate 9) and, on a larger scale, the older, harder rocks of the north and west of Britain form higher ground than the younger, softer rocks of the south and east.

(b) PERMEABILITY

The principal erosion of the land surface is caused by running water, so that rocks which naturally reduce runoff and allow ready percolation often possess considerable resistance to erosion. Permeability may often be of more significance than hardness, for even soft limestones, such as chalk, often form high ground. A probable example of this effect is the Carboniferous Limestone escarpment near Shap on the eastern side of the Lake District. The Carboniferous Limestone lies over folded Ordovician rocks, in places the Borrowdale Volcanics, which are resistant but impermeable (Fig. 7.2) and, therefore, support a surface drainage which erodes them, while the Carboniferous Limestone, having little surface drainage, suffers little surface erosion.

The significance of permeability is also well illustrated by the Chalk in the Cambridge district. The Chalk Marl, which forms the lower half of the Lower Chalk, contains such a high proportion of clayey impurities that it is virtually an impermeable rock. The main spring line is at the level of the Totternhoe Stone, the underlying Chalk Marl together with the Gault forming the strike vale of the river Cam (Fig. 7.3). In fact, from a geomorphological point of view it is more satisfactory to group the Chalk

PLATE 9. The Ochils from Stirling Castle. The Ochils, formed of Devonian lavas and ashes, are separated from lowlying Carboniferous rocks to the right by a magnificent fault line escarpment. In the left centre the wooded hill is developed on a sill

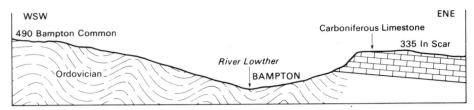

FIG. 7.2. Relations between Carboniferous Limestone and Ordovician rocks near Shap

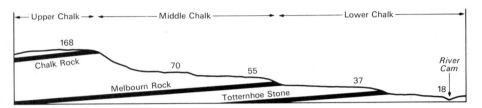

FIG. 7.3. Chalk escarpment of south-west Cambridgeshire

Marl with the Gault than with the main mass of the Chalk above. A second but less important spring line occurs at the level of the Melbourn Rock, where the water is thrown out by a metre or so of marly chalk beneath. Spring sapping at these two levels is controlling the recession of the erosion terraces in front of the escarpment. Thus, the permeable rocks are little affected by surface erosion, but are cut back from their margins where springs are thrown out by underlying impermeable beds.

Although limestones are among the most permeable of rocks, they may occasionally be in juxtaposition with even more permeable beds. On certain parts of the dipslope of the North Downs the lower Eocene beds appear to be even more permeable than the Chalk, as is suggested by the way in which the Eocene beds form a minor escarpment just east of Croydon (Fig. 7.4). Presumably, water issuing at the base of the Eocene as springs

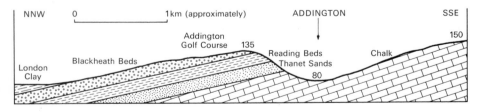

FIG. 7.4. Sketch section across the Tertiary-Chalk junction east of Croydon

tended to flow along the Eocene-Chalk junction and to erode there a strike valley, the northern side of which is still capped by the permeable sands and gravels of the lower Eocene.

(c) Joints, cleavage, schistosity, bedding planes.

Minor rock structures of these types have been considered in Chapter 3, as they have an important effect on weathering. The rate of weathering obviously influences the resistance to erosion, provided that the products of weathering can be easily removed. In addition, blocks of material may be pulled away by erosive agents, such as water and ice, when the joints have been opened up by weathering. Usually, then, joints tend to facilitate erosion, but the presence of an open joint system also increases the permeability of a rock and thus increases its resistance to erosion. Carboniferous Limestone, for example, probably owes its permeability largely to its joints.

The effect of resistance on the relief

(a) The importance of adjacent beds

If a layer of moderately resistant rock could be sandwiched between two layers of non-resistant rocks in one place and between very resistant rocks in another, the rock concerned should form a ridge in one place and a valley in the other. Although few rock types remain lithologically constant over large areas, a point which is referred to below, examples illustrating this effect can be drawn from Carboniferous Limestone areas.

One of the best Carboniferous Limestone escarpments in the country is that on the west of the Pennines overlooking the Vale of Eden (Fig. 7.5).

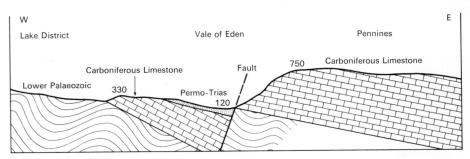

Fig. 7.5. Diagrammatic section through the vale of the Eden

The juxtaposition of resistant, permeable Carboniferous Limestone against weaker Permo-Trias rocks by faulting is probably the main factor controlling the relief.

In the northern part of the South Wales coalfield, however, the Carboniferous Limestone is not a great relief former. It is comparatively thin and, being wedged between the resistant Old Red Sandstone of the Brecon Beacons and the Millstone Grit, which contains large amounts of quartz conglomerate here, it forms only a minor relief feature (Fig. 7.6). Although the limestone is thinner than it is in the Pennines, it is probably the resistance

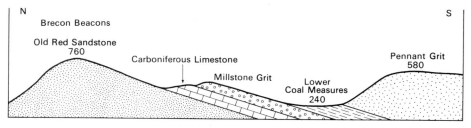

FIG. 7.6. Sketch section through the northern part of South Wales coalfield

of the adjacent rocks which is significant in suppressing the influence of the Carboniferous Limestone on the relief.

(b) THE EFFECT OF DIP

The area covered by a bed of given thickness will depend largely on the dip. A bed 300 m (1 000 ft) thick may cover many square kilometres if horizontal, e.g. the Chalk of Salisbury Plain, but, if vertical, its outcrop will be only 300 m wide, e.g. the Chalk in parts of the Isle of Wight. As the agents of erosion have to reduce a larger area in the first instance, it is probable that any bed will retain higher elevations where its dip is slight than where it approaches the vertical. The effects are admirably illustrated by some of the Chalk outcrops of southern England.

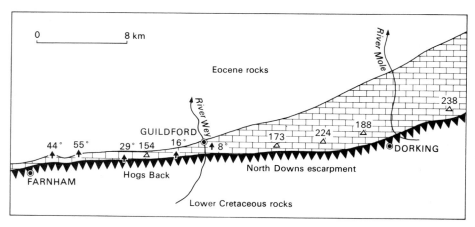

FIG. 7.7. Relations between angle of dip and elevation in the western part of North Downs

The Chalk outcrop of the North Downs (Fig. 7.7) narrows steadily westwards from the Mole gap to the Hog's Back west of Guildford, while the dip increases westwards from a few degrees to 40 to 50 degrees near Farnham. The crest of the escarpment in the east is above 210 m (700 ft) OD, but it falls westwards and, on the Hog's Back, barely exceeds 150 m

(500 ft) OD. West of Farnham, where the dip decreases, the outcrop widens and the crest elevations increase again.

The central Chalk outcrop of the Isle of Wight shows comparable features. In many places the beds approach the vertical (Fig. 7.8), as they are

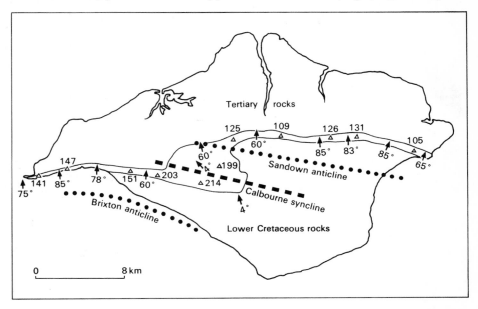

FIG. 7.8. Relations between angles of dip and elevation on the main Chalk outcrop of the Isle of Wight

involved here in the Isle of Wight monocline, but in the centre of the island the Chalk outcrop widens, due to the influence of the Calbourne syncline. Away from the central area the maximum elevations are everywhere below 150 m (500 ft) OD, but in the centre, where the dip is slight, the land rises to 210 m (700 ft) OD.

(c) THE EFFECT OF THE LENGTH OF THE PERIOD OF EROSION

In discussing the influence of structural patterns on relief it was shown that the length of the period of erosion is very important. The same is true when one particular bed is being considered: it may form no relief on an uplifted plain of marine erosion, but, given sufficient time, it will probably be etched out as an area of high ground.

Effects of this sort are seen on the Carboniferous Limestone near Llangollen. Immediately north of the town Eglwyseg Mountain is a very impressive escarpment (Fig. 7.9B), as the Carboniferous Limestone is probably harder and certainly more permeable than the underlying Ludlow Shales. The junction is being etched out by the little Eglwyseg river, which flows at the foot of the scarp on the impermeable Ludlow Shales. A few

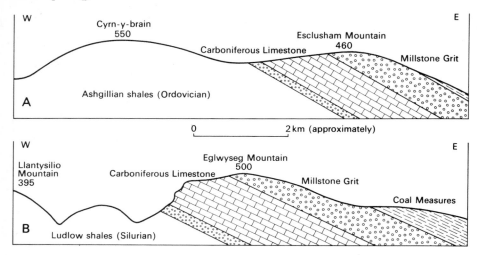

FIG. 7.9. Sections north of Llangollen

kilometres to the north, however, Cyrn-y-brain on Ordovician shales and
Esclusham Mountain on the Carboniferous are separated by no more than
a shallow depression (Fig. 7.9A). The Ordovician rocks seem to be similar
lithologically to the Ludlow Shales to the south, so that the absence of an
escarpment seems to be due, not to changes of lithology, but merely to the
fact that the Eglwyseg river has had insufficient time to erode northwards
and to expose the lithological differences. Whether the Carboniferous
Limestone forms an escarpment in this district seems to depend mainly on
the stage of erosion reached.

All the factors discussed above are affected by changes in climate, as
climatic changes involve a shift in the assembly of processes, but a fuller
discussion of this point must be left to the chapters on arid and glacial landforms.

Practical considerations

In practice, the various ways in which rocks affect relief may rarely be
isolated. As no single rock type remains homogeneous over considerable
distances and many of them change to a marked degree, it is difficult to
generalise about the effects of rocks on relief. That such generalisations
have many exceptions can be seen from a consideration of the way in which
some of the British granites affect the relief.

The common conception, that granites form upland areas of rounded
relief in this country, holds for many intrusions but not for all (Plate 10).
A typical example of granite forming high ground is found in Arran (Plate

PLATE 10. Rannoch Moor, Argyllshire. The heavily glaciated granite does not form
outstanding relief

11), where the main mass in the northern part of the island exceeds 840 m (2 800 ft) OD, while the mixed sediments and igneous rocks of the southern part of the island hardly reach 450 m (1 500 ft) OD, this maximum elevation being on a small intrusion. A comparable upland granite area is Dartmoor, which, with a maximum elevation of just over 600 m (2 000 ft) OD, forms the highest land in south-western England. In both examples the superior resistance of the granite appears to be the obvious cause.

Much more subdued granite relief is found on the eastern side of Bodmin Moor, where the granite rises about 180 m (600 ft) above the surrounding Devonian sediments (Fig. 7.10), which form a marked plateau at 150–195 m (500–650 ft) OD. At the present time the river Lynher does not seem to be actively eroding the escarpment, as in two places, to the north-west of

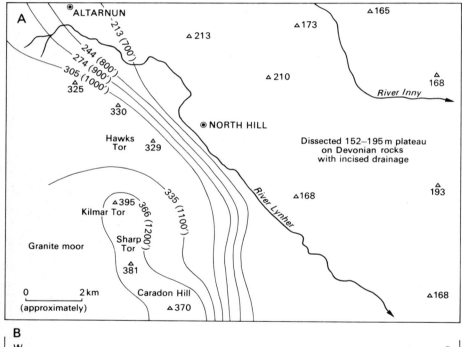

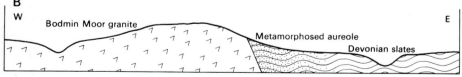

FIG. 7.10. Sketch map and section of the eastern side of the Bodmin Moor granite. Generalised contours are shown on the granite

PLATE 11. Goat Fell from Brodick, Arran. The granite of Goat Fell rises sharply above the surrounding plateau (AA′) developed on Dalradian and Devonian rocks

North Hill and in the south of the area mapped the plateau continues west of the river, which is incised some distance away from the edge of the granite. The steep eastern slopes of the granite may be relics from an earlier period of erosion, when the plateau on the Devonian rocks was cut by rivers or by the sea, which was unable to wear the granite down to the same extent. The escarpment, on this assumption, would still reflect the relative resistance of the rocks in relation not to the present cycle, but to an earlier one.

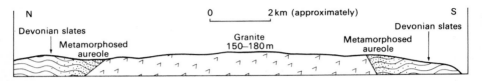

FIG. 7.11. Sketch section across the eastern part of the St Austell granite

The eastern part of the St Austell granite (Fig. 7.11) affects the relief to a very limited extent. Whether this is due to there being only a slight difference between the resistance of the granite and that of the surrounding rocks is uncertain, as it appears from a map that the area may have been planed off by some agency and that there has been too little time since the planation for subaerial erosion to work out the lithological differences.

The western part of the Shap granite affords an example where the intrusion appears to be slightly less resistant than the surrounding rocks. The granite area is to a considerable extent covered with peat bog, while a slight rise and a change to a drier type of vegetation denote the onset of the Ordovician volcanic rocks to the west.

One of the finest examples of an intrusion forming lower ground than the surrounding sediments occurs in the Southern Uplands of Scotland. The Loch Dee intrusion (Fig. 7.12) is complex, but consists essentially of a central mass of granite surrounded by tonalite, an intermediate rock containing considerable amounts of plagioclase felspar and quartz. The surrounding Ordovician sediments have been metamorphosed at their contact with the intrusion. The highest ground is found on the altered sedimentary rim, the Merrick range in the west reaching 830 m (2 764 ft) OD and the Rhinns of Kells in the east 800 m (2 668 ft) OD. Within the igneous area the central granite rises to 680 m (2 270 ft) OD at Mullwharchar and stands above the surrounding tonalite. It is interesting to find, not only that the igneous rocks have proved less resistant than the sediments, but also that the more basic igneous rock, the tonalite, has been more readily eroded than the granite, a state of affairs to be expected had chemical weathering been the main agent (see Chapter 3).

Finally, in Skye the granites of the Red Hills form considerably higher ground than the basalts but are lower than the gabbro mass of the Cuillin Hills. Unlike the Loch Dee area, the acid rocks of Skye have been more

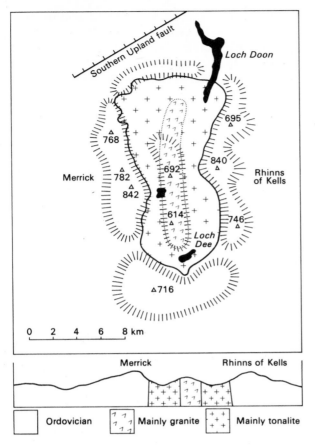

Fig. 7.12. Diagrammatic map and section of the
Loch Dee intrusion

eroded than the basic rocks, an apt illustration of the complexity of
weathering processes and the dangers of generalisation.

The significance of facies

If the British Isles and the adjacent areas of sea floor were to be uplifted and
the marine deposits so exposed consolidated, a variety of rock types would
be formed. Near the old coastline conglomerates would represent the old
beach pebbles, sandstones would be formed from the shallow water deposits, and clays or shales from the finer deposits of deeper water. These
rocks would be very different although of the same age: they are termed
facies and the general aspect of the rock changes from one facies to another.
Certain geological beds show many facies, a notorious example being the
Carboniferous Limestone. In south and central Britain it is predominantly

a limestone formation, but in north-east England, Scotland and Ireland quite different facies occur. In the Central Lowlands of Scotland coals and oil shales are important rocks in the Carboniferous Limestone and changes of this sort are bound to affect the relief.

In Arran Carboniferous Limestone appears on the north-east coast, where it is faulted against Dalradian schists (Fig. 7.13). It is not wholly a

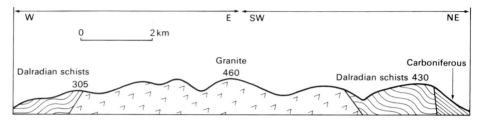

FIG. 7.13. Sketch section across the northern part of Arran

limestone formation. Partly for this reason and partly because of the resistance of the schists, it does not form high ground but merely a steep slope down to the sea.

In southern Ireland yet another facies of the Carboniferous Limestone occurs: in the far south-west it is a shale formation largely metamorphosed to slates in the Hercynian period of folding. The synclines in which the Carboniferous Limestone is preserved form the long bays and valleys of the area, while the intervening ridges and headlands are formed of Old Red Sandstone exposed in the anticlines. Even around Cork, where limestone becomes more important in the Carboniferous Limestone, the Old Red Sandstone is still more resistant as can be seen from the outline of Cork Harbour, which represents a drowning of subaerial relief (Fig. 7.14).

Another bed subject to marked lithological variations is the Lower Greensand of the Weald, the main divisions of which are shown in Fig. 7.15. The principal variations occur in the Hythe beds, which in East Kent are alternations of sands and sandy limestones giving rise to a marked but not outstanding escarpment. In west Surrey, however, they are reinforced by cherts and hard sandstones, which cause the Lower Greensand to dominate the local relief in such places as Leith Hill (290 m: 965 ft OD), Hindhead (268 m: 895 ft OD) and Blackdown (275 m: 918 ft OD). Such relief continues round the western end of the Weald into West Sussex, where there is a marked escarpment on the Hythe Beds, the strike vale of the Rother on the Sandgate Beds and a series of low plateaus on the Folkestone Beds (Fig. 7.15). At the river Arun the lithology changes abruptly, the hard beds disappearing, so that to the east the whole of the Lower Greensand makes a feature of minor importance easily missed in the field.

Changes of facies, such as occur in the Carboniferous Limestone and the Lower Greensand and to an even greater extent in the English Jurassic,

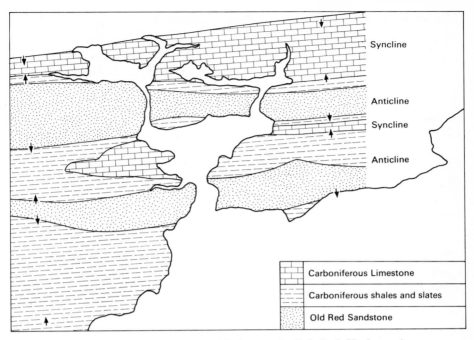

Fig. 7.14. Relations between structure, lithology and relief, Cork Harbour. Arrows indicate dip (*after Charlesworth*)

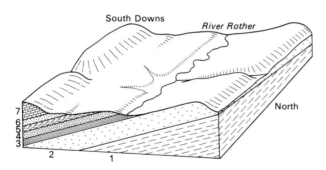

Fig. 7.15. Relations between rocks and relief round Midhurst, Sussex

        7. Chalk
        6. Upper Greensand
        5. Gault
        4. Folkestone Beds ⎫
        3. Sandgate Beds   ⎬ Lower Greensand
        2. Hythe Beds
        1. Weald Clay

must be considered in any analysis of relief. They add greatly to the difficulty of generalising about the effects of rocks on relief.

The importance of detailed lithological studies

Although the study of facies is really a study of lithological detail, yet another problem must occasionally be considered: it is sometimes found that beds which appear to be hard form no significant relief features, while in other places appreciable relief appears to occur without corresponding lithological differences.

The Melbourn Rock, a nodular rubbly bed at the base of the Middle Chalk, has been said to cause a ledge in the Chalk escarpments of parts of south-east England. It has been shown, however, that the ledges in the face of the South Downs escarpment are probably erosion terraces, as they do not coincide with the Melbourn Rock (Bull, 1936), and the same appears to be true of parts of south Cambridgeshire (Sparks, 1957), where comparable ledges stand out in front of the escarpment (Fig. 7.3).

While the obvious harder beds appear to affect the relief of chalk areas very slightly, there are marked effects in places where it is very difficult to distinguish any significant lithological difference. In parts of the Hampshire Basin and South Downs, a second escarpment is sometimes found half way down the dipslope (Fig. 7.16), but it is very difficult to find

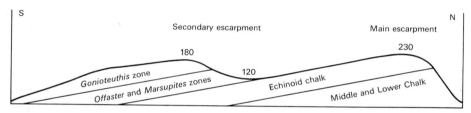

FIG. 7.16. The South Downs escarpments in relation to the divisions of the Chalk

the lithological cause. The arrangement of the Chalk zones involved is shown in Fig. 7.16, and from this disposition it seemed that there must be some factor giving superior resistance to the Chalk of the *Actinocamax*, or, as it is now called, *Gonioteuthis* zone, when compared with the *Offaster* and *Marsupites* zones below. But the only differences detected in an earlier study (Sparks, 1949) were the presence of a greater number of thin marl seams in the latter two zones, and an apparently greater percentage of insoluble residue (3·3–4·1 compared with 1·4–1·9 in the Chalk above and below). It was suspected that this might have slightly decreased the permeability when the water table was high and so allowed more surface erosion, although the amount of evidence for differences in insoluble residue percentages was too slight to be at all certain.

Much later Williams (1969) performed laboratory freeze-thaw experiments on rocks from these zones, and, although such experiments are difficult to set up and the interpretation of their results is not always clear, the Chalk of the *Offaster* and *Marsupites* zones did seem to be more susceptible to freeze-thaw as a result of a higher marl content.

Recently the question has been discussed again by Small and Fisher (1970). On the basis of more insoluble residue analyses it was shown that the percentages were more variable than shown in earlier work, though the mean for the *Actinocamax* zone is little more than a half of the mean for the *Offaster* and *Marsupites* zones combined. However, the authors rightly stress that it is fracture permeabilty which really counts in the Chalk and not even marl seams are impermeable if they are fractured. The authors incline to put their faith in the greater frequency of nodular flint seams in the *Actinocamax* zone and perhaps also in more effective jointing of that zone: the latter of these two suggestions seems the more likely.

The whole question is a local one but it well illustrates the great difficulty of locating the lithological reasons for a marked relief feature. No more progress in the determination of the average percentage of marl seems possible without a massive sampling exercise, for, even though intrazone variations may occasionally exceed interzone variations, there may still be significant differences between the zones. These seem to be more likely to have affected freeze-thaw susceptibility than permeability, and it may be that periglacial processes contributed to the lowering of divides between valleys and the formation of the escarpment in its present form. At the same time the characteristics of the *Actinocamax* zone stressed by Small and Fisher may well have contributed in the other direction by protecting the chalk of that zone from denudation.

Igneous relief

While every rock produces relief features with a certain degree of individuality, certain rocks, notably igneous rocks and limestones, give rise to landforms that are so distinctive as to deserve special treatment.

A distinction may be made between modern extrusive rocks on the one hand and intrusions and ancient extrusions on the other. The modern extrusives, the lava flows, add to a landscape an area which may be in a strikingly different stage of development from the rest of the landscape. The intrusions and the ancient extrusions have to be exhumed from beneath a cover of sediments and their general shape may be reflected in the landscape. The jointing, characteristic of many igneous rocks, leads to distinctiveness in detail in some cases, though it must be remembered that many sedimentary rocks also have joint patterns. A schematic illustration of the effects of various types of intrusions and extrusions on relief is given in Fig. 7.17.

Many intrusions are roughly circular or oval in plan, and if they are intruded into sedimentary rocks in which the relief trends are linear, they may give rise to areas in which the general pattern of the relief is distinct. The domed form of such batholiths or bosses may also be reflected in the general elevation of the surface, as if the igneous rocks are more resistant, their surface may be approximately exhumed by erosion. The exhumation

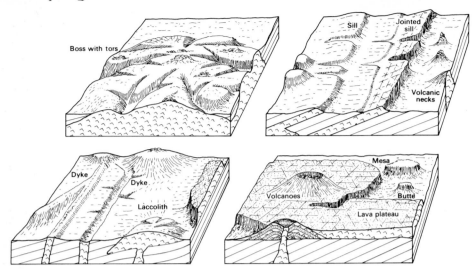

FIG. 7.17. Diagram of possible relations between igneous rocks and relief

is far from perfect and the actual surface of the intrusion probably rarely survives over large areas, as can be seen from the example of the Wicklow Mountains (Fig. 7.18). The detail of the relief, especially when the intrusions are granites, may be considerably influenced by the jointing: masses of weathered jointed granite may be left as tors on the higher parts of the intrusion (Figs. 3.5 and 7.17), while many of the smaller valleys may be locally guided by the joints.

Smaller intrusions such as laccoliths and volcanic necks may produce distinctive if smaller effects on the general relief. Laccoliths, which are closely related to sills but which were formed from a magma too viscous to spread far, may form local domelike features when exposed by erosion. The type laccoliths are those of the Henry Mountains of southern Utah (Gilbert, 1880; Hunt, 1953), where porphyritic trachyte, a viscous magma responsible in other areas for such bulging features as the Puy de Dôme in central France, has been intruded into Mesozoic strata. The effects of the

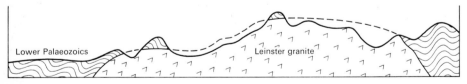

FIG. 7.18. Section across the Wicklow Mountains (*after Charlesworth*)

PLATE 12. The south coast of Arran from Bennan Head. Numerous basic dykes project like groynes into the sea. The '25-foot' (7·5 m) raised beach and abandoned cliff are clearly seen. In the left foreground is the edge of the quartz-porphyry sill of Bennan Head

intrusions on the relief are clearly seen in this semi-desert area, as erosion is at present engaged in exposing them (Fig. 7.19).

Consolidated magma in the vents of ancient volcanoes is often considerably more resistant than the ashes of the cones and the surrounding

FIG. 7.19. An eroded laccolith in the Henry Mts, Utah (*after Gilbert*)

rocks, both of which may be removed by erosion to leave the volcanic neck as a conspicuous feature of the relief (Fig. 7.17). Many examples occur in the Central Lowlands of Scotland and in the Puy district in the upper Loire valley, central France.

Dykes (Plates 12–14) may form linear features running across country at an angle to the general grain of the relief, while sills (Plate 3) often form especially resistant members of the series into which they are intruded. Not only may they form escarpments or ledges, but their jointing may add a distinctive feature to the relief: the quartz dolerite Whin Sill of the north of England affects the coastal scenery in this manner, while inland it causes waterfalls on some streams and in some places a marked feature, as for instance where Hadrian's Wall is built upon it. Sills and dykes are defined geologically by their relation to the rocks into which they are intruded, a sill being concordant with the bedding and a dyke discordant. From a geomorphological point of view the present attitude of the intrusion is the factor of the greatest importance: a vertical sill, for instance, may have roughly the same effect on relief as a dyke.

Volcanic rocks, except those which have been exhumed from beneath a great thickness of sediments, form distinctive relief. Volcanoes themselves vary from the enormous gently sloping cones of the Hawaiian Islands to

PLATE 13. Ceum na Caillich, Arran. An eroded basic dyke in massive, jointed granite

composite ash and lava cones, such as Vesuvius, and to the bulging trachytic puys of central France. Generally speaking, volcanoes emitting basalts produce streams of fluid lava, while more acid magma is usually accompanied by explosive activity with the formation of ash and fragmented material. This new constructional volcanic relief may be added to an old eroded landscape: the puys of central France are of late Tertiary or Quaternary age and they surmount a landscape formed essentially of an uplifted, faulted and tilted surface of erosion of early Tertiary age.

Although volcanoes are spectacular, they are not as important regionally as the lava plateaus of the world. Whether such plateaus were formed by fissure eruptions or by the spreading of very fluid lava from a series of central volcanoes is a geological rather than a geomorphological problem. Whatever the origin the result is a plain of lava, a constructional form in the midst of an erosional landscape in many instances. The western part of the Deccan of India, parts of central Arabia, the Snake–Columbia basin of the north-west of the United States, and the Inner Hebrides and parts of the adjacent mainland are all extensive areas of lava. Typically the relief is plateaulike, abrupt edges occurring either at the coast or where erosion has cut through a lava bed into less resistant rock such as an ash bed. On the scarp edges of the plateaus the detail of the relief is sometimes strongly affected by the vertical polygonal jointing typical of basalts and various other lavas. Unlike major intrusions, which usually have rectangular joints with an important set parallel to the surface of the intrusion, the lava flows have dominantly vertical joints (Fig. 3.6). The dissection of lava plateaus, especially in semi-arid regions, leads to the formation of detached portions or outliers, the jointing strongly influencing these mesas and buttes as larger and smaller outliers are termed (Fig. 7.17).

A comparable type of plateau, though of rather different origin, is the ignimbrite sheet of North Island, New Zealand (Cotton, 1944). Ignimbrite, or more expressively welded ash, is the name given to the product formed by the deposition of great thicknesses of incandescent ash. If the temperature of the material is sufficiently high, the particles of ash may fuse together forming a rock of cellular texture, which, if buried, often becomes firmly welded to a strong rock resembling rhyolite or even obsidian. Like basalts, ignimbrite may become vertically jointed, though usually less regularly than basalt, and steep, jointed scarps may result from erosion (Fig. 7.20).

Both basalts and ignimbrites, then, may form plateau relief with scarped edges; the valleys which cross them will tend to be steepsided and, if layers of alternate harder lavas and softer ashes occur, the harder beds will tend to form ledges on the valley sides.

PLATE 14. An eroded basic dyke in Dalradian metamorphic rocks on the north coast of Arran

Fig. 7.20. Escarpment of ignimbrite plateau, North Island, New Zealand (*after Cotton*)

Limestone relief

Limestones develop more distinctive relief features than any other type of rock, primarily as the result of the jointing of the rock, its permeability and its solubility in water containing carbon dioxide or humus acids. The lithology of the limestone is also significant, as hard, well-jointed limestones possess different features from those of soft limestones such as chalk (Plates 4, 15–17).

Landforms on massive limestones

Rocks such as the Carboniferous Limestone of parts of England, the resistant Mesozoic limestones of the Causses of central France and similar rocks in the Karst region of Yugoslavia, are often said to possess typical limestone landforms. The main process involved is the widening of fissures, joints and faults by solution, and a prerequisite for its action is that the water table must be well below the surface to allow water to percolate continually downwards through the rock. The vocabulary used to describe the features so produced is a compound of terms from several languages and dialects: the synonyms are many, the meanings of the terms not immediately apparent, and their pronunciation often difficult, so that it is tempting to ignore them completely. The words are not technical terms describing specific geomorphological concepts, but local names for physical features, which can be readily described in everyday words. For this reason the following descriptive account makes use of common words, the local terms being inserted in brackets at suitable positions.

PLATE 15. The Chalk escarpment of the South Downs near Brighton, Sussex. This soft limestone escarpment should be compared with the hard limestone escarpment shown in Plate 16

The surfaces of many limestones are minutely sculptured. In the Ingleborough district of Yorkshire and in the western parts of Eire the Carboniferous Limestones usually have the joint planes enlarged as grykes, while the joint blocks remain as clints (Plate 18). Of different origin (see below) are the finely dissected limestones of other parts of the world whose surfaces are fretted with systems of minute channels, for which the usual modern term is Karren, although lapiés has been commonly used in the past. Some of these features are due to joint enlargement, but others are smaller-scale runnels on the blocks bounded by the joints.

Provided the water table is below the surface rain will percolate downwards along the joints and at certain positions, probably the intersections of joints, there may be naturally easier channels. These will be slowly enlarged by solution into holes, the shape of which will depend largely on the control exercised by the minor structural features of the rock. Two main types of solution holes have been distinguished; funnelshaped depressions with a hole at the centre (doline, sotch, creux, sink hole, swallow hole, swallet) and shaftlike holes (ponor, avens, gouffre, puits). With continued solution such holes may enlarge and in places several may coalesce to form larger, compound solution holes (uvala). It is also possible to differentiate between solution holes formed mainly from the surface downwards and those caused by underground solution followed by surface collapse, a point which will be discussed below.

The largest depressions of Yugoslavia, the poljes, are probably not solution forms at all but tectonic depressions modified by solution of the limestone preserved in them.

The surface drainage, which sinks through these various types and sizes of solution hole, escapes laterally underground. The actual form of the water table will depend on the degree of permeability of the rock. If the joints are small and irregular, a head of water must be built up before lateral movement can take place and this will result in the water table being domed up. If the joints are open, especially if they have been enlarged by solution, the water escapes rapidly and the water table may be almost flat.

One may distinguish, after Cvijić, between parts where the underground drainage is mainly downwards and parts where it is mainly horizontal. In reality there will be three zones, an upper or vadose zone in which the movement is always downwards, a lower phreatic zone of permanent saturation where the gradient of the water table ensures that the drainage is laterally directed towards the margins of the limestone area, and a middle zone which is occupied by streams only when the water table is high. The lateral movement of the underground streams results in the frequent formation of caverns in the lower two zones, probably near the junction of the two, as lateral flow will be most frequent at this level. Certain problems of cavern formation are discussed below.

PLATE 16. Durness Limestone (Cambrian) escarpment, near Inchnadamph, Sutherland, in late afternoon sun

Thus it can be seen that the action of chemical erosion on limestones results in a fretting of the surface, the disappearance of surface drainage down solution holes and its escape underground through a series of channels, largely joint-guided but opened up by solution.

If for some reason, such as the erosion of adjacent areas of impermeable rocks, the water table in a mass of limestone becomes lowered, the main underground channels will be displaced to successively lower levels. At the same time general lowering of the limestone surface is said to thin the rock above the underground caverns so much that eventually the roofs collapse and the drainage reappears at the surface in deep narrow gorges. Certain narrow valleys in Yugoslavia have been attributed to such cavern collapse and the same hypothesis has been applied to Cheddar Gorge. It must not be thought, however, that every narrow limestone valley is a collapsed cavern. In fact, the general lowering of the whole surface by solution is probably such an extremely slow process that one has doubts whether any narrow limestone valleys are collapsed caverns. The lowering of the water table in a limestone region is effected largely by the entrenchment of the valleys. During the process some rivers cut down their valleys more rapidly than others and, by the underground abstraction of drainage, the majority of the valleys become dry. The rivers which survive receive no surface tributaries but, instead, a supply of water from springs at river level. Streams having headwaters outside the limestone region will be assured of a supply of water, and consequently may survive as the main surface streams. Very good examples are provided by the Tarn and the Jonte in the Causses of France. The valleys of the few surviving streams are steep sided and deep because there is no surface drainage nor tributary streams to erode back the valley sides. In fact it is probably more usual for the water table to be lowered by the main streams than for some unspecified external cause to effect such a rapid lowering of the water table that all streams dry up.

Some streams in limestone districts may be partly underground and partly on the surface, one example being the stream from Malham Tarn, Yorkshire, which disappears into a plateau and reappears lower down in the headwaters of the river Aire. A somewhat different disappearance and reappearance of a stream takes place in the headwaters of the river Neath near Ystradfellte in South Wales. The upper course of the stream follows the dipslope of the Brecon Beacons on Old Red Sandstone, from which it passes on to the Carboniferous Limestone, where for about 250 m its course is underground (North, 1949). The stream disappears into a cave, Porth-yr-Ogof, at the base of a cliff, 9–12 m (30–40 ft) in height, and reappears from a similar cave lower down. A possible explanation for such behaviour is that, during a period of rapid downcutting, the drainage across the

PLATE 17. Chalk valleys in the South Downs, north of Goodwood Racecourse, Sussex

Carboniferous Limestone was diverted to joints underground and has re-mained there ever since.

In the Yugoslavian Karst region, which is probably unique both on account of its area and because of the great thickness of its limestones, it has been found possible to formulate a Karst cycle of erosion (Saunders, 1921). The cycle, which was developed by Cvijić, includes three important assumptions: a thick and extensive mass of limestone, an underlying im-permeable stratum, and a surface layer of impermeable rocks for the initia-tion of a stream pattern.

In the youth of the cycle the upper impermeable layer is removed by the streams, which then proceed to disappear underground through enlarged joints and fissures (Fig. 7.21A). It is really a period of disintegration of the drainage pattern: streams flow down normal valleys only to disappear into solution holes at blind ends. Throughout the youth of the cycle the under-ground drainage develops slowly and when it is all underground maturity is reached (Fig. 7.21B). At this stage it will possess the three zones mentioned above: the upper zone of downward movement, the lower zone of lateral movement, and the middle zone which, according to the level of the water table, is alternately one of lateral and downward movement. Late maturity sees the underground streams reaching the impermeable underlying stra-tum and the collapse of the cavern roofs due to the lowering of the surface (Fig. 7.21C). In old age all the roofs disappear and the drainage reappears at the surface on the exposed underlying impermeable bed (Fig. 7.21D). The limestone cover is reduced to a few outliers (hums), honeycombed with caves, and is finally removed.

The cycle of Karst erosion has so far been applied only to the type region, where limestones of a great thickness cover a wide area. Where the lime-stones are thinner and occupy smaller areas, as they do in Britain, they come under the influence of the normal cycle of erosion and form areas of specialised relief in regions of normal landscape. Although the drainage may be locally underground in such areas, it is very unlikely that many of the streams will disappear in maturity and reappear in old age. Many features of Karst type may occur, but it is doubtful whether it is proper to refer to the landscape as a whole in terms of the Karst cycle. As a concept, the cycle is much more rigid than Davis's general cycle of erosion, which is so flexible that it fits a great variety of conditions, though it is not readily adaptable to limestones—hence Cvijić's special cycle.

In fact the formation of well-developed karst demands that a number of conditions be satisfied.

(*a*) There must be a large area of limestone, otherwise the relief developed will be constrained by the hydrological conditions on adjacent outcrops

PLATE 18. Limestone pavement near Ingleborough, Yorkshire, with joints etched out by solution

Geomorphology

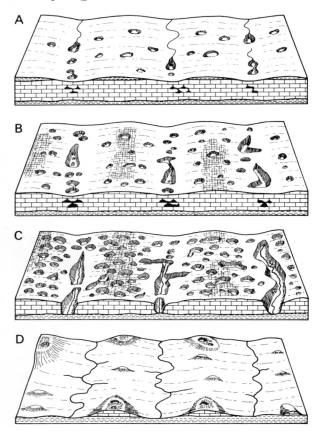

FIG. 7.21. Simplified representation of the Karst cycle of
erosion (for explanation see text). The depth and size of the
solution holes is greatly exaggerated in relation to the
thickness of the limestone

and, although some karst features may be present, a full karst scenery is
unlikely to develop.

(*b*) There should be a considerable thickness of carbonate rocks, otherwise
intercalated lithologies of other types will hinder karst development.

(*c*) The rock must be massive, crystalline, well bedded and well jointed to
ensure that the drainage will be mainly along fractures, and that karst
features are not denuded almost as fast as they appear, which is the case on
soft limestones such as the Chalk.

(*d*) The height of the area above sea level or above the level of through-
flowing rivers must be great enough for a full circulation of underground
water to develop.

(*e*) There should be sufficient rainfall to promote solution. Karsts do occur

198

in deserts but they may be relict forms, or alternatively, if found in semi-deserts, they may have taken a long time to form.

General limestone problems

In recent years there has been a revival of interest in karst landforms and several major points of discussion and controversy have emerged, so that some doubt has been cast on certain of the hypotheses stated above.

Limestone solution is a more complex process than a simple statement of the equation denoting the interaction of calcium carbonate, carbon dioxide and water to form calcium bicarbonate would seem to suggest. A number of equilibria have to be maintained, not the least of which is equilibrium between the carbon dioxide content of the water and the partial pressure of carbon dioxide in the gas phase of the system, whether this is the free atmosphere, the soil atmosphere or some underground atmosphere. The higher the partial pressure the greater the amount dissolved, but it should also be noted that solubility varies inversely with temperature. A simple account of the equilibria involved can be obtained from Sparks (1971), though a book such as Garrels and Christ (1965) should be consulted for a precise treatment.

Bögli (1960) attempted to relate certain aspects and features of limestone surface morphology to an intricate pattern of chemical phases—a summary is given by Sparks (1971). Broadly speaking, he considered that his first two phases involved rapid solution at a high rate, the whole reaction being completed in a minute. In the first of these two phases a rise of temperature of 10 degrees C (18 degrees F) approximately doubles the rate of reaction i.e. there should be a fourfold increase in the amount of solution due to this cause between the Arctic and the Tropics. His third phase, involving the diffusion of atmospheric carbon dioxide into the water, was slow, twenty-four hours at least, and the general rate of solution, though not necessarily the total amount, low.

Because of the short time involved in the first two phases solution during them would be largely confined to the surface outcrops of limestone. The first phase is held to cause Rillenkarren, i.e. very small-scale, closely spaced fluting, and the second to form Trittkarren, which are minute, flat-floored depressions. Such solution would not occur in channels, nor beneath the soil cover, nor underground, because the reactions would be virtually complete before the runoff had time to reach such positions. In these positions phase three would be in operation and the main part of channel, subsoil and underground solution would be attributable to this phase.

The importance of soil carbon dioxide is very great. The average atmospheric content of carbon dioxide is 0·03 per cent. In the soil it may be one hundred times greater and figures as high as 25 per cent have been recorded. As most humic soils have low permeability, the water slowly passing through them would have ample time for its carbon dioxide content to come into equilibrium with the high partial pressure of carbon dioxide in the soil atmosphere. Thus we might expect large amounts of solution in situations immediately below rich vegetation and humic soils. Because some of these processes (e.g. Bögli's first phase and the rate of diffusion of carbon dioxide from atmosphere into water) are increased with temperature, there is available a possible explanation of the intense surface karstification of tropical limestone areas.

Intense surface karstification with Karren and lapiés does not play a great part in the development of limestone landscapes in the British Isles, possibly because of the cool climate. Instead, we have limestone pavements with the joints enlarged into grykes, which may be up to 0·6 m (2 ft) wide and 3–4 m (12–15 ft) deep, separating clints of approximately 2 m by 1 m 6 by 4 ft) (Sweeting, 1966). These have long been recognised to be confined to glaciated massive limestones, and are absent in unglaciated areas of similar lithology, for example the Mendips. They only occur in massive rocks for the beds have to be strong enough to resist general glacial destruction and yet fissile enough along major bedding planes to be stripped off. P. W. Williams (1966) has attributed the formation of clints and grykes to a phase in the Post-glacial period, after the drift has been stripped from the limestone surface, when the joints are enlarged by weathering as the first phase in the destruction of the surface into rubble and ultimately soil. Thus, in this view, clints and grykes are essentially transient features.

A second major karst problem is the origin of solution holes. Are they due to solution from the surface or to underground solution followed by collapse? There are many types of solution hole ranging from the craterlike forms of the northern part of the South Wales coalfield syncline to the broad, shallow depressions, sometimes several kilometres broad and often filled with terra rossa, that occur widely in the Causses of France and the Karst of Yugoslavia. It is easier to prove the solution collapse hypothesis than the surface solution idea, and no better case has been made out than that for the South Wales coalfield (Thomas, 1954). Here, the Basal Grit of the Millstone Grit laps on to the Carboniferous Limestone along the northern outcrops, for example north of Penderyn. Of the thousands of solution holes in this area, the best by far are on the Millstone Grit where a thickness of up to 60 m (200 ft) overlies the Carboniferous Limestone. Their size is such that Thomas reckons that the major solution holes would each have required the collapse of a cavern as large as any known in Britain. Solution from the surface is out of the question as the overlying Basal Grit is an insoluble quartzose rock.

On the other hand Ford and Stanton (1968) in a study of 600 closed depressions on the Mendips maintain that 90 per cent of the depressions lie in the floors of dry valleys, where one would expect to find solution forms generated from the surface, and that their frequency increases where the gradient slackens, i.e. where slower flow gives the water a greater time to find its way underground. It should be noted that these suggestions are at variance with some earlier work on the area, but it seems difficult not to accept them.

In the tropics much solution seems to be initiated from the surface. Aub has presented strong evidence from Jamaica of the lack of collapse in the bottom of cockpits in the karst there. In fact he was prepared to say that 60 per cent of the cockpits or depressions certainly were initiated and developed by surface solution.

In short, this question may be another example of the false problems geomorphologists continually set themselves by trying to find a common explanation for phenomena which are not identical but at best vaguely similar.

Thirdly, we must consider briefly the question whether climate or lithology is the more important in karst formation. Over the years Corbel (1959) has been the main protagonist of the climatic view, which has an appealing overall simplicity. Because of the greater solubility of carbon dioxide at low temperatures (see Chapter 3) and because of a series of figures which showed high amounts of dissolved calcium carbonate in the drainage waters of cold regions, Corbel strongly advocated maximum karst development under cold conditions. These observations conflict with the well-known fact that tropical limestones are often intensely karstified at the surface (for further discussion see Chapter 12). This may be due to aspects of carbonate equilibria mentioned above, but Corbel suggested that it could be caused by the fact that there had been a tremendous length of time with unchanged climate available for the karstification of the older limestones, many of them Mesozoic, in the tropics. On younger Tertiary limestones, Corbel suggested that there was a lower degree of karst development because of the shorter time available. On the other hand, workers in the West Indies and New Guinea have pointed to highly developed karst on limestones as young as Miocene and Pliocene.

In fact, the conflict suggested by the last antithesis might be resolved in terms of lithological differences, which have been shown to be important in tropical areas (e.g. Sweeting, 1958; Monroe, 1964). We have already seen how Williams and Sweeting have indicated the importance of lithological control in the glacial stripping of British limestone pavements. Lithological differences obviously affect limestone landforms in places as unlike as Britain and the southern French Alps. Even in the Dolomites de Smet and Souchez (1964) have pointed to the lithological effects of the local Triassic dolomites in the relief. In this area the Sciliar Dolomite, a reef formation strongly affected by vertical fracturing, was intensely karstified in preglacial times to give rise to a tortured relief, whereas the Main Dolomite, a massive near-horizontal formation, gives a slab-like, massive plateau relief, bounded by vertical cliffs. In this particular example we touch on the interaction of lithology and structure with both present and past landforming processes.

To approach a fourth set of problems let us consider underground solution and cavern formation. It might seem that all the dissolving powers of carbon dioxide solutions would have been absorbed at the surface of the rocks. That this is not so is obvious from the presence of enlarged passages and caverns. If Bögli's third phase takes a day or more for equilibrium to be reached, then it is highly likely that there will be surplus capacity for attack on limestone underground, especially in cold regions, where the time taken for equilibrium to be reached between the gas and liquid phases is at a maximum. Again, equilibrium can be upset undergound. Let us assume that it is colder underground than at the surface and that the partial pressure of carbon dioxide remains unchanged. In this case, more carbon dioxide can be dissolved and the attack on the limestone renewed. On the other hand, if the temperature remains the same and the partial pressure of carbon dioxide underground is less than in the soil atmosphere, carbon dioxide will be transferred from liquid to gas phases and precipitation of calcium carbonate will take place. Under hydrostatic pressure below the 'water table' when the liquid completely fills the space and there is no gas phase, no further interaction can take place, because there can be no restoration of equilibrium in the absence of one phase. Yet solution seems to take place at these levels and a reason for this has been suggested by Bögli (1964). He pointed out that the relation between calcium carbonate in solution and carbon dioxide is not arithmetical and that if two water masses of different content mix, the form of the curve ensures that the mixture will always be aggressive, i.e. capable of further calcium carbonate solution. This phenomenon can, of course, occur at any level.

It is often said that limestone caverns occur roughly at the level of the water table between the vadose and phreatic zones. Not all observers believe that there is necessarily a simple level in limestones below which the rock is completely saturated. Pockets of air have been found well below the alleged water table level, and some underground forms appear to have been produced by pressure flow. Yet there is a fair amount of evidence consistent with the formation of caves at water table level. According to some authors both chemical attack and mechanical corrasion caused by the load in the underground streams may contribute at this level. The type of stratification of caves and passages at certain heights found by Sweeting (1950) in Yorkshire suggests a development along a flattish water table surface. It has been said that many cave systems are three-dimensional and that this is not consistent with development at one level, but it might be consistent with development under the influence of a falling water table, a very possible state of affairs. On the other hand tubular passages are known with blind pockets and with evidence of flow in both ceiling and floor, facts more consistent with a phreatic origin, while the up and down profiles of some cave passages, by analogy with eskers (see Chapter 14), points to an origin under the influence of pressure flow. There are important differences between caves: some are free from secondary deposition, others have magnificent development of dripstone in the form of stalactites and stalagmites, features which might be attributed in the first case to present enlargement and in the second to present choking-up with deposition. It is probably true that caverns can be formed in either zone, and that they may be modified in the other.

Even limestone caves may be strongly affected by rock type if some suggestions made by Sweeting (1968) prove to be generally applicable. These related the shape of the cave to the detailed lithology and structure of the rock.

Finally, the deduced rates of overall solution of limestone give one cause for thought. Most of these have been arrived at by a formula proposed by Corbel which gives annual denudation loss in millimetres per 1 000 years in terms of the runoff and the average calcium carbonate content of that runoff in mg per litre. Later authors have introduced terms for magnesium carbonate solution and for specific gravity of limestone different from Corbel's standard 2·5. Corbel himself introduced a term to cover the proportion of the drainage basin made up of limestone. There is a convergence of thought from these calculations, even discounting some of Corbel's enthusiasm for cold region limestone solution, from comparisons of the level of limestones beneath datable features such as glacial erratics and barrows and the level nearby where the limestone has not been protected, and from the obliteration of dated inscriptions on tombstones, of limestone solution rates equal to 50–100 mm (2–4 in) per 1 000 years. Not all of this need be surface solution, but Pitty (1968) has argued that most of the solution in the southern Pennines is probably surface solution. He pointed to the generally saturated state of underground water, and made the telling point that the annual solution loss was greater than the total volume of known caverns, a fact which seems to put underground solution into the background as a minor contributor to total solution loss.

If we assume a loss of 50 mm (2 in) per 1 000 years it means a lowering of 50 m (approx. 170 ft) per million years, i.e. since early in the Pleistocene. Yet other geomorphologists have postulated arrested surface dissection and the consequent preservation of terraces as characteristic limestone features. The south of England has probably been exposed to subaerial denudation for about 25 million years (since the beginning of the Miocene at a conservative estimate). In that time the surface of the Chalk could have been lowered 1 250 m (4 000 ft). Where none is missing the Chalk reaches a maximum thickness of about 450 m (1 500 ft). It may not all have been exposed to denudation immediately, and a lot of time may have been needed for stripping off the overlying Tertiary rocks. It could be said that a constant extrapolation backwards in time is not justified. But it is, because the solution of limestone is one of the few processes which probably do not vary with time. Obviously this is a case of head-on conflict, where both rethinking and more observation are needed.

The discussion of tropical karst is deferred until Chapter 12. Many of the questions summarised here are dealt with more fully elsewhere (Sparks, 1971).

Chalk relief

Chalk and other soft limestones form relief which differs greatly from that developed on Carboniferous and similar massive limestones. Chalk does not usually possess such a neat series of joints but a multitude of irregular cracks, as any inspection of a chalk pit will show. Thus there is usually no joint control of the relief comparable with that of Carboniferous Limestone districts, nor is the rock hard enough to be fretted at the surface by solution, nor is it strong enough to allow a great development of caves. A few caves and gaping fissures have been discovered, but they are not common, presumably because the weight of fractured chalk above tends to close up any fissures widened by solution.

The general form of chalk landscapes, dominated by smooth convexo-concave curves (Plates 15 and 17), is probably caused by the permeability of the rock and its waste mantle. As was seen in Chapter 4, a high degree of permeability is thought to lead to the convex upper part of the slopes attaining a considerable development. Owing to the bareness of many chalk areas, especially those of Wessex and the eastern part of the South Downs, the slope forms are open to view, whereas the slopes on other rocks are often obscured by hedges and patches of woodland. Whether the form of chalk slopes is unique, as is often maintained, is doubtful, for some other rocks, such as the soft Jurassic limestones and some permeable sandstones, if stripped of their vegetation, might well appear very similar. The typical rolling downland of chalk regions probably owes as much of its quality to its vegetation and land use as to its slope forms. Where the chalk is mantled by superficial deposits, such as Clay-with-flints or boulder clay, its vegetation and land use change and it appears to lose many of its typical landscape qualities. The typical chalk landscape consists of convex divides, fairly steep valley sides and trough-shaped valleys, but there are all sorts of variations on this form. In many places the valleys are wider and shallow, in other places they are asymmetric in cross-section, while in certain localities the valleys may be steep gashes with the appearance of railway cuttings.

The role played by solution in the formation of chalk landscapes is difficult to assess. The number of caves is, as noted above, small and the same is true of active solution holes. Exception must be made, however, of the valley of the river Mole across the North Downs, as in dry seasons the river goes underground in places through numbers of half-choked swallow holes in its bed (Fagg, 1958). Even more spectacular solution holes are those at North Mimms, Hertfordshire (Wooldridge and Kirkaldy, 1937); dozens of choked holes occur in the district, many of them dimples in the fields, others pools with a considerable thickness of mud in the bottoms, and a few active holes in which small streams can be seen to pour into the chalk through a manmade debris of old tyres and detergent foam.

More widespread are the gravel-filled holes often exposed in chalk pits and railway cuttings. Some of these are small narrow pipes, others irregular depressions, while a few, notably those at South Mimms, Hertfordshire (Kirkaldy, 1950), are deep cylindrical holes with tributary 'sheet pipes' developed approximately horizontally along the bedding of the chalk. The holes are usually filled in with a rubble of iron-stained flints, sand and clay, derived from overlying Tertiary or glacial deposits. The holes are scattered in valleys, on slopes and on the tops of the hills. They are unlikely to represent the courses of former streams. Usually they appear to be absent from bare chalk areas long stripped of the overlying Tertiary beds such as Salisbury Plain and the South Downs. They are especially abundant at the margins of and beneath thin remnants of Tertiary beds, such as Reading Beds, Blackheath Beds and Bagshot Beds and similarly gravelly or sandy Quaternary beds. This coincidence was noted many years ago by Prestwich (1855), who considered that water passing through the more abundant vegetation on the Tertiary outliers would have an increased carbon dioxide content and that such water would be held up at the Tertiary–Chalk junction due to the fact that the latter is less permeable than the Tertiary sands and gravels. The Tertiary beds would act, then, like a sponge soaked in weak acid applied to the chalk. The possible significance of humus acids in assisting the solution of the chalk appears to have been first recognised by Reid (1899) in describing Puddletown Heath, Dorset, where the Reading Beds have collapsed, allegedly through solution of the underlying chalk. The humus derived from the heath vegetation characteristic of many of the Tertiary sands and gravels is acid, and the presence of this acid, far stronger than the available concentrations of carbon dioxide, probably explains the relations so often observable between solution holes and Tertiary outliers. Where the former exist without the latter it is often possible to show that the Tertiary beds have been stripped away in geologically recent times. Although many of the solution holes appear to be fossil forms, some appear to be still active, as is suggested by the descriptions of broken ground and inclined and prostrate trees at Lane End on the Chilterns (Jukes-Browne and White, 1908). In addition, the Puddletown Heath hollows affect Roman tracks, banks which are probably in a general sense medieval, and even a conifer planted in 1927.

The Clay-with-flints, strictly speaking a mass of rusty clay with fractured flints but in practice an omnibus term covering a variety of deposits of different types, has sometimes been attributed to solution of the chalk. This appears to be unlikely. As long ago as 1906 Jukes-Browne showed that in order to explain the existing thickness of Clay-with-flints it would be necessary to assume that 30–60 m (100–200 ft) of chalk had been removed by solution. Furthermore, the residue, at least from the Upper Chalk, would be a flints-with-clay, the former greatly exceeding the latter in bulk. In places it can be shown that the maximum thickness of chalk which can have been eroded is insufficient to provide the known thickness of Clay-with-flints. Much of the mapped Clay-with-flints is probably a mixture of Tertiary beds resorted by frost heaving or cryoturbation in periglacial conditions, a process to be described in a later chapter. These conclusions have been confirmed in detail for a part of the South Downs Clay-with-flints, in which the habit of the clay mineral, montmorillonite, shows it to be derived from the Reading Beds above and not from the Chalk, while fluidal structures in clay in cavities and cracks in the Chalk suggest translocation downwards from higher levels (Hodgson, Catt and Weir, 1967). Thus, the presence of Clay-with-flints cannot be used as evidence of chalk solution on a vast scale.

Solution must take place in the Chalk, for the hardness of the water derived from that rock is well known. It would be tempting to conclude that most solution derives from the enlargement of underground joints and cracks. But, in view of the problems associated with the amount of surface lowering suggested by solution on limestones and the opposing evidence of the widespread preservation of terraces dating back to the early Pleistocene on the Chalk in south-east England, the effects of solution over a long period must be considered an open question. One final observation must be made: where solution obviously has occurred, i.e. in the piped and pitted areas or in places like the Mimms depression, the surface is highly irregular, which is precisely what one would expect of solution. But over most of the Chalk the surface is smooth, so that, if it has been lowered greatly by solution, it must have been lowered remarkably evenly.

Probably the most discussed feature of the chalk is the dry valley. For some reason most discussions on dry valleys have been based upon chalk dry valleys to the exclusion of similar features occurring on other soft limestones such as the Jurassic and on some permeable sandstones. There are two plausible theories of dry valley formation, one based upon the steady operation of processes working very slowly and the other on catastrophic processes during the glacial period.

The first idea, championed by Chandler (1909) and Fagg (1923, 1939), is based upon the fact that, as the chalk escarpment recedes, the level of the escarpment springs at the Chalk–Gault junction, which controls the level of the water table within the chalk, is lowered. If the water table falls at a rate faster than the dipslope valleys are able to cut down, the valleys will probably dry up. The hypothesis is illustrated in Fig. 7.22. At present the

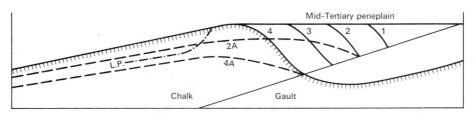

F IG. 7.22. Fagg's hypothesis of dry valley formation (for explanation see text) (*after Fagg*)

chalk escarpment is in such a position (4) that the water table (4A) lies below the valley floor (LP). But, at an earlier stage, when the escarpment was at position 2, the water table was at the level indicated by the line 2A and thus springs would have occurred in the head of the valley. This hypothesis was based mainly on work on the North Downs, where it is very likely that the escarpment has receded in the manner indicated. A similar recession has probably taken place in most other chalk areas. There may well have been phases of more rapid recession and phases of stillstand corresponding with periods of rejuvenation and periods of little erosion. During the stillstands the springs may have tended to sap headwards forming the steep rounded ends common to many dry valleys. The periods of rejuvenation and rapid recession of the escarpment probably involved comparatively rapid falls in the water table when the valleys were dried up either wholly or in their upper reaches.

The operative mechanism in Fagg's hypothesis is the level of the Chalk–Gault junction at which the springs occur. If the escarpment remains stationary it is quite possible to extend the hypothesis to apply to valleys in the face of the escarpment (Fig. 7.23). A phase of planation may be

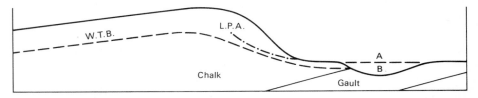

FIG. 7.23. Possible application of Fagg's hypothesis to scarp valleys (for explanation see text)

imagined to have produced a flat surface cut across the Gault and into the lower part of the Chalk (A on Fig. 7.23). Scarp valleys such as that shown in profile (LPA) might well be graded to the plain in front of the escarpment. If the stream following the outcrop of the Gault were to be rejuvenated it might lower its valley to the level B and with it the level of the water table in the Chalk to the position WTB, thus drying out the valley in the escarpment.

These hypotheses have both involved spring erosion, but it is quite possible to attribute the valleys to meltwater flow during the Pleistocene period. As pointed out by Reid (1887), the valleys might have been formed under tundra conditions just beyond the margin of the ice sheet. Although chalk is normally permeable, it would be rendered impermeable by being frozen to great depths, while the spring thaws might affect only the surface layers. The result would be a very rapid spring runoff and the violent, though temporary, streams so formed would have very great erosive power. In Sussex, to which area the hypothesis was originally applied, the material scoured from the valleys is alleged to have accumulated as the spread of coombe rock at the foot of the dipslope of the South Downs. It is a rather curious fact that the meltwater hypothesis has been championed by workers on the South Downs (Reid, 1887, and Bull, 1940), while Fagg's hypothesis was derived mainly from work on the North Downs. The latter, being nearer to the former position of the ice front, should show the effects of meltwater at least as strongly as the South Downs.

Bull (1940) has argued specifically against the application of Fagg's hypothesis to the South Downs valleys. The first argument is that the presence of 100 m (320 ft) and 60 m (200 ft) terraces cut into the face of the escarpment shows that there has been little recession of the escarpment since the terraces were cut. But if the former position of the Gault outcrop at these levels is reconstructed (Fig. 7.24), it can be seen that the Gault has

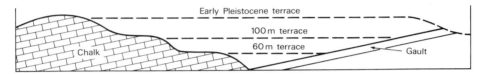

FIG. 7.24. The recession of the South Downs escarpment near Eastbourne (*after Bull*)

been considerably lowered and that the lower part of the chalk escarpment has receded some distance since the cutting of the terraces. The lowering of the Gault has allowed a fall in the level of the water table within the Chalk without the recession of the upper part of the chalk escarpment. Bull's second argument was that if the escarpment had receded the heads of the dipslope valleys ought to cause notches in the crest of the Downs. It was once thought (Sparks, 1949) that the relations between the valley pattern near Beachy Head and the 145 m (475 ft) (Fig. 7.25) shoreline suggested that some of the dry valleys were initiated on sea floors below the level of the early Pleistocene sea, the shoreline of which ran east–west north of the present escarpment and on the emerged floor of which the main valleys were presumably generated. But although this hypothesis might apply locally, it certainly cannot be used elsewhere to explain the lack of breaching. The problem of the lack of escarpment breaching is very difficult to explain if recession has taken place. Small (1961) measured 520 km (325 miles) of Chalk crestline and found 103 cols, by his definition, therein: if the escarpment retreated another 1·6 km (1 mile) there would be 417 cols, by the same definition. This seems to suggest that most of the dry valleys have been formed with the escarpment, or at least a surface and sub-surface drainage divide, roughly in its present position. After rejecting a number of hypotheses, Small advocated a two-stage development, which is a little complex to go into fully here, but the essence of which in terms of sections is illustrated in Fig. 7.26. It will be noted that this involves a general coincidence, though not everywhere of the same degree if one may judge by the varying degree of breaching of the escarpment crest, between the

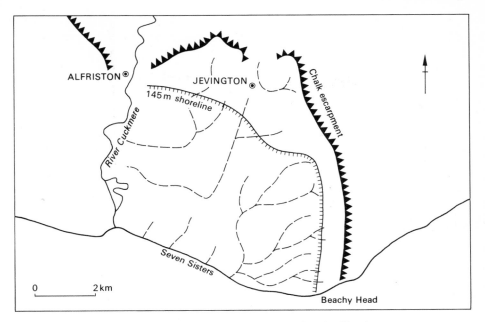

FIG. 7.25. Dry valley pattern on the South Downs near Beachy Head

drainage divide at the end of the first cycle of erosion and the present position of the escarpment. But coincidences must occur sometimes.

In the last twenty years attention has been focused primarily on escarpment valleys and not on the dipslope valleys which had concentrated previous attention. Interest was started by Lewis, who followed up a series of earlier isolated comments on Chiltern escarpment valleys, with a detailed investigation of the scarp valleys near Pegsdon, Hertfordshire (Sparks and Lewis, 1957). The type of valley in question is of the order of 1·5 km (1 mile) long and the distance between valleys is also about 1·5 km. They occur in parts of the Berkshire and Marlborough Downs and are again well developed on the southern side of the Vale of Pewsey and in the western Weald escarpment.

The valleys are steep-sided, often bluntended and sometimes possessed of angular bends. They are notably fresher in appearance than the majority of dipslope valleys. Lewis's Pegsdon studies led to the conclusions that the sharp breaks of slope at the top and the bottom of the valley sides as well as the flat floors were accentuated if not entirely caused by cultivation, that spring-sapping was the basic mechanism of formation and that latent joint-control was the cause of the rightangle bends. Deposits in the valley floors showed the presence of a stream in one valley in Post-glacial times, so that there may have been a sharpening of valley forms with higher water tables early in that period, but both Lewis and Sparks considered it very doubtful whether the whole of the valley forms had been cut at that time.

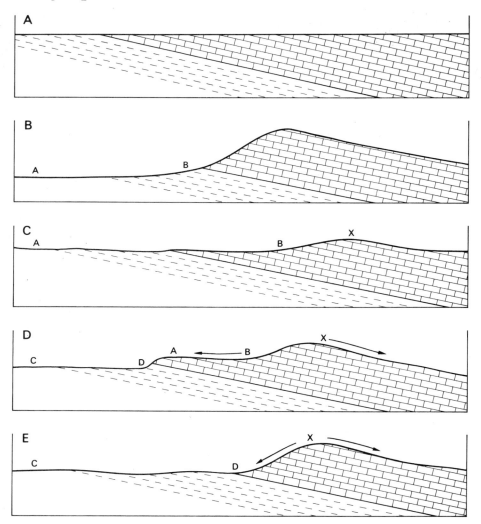

FIG. 7.26. Small's explanation of non-breaching of Chalk escarpments
A, B. Initiation and development of scarp in first cycle
C. Subdued, unbreached divide (X) at end of first cycle
D. Rejuvenation and development of new scarp in second cycle
E. New scarp pushed back to position of unbreached first cycle divide (X)

Small (1964) greatly extended and amplified these studies and used an idea put forward, along with a number of less acceptable notions about escarpment dry valley formation, by Fagg (1954). The idea was that, as the escarpment recedes, scarp-foot springs are liable to bite into and divert scarpwards sections of the heads of former dipslope dry valley systems. Thus, the steep, fresh escarpment valleys may represent the rejuvenated and diverted valley heads of dipslope systems. The essence of the idea may be derived from Fig. 7.27. Here, the scarpfoot spring has bitten into a recession col (i.e. a place where the scarp crest has been lowered by scarp retreat) and sapped back into valleys A and B, which formerly paralleled valleys C and D respectively in being tributaries of the dipslope dry valley system shown.

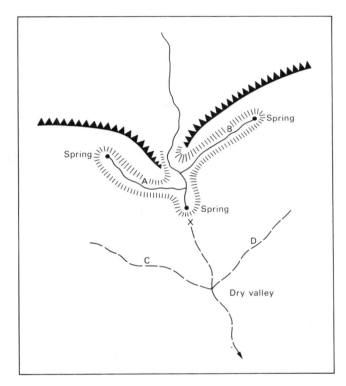

FIG. 7.27. Formation of scarp valleys by capture and rejuvenation of dipslope valleys

There are a number of conditions governing the likelihood of such diversions happening. A revival in vigour of the scarpfoot drainage is necessary. Thus, on the southern side of the Vale of Pewsey, examples of such captures at Bratton and Tinhead are in part due to the rejuvenation of the Bristol Avon drainage system. In the western Weald more examples are found associated with the headwaters of the West Sussex Rother than with the Wey system: the Rother is a well-known example of an active, revived stream.

In a number of examples the springs have obviously taken advantage of recession cols, probably for the following reasons. The height of Chalk headwall to be removed per unit of recession is less than on the face of the escarpment. As the spring cuts back the headwall height is liable to be further reduced for two reasons: the spring must rise in elevation to maintain flow, and also once the divide in the col has been passed the height of the backwall must be progressively reduced. Very often Middle and Lower Chalk are exposed in the cols and these are both less resistant than the Upper Chalk. Cols and dry valleys may also be zones of greater water transmissibility because subsurface flow has widened joints and fissures along these lines.

The general form of the water table, if it represents a subdued reflection of the relief as is usually believed, must result in convergence at valley heads and hence the development of the strongest springs there. A col in an escarpment is virtually a valley head, and hence the more powerful scarp-foot springs should be located there. In fact, as the col is sapped into, there should be an increase in spring flow which should further accelerate sapping.

Actual evidence of the reversal of drainage is also sometimes present. Streams in position such as those of A and B (Fig. 7.27) could never have sapped back *ab initio* along the strike of the escarpment: they must represent valleys sapped back well within the escarpment and now almost removed by scarp recession. Occasionally high-level valley floors, representing former stages of dissection, run along valleys such as A and B but then lead over the col at X into the dipslope valley system.

In places the latent joint or fault control suggested by Lewis for the right-angle bends in the Ravensburgh valley at Pegsdon has been confirmed by Small, especially in some of the Berkshire and Marlborough Downs valleys. Gullying and fluting on the Chalk escarpment of the southern side of the Vale of Pewsey also may well reflect joint control.

Some very different scarp valleys have been investigated by Kerney, Brown and Chandler (1964) on the North Downs near Brook in Kent. These are roughly 400 m (440 yards) long and 400 m apart and the dissection is therefore much nearer the texture produced by the periglacial valleys on the north side of the Cromer moraine in Norfolk than the texture of the Wessex and Pegsdon valleys, apart from some of their minor tributaries. A fan of chalk debris in front of the Brook set of some half a dozen valleys yielded a radiocarbon date near its base of about 12 000 years ago, that is in Zone II of the Weichselian Late-glacial (see Chapter 9). Much of the debris above, which is mainly solifluxion, must therefore be attributed to Zone III. As this debris would go far towards filling the valleys, a figure of one third seems to emerge from the description, it seems that the valleys were completely formed or at least greatly modified in Zone III, the last periglacial phase to affect Britain, at the very end of the Weichselian glacial. It is unlikely that these valleys, so different in their scale, are comparable with the other scarp valleys, so that we must avoid creating yet another false problem by assuming that the origin of both sets must be the same.

Summing up, there is much to be said, apart from the joint control, for the slow development of most valleys over a long period by normal processes. The apparent relation between valleys and shorelines near Eastbourne, the relation of certain South Downs valleys to structure, the correspondence between certain strike valleys near Mickleham (Fig. 7.28)

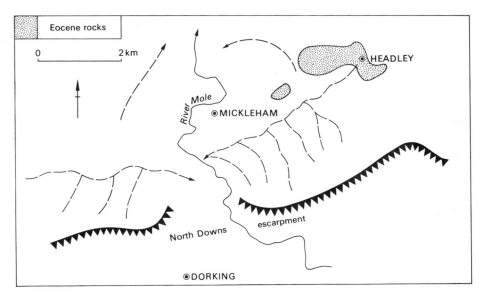

FIG. 7.28. Dry valley pattern near Mickleham, Surrey

and a probable former position of the Tertiary escarpment and the possible joint control of the Ravensburgh and Berkshire Downs valleys all suggest that the valleys have behaved like stream valleys in their adaptation to structural and relief control, and not like meltwater valleys which one would expect to be largely independent of structural guidance. The Mickleham valleys are especially interesting as they have the appearance of having originated at the foot of a slight Tertiary escarpment the position of which is indicated by the outliers near Headley. On the other hand valleys such as those at Brook have either been originated or modified by periglacial conditions. Yet others may have been modified by accelerated spring flow or meltwater in similar conditions, for they possess fresh undercut slopes not all of which can be explained away by cultivation. In the South Downs some valleys are graded to levels below the river alluvium of the gaps, indicating erosion under cold conditions accompanied by a low sea level. In periglacial conditions meltwater activity and accelerated spring flow, due to saturation of the rocks, could have gone hand in hand. It could be that maxima of spring flow occurred soon after the periglacial phases, so that one might speculate whether the two main rival hypotheses of periglacial and spring sapping origins present such a strong antithesis as they seem to do at first sight.

8
Coastal features

Although normal subaerial erosion may act with greater power on some parts of the earth's surface than on others, an obvious example being found in the contrast between rates of denudation in river valleys and on plateau surfaces, the general result of such erosion is an overall reduction of the surface. The action of the sea, however, is concentrated within comparatively narrow limits. In plan the length of the coastline obviously limits the area which waves can attack, so that the more highly indented coastlines offer greater lengths to wave action. In section the range over which erosion is operative is equally restricted, although there is as yet no precise agreement as to the exact limits. Waves certainly erode a metre or so (several feet) above the level of the highest spring tides, but the depth below the lowest neap tides at which erosion occurs appears to be much more variable.

A second contrast between subaerial and marine erosion is to be found in the relative importance of erosion and deposition. Although deposition may be an important feature in the process of stream and slope grading, it is usually temporary and the products of erosion are finally reduced in size and transported to the sea. Thus the coastline receives not only the products of marine erosion but also the waste derived from subaerial erosion. It is not surprising, therefore, that on the coast forms due to erosion and to deposition are more equally balanced than they usually are inland under the influence of subaerial erosion.

Finally, it must be pointed out that the recession of a coastline is not merely the result of marine erosion. Where the cliffs are high they may be undercut by the sea with the result that most of the material may be removed by the subsequent collapse of the upper undermined section. Alternatively, the rate of marine erosion at the base of the cliffs may be less than the general rate of subaerial denudation over the whole cliff. Such a condition leads to a degraded cliff, one that is principally the product of subaerial erosion and only locally sharpened at the base by marine erosion.

Factors affecting marine erosion

The factors affecting erosion on the coast may be divided into two broad groups. On the one hand there are the waves themselves and the various factors which control their size, shape and approach to the coast. On the other hand are the characteristics of the coast: the height of the cliffs, the lithology of the rocks, the orientation of the coast with respect to the waves and the supply of debris available on the beach.

The effect of waves

The erosive activity of waves is partly due to the effect of impacts of water against the coast and partly to the action of stones and boulders moved by the waves.

The nature of the impact has a considerable effect on the pressure exerted on the cliff face and a number of different types have been examined by Russell and Macmillan (1952).

If the sea is deep right up to the coast the waves may reach the coast and be reflected back without breaking. The reflected waves interfere with the incoming waves to produce a pattern of standing waves or clapotis, the crests of which always appear in the same places (Fig. 8.1A). Such waves

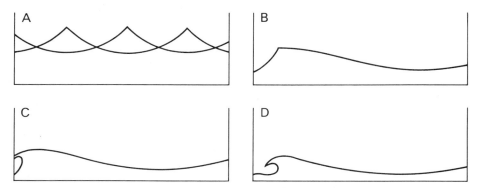

FIG. 8.1. Types of wave action to which cliff may be subjected (for explanation see text) (*after Russell and Macmillan*)

are said to be capable of setting up oscillations which may damage a jointed structure, as the increases and decreases of pressure tend to pull jointed blocks away. In designing artificial defences the structures are made sufficiently strong to resist this type of wave, so that one might, therefore, expect closely jointed rocks to be susceptible to damage by clapotis as such rocks are the nearest approach to natural walls. Yet plunging cliffs, i.e. those whose profiles are continued straight down below the sea, are free from notches, benches and other forms of wave action. Some plunging cliffs may be due to faulting, downwarping and volcanic flows of very recent date in tectonically mobile areas such as New Zealand. The lack of notching of these, therefore, may be due to lack of time. Other plunging cliffs, however, have been formed by the Post-glacial rise of sea level, for example the sheer, gaunt, limestone cliffs that form Cape St Vincent at the south-western tip of Portugal. These cliffs fall from a monotonous, limestone, coastal plateau straight down almost vertically into deep water. Their steepness leads one to infer modern erosion due to their great exposure, but there is no wavecut bench to indicate any erosion. Any debris dislodged from them by hydraulic action sinks to the depths and cannot be used by the waves to attack the cliffs. Thus, plunging cliffs seem to suggest that, at least in certain areas, reflected waves and clapotis are not very effective.

The highest pressures exerted on the rock are connected with breaking waves, but the pressure varies with the distance between the point where the waves break and the cliff. When waves break it has been observed that a jet of water at the crest of the wave may move forward with twice the velocity of the wave as a whole, so that high pressures may be exerted for a period of a second or two on the cliff face. Such conditions, shown in Fig. 8.1B and D, have been shown to be the main ones responsible for damaging marine structures.

A pocket of air may be trapped between the breaking wave and the cliff, if the wave breaks in the right position relative to the cliff (Fig. 8.1C). If this happens very high pressures may result due, not to the wave itself, but to the pressure of the air compressed between the wave and the cliff. Work by Bagnold has shown that the pressures are greatest when the pocket of trapped air is thinnest and that when the dimension of the pocket at right angles to the cliff reaches half that of the depth of the pocket the pressure is negligible. The possibility that these momentarily high pressures may damage jointed rocks cannot be ignored, but there is, curiously enough, no direct evidence as yet that actual damage has been caused by waves of this type. The reason may be found in the fact that the air tends to burst upwards through the water along the path of least resistance. Very high velocities have been observed in the movement of spray rising from impacts of this type. Large masses of spray have been recorded moving upwards at above 110 km/h (70 mile/h), while small jets have been observed with the astounding speed of over 270 km/h (170 mile/h).

The amount of erosion of any rock cliff will depend not only on the way in which the waves break against the cliff but also on the size of the waves. The size of the waves is governed largely by the time during which they have been moving under the influence of the wind. The time may be limited by the proximity of another shore from which the wind is blowing or by the duration of the wind. Where the fetch, or distance at the disposal of the wind for the generation of waves, is very large the factor governing the height of the waves is the time during which a wind of a given strength has been blowing. As an example Russell and Macmillan (1952) quote the fact that with a westerly gale of twelve hours' duration in the Atlantic the size of the waves reaching Cornwall would be limited by the time and not by the fetch, i.e. the distance between Cornwall and America. On the other hand fetch is the limiting factor in the height of waves generated by easterly winds over the North Sea. As a more extreme example may be quoted the case of a south-easterly gale in the Straits of Dover: however long the gale blew the waves could never exceed a certain height determined by the distance across the Straits.

Each sea and ocean tends to have a maximum wave size (Zenkovich, 1967) depending upon the size and shape of the basin and the weather conditions. In mid-ocean storm waves are typically 15–20 m (50–65 ft) high and 300 m (1 000 ft) long at a maximum. In the Mediterranean waves 9 m (30 ft) high and 200 m (650 ft) long have been reported, while in the North Sea, the Bering Sea, the Barents Sea and the Sea of Okhotsk waves 8 m (25 ft) high and 150 m (500 ft) long have been observed. The Baltic and Black Seas, which are smaller, seem to have maximum waves about 6–7 m (20–23 ft) high and 60–100 m (200–330 ft) long.

These figures refer to the size of free waves in the open ocean. When waves approach the shore they suffer a number of changes. Storm waves produced in open oceans travel on far from the storm centres with velocities which depend on wave length, the longer the wave length the higher the velocity, and on the height of the waves, the lower the waves the less readily they are damped out. Such waves are known as swell, and they are often the main waves affecting coasts in the southern oceans, for example South Africa and Australia.

As swell waves or wind-driven waves run into shallow water the velocity and wave length both decrease, as also does the wave height at first, but when the ratio of depth to wave length falls below 0·06 the waves increase in height immediately prior to breaking (Bird, 1968). At the same time changes of wave direction are imposed on waves by the sea bottom interfering with the flow pattern in the waves, but these directional changes are considered later on.

In connection with wave attack it is useful to distinguish between prevalent and dominant winds. Prevalent winds are those which blow with the greatest frequency at any place. Over the British Isles the prevalent wind at all stations is the westerly wind. Dominant winds are those which produce the largest and most damaging waves at any station. The prevalent and dominant winds may blow from the same direction or they may blow from almost opposite directions. On almost the whole of the exposed parts of the western coasts of Britain both the prevalent and dominant winds are the westerlies, but on the east coast, where the prevalent winds are still the westerlies, the dominant winds are the easterlies, as the westerly winds are blowing offshore and are incapable of producing waves affecting the coast.

A final factor affecting the erosion of cliff coasts composed of hard rock is the load of stones and boulders stirred up and moved about by the waves. The effect of the load is probably somewhat different from the variations of pressure set up by breaking waves, which may dislodge jointed blocks. Boulders flung against the cliff by the waves are much more likely to knock off projecting corners and become themselves rounded in the process than to dislodge rock masses. The amount of attrition can be readily observed in the rounded nature of many cliffs below the high tide level and in the rounding of the beach material itself. Care must be exercised in introducing the latter as evidence of marine abrasion, as in many parts of the coast the beach material is derived from the erosion of cliffs of incoherent material already containing rounded or partially rounded stones. Well-known examples are the cliffs of Pleistocene deposits on the east coast and the cliffs of Tertiary beds and coombe rock on the south coast, all of which may contain rounded flint material.

The purely surface nature of the abrasion of beach pebbles may be deduced from flint beaches. Flint is comparatively brittle but the number of flints with fresh fractured surfaces is usually small. On the other hand the roughening due to superficial abrasion is quite apparent.

Most of the remarks so far apply particularly to coasts of hard rocks. Where the rocks exposed in the cliffs are incoherent, pressures produced by the waves may be of less importance and much of the erosion accomplished by the load flung against the cliffs and by the simple swilling action of the waves, which removes the finer matrix from gravel beds. The beds are quickly eroded, the cliff oversteepened, and collapse follows due to the inability of the material to maintain very steep faces. Such coasts are often fronted by large shingle beaches, derived from the destruction of the cliffs, and the erosion of the cliff depends to a large extent on the protection afforded to it by the beach.

On shores composed mainly of shingle the amount of erosion performed by waves depends to a large extent on whether they are constructive or destructive. The differences between these waves have been examined by Lewis (1931). Broadly speaking, waves with a high frequency (13–15 per minute) are destructive, while waves with a lower frequency (6–8 per minute) are constructive. The difference appears to be caused by the direction of the plunge of the mass of water when the wave breaks. In high-frequency waves the water motion appears to be nearly circular so that, when the waves break, the mass of water is directed downwards at the beach (Fig. 8.2A). With this type of wave the swash, or rush of water up the beach from the breaking wave, is weak as the water plunges directly on to the beach. In addition the backwash from the last wave is still moving down the beach when the next wave arrives due to the high frequency of the waves. Thus, an already weak swash is further impeded in its movement up the beach by interference from the backwash of the wave before. The result is that this type of wave combs material down from the top of the beach giving a net erosive effect.

Constructive waves, on the other hand, have a lower frequency and the particle movement within them is more elliptical (Fig. 8.2B). The result is a powerful swash which moves material up the beach. The backwash is reduced in energy by the percolation of water into the shingle, so that the backwash is not necessarily able to return the material carried forward

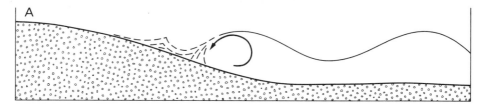

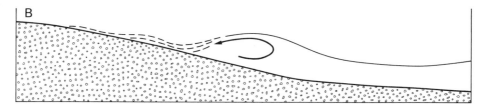

FIG. 8.2. A. Destructive waves. B. Constructive waves (*after Lewis*)

by the swash in spite of the fact that it is combined with the effect of gravity, whereas the swash is acting against gravity.

It is usual today to refer to the two types of breaker described above as plunging and spilling breakers respectively. The simple association of degradation and aggradation of beaches with wave type needs some modification. Obviously waves may have frequencies of intermediate magnitude and so be difficult to classify as either destructive or constructive. Further, large waves of destructive type, although combing the beach down, may at the same time fling shingle up above high water mark and so form beach ridges. In fact, very large waves, whatever their frequency, tend to erode the beach, while very small waves tend to build the beach up. Finally, it is not entirely true that the wave is the independent variable and the beach the dependent one, because the form of the wave, at least when it nears the coast, is to some extent dictated by the form of the beach and the immediate offshore bottom. An interaction exists just as it does in river processes.

A factor contributing to wave action is the wind. High-frequency destructive waves are usually wind-driven waves, developed by strong onshore winds. The wind leads to the piling up of water near the shore, and this water has to escape seawards in one way or another. It used to be thought that there was a seaward movement of water on the bottom, so producing the undertow which was considered so dangerous to bathers. If such a flow exists it may assist the seaward movement of debris and hence contribute to the erosion of the coast. It would not necessarily be capable of moving debris, but any debris lifted by wave action would have its direction of fall influenced by the direction of the bottom current. In recent years it has been generally held that the backwash down the beach contributes to the breaking of the next wave, becomes involved in it and is returned with the swash, so that there is no continuous undertow transfering water out beyond the breaker zone. Whether there is a percolation of water seawards in the beach material is another possibility, which does not seem to have received much attention, but it is unlikely that such a form of movement would be capable of dealing with the total volume of water involved.

Starting with Shepard's work in the late 1930s it has become increasingly normal to attribute the return flow to local rip currents. There must come a time when so much head of water is piled against the beach that at certain points it is enough to initiate return seaward flows, which are the rip currents. Once a current is developed and the head of water is lowered a hydraulic gradient is developed along the beach and a circulation pattern develops of longshore currents feeding seaward rip currents (Fig. 8.3).

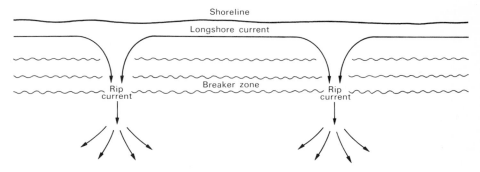

FIG. 8.3. Formation of rip currents

Speeds of 8 km/h (5 mile/h) with waves 3 m (10 ft) high are normal, while speeds of double this may occur in severe storms. These currents show on aerial photographs because of the colouring given by the amount of sand being carried by them. Rip currents also occur with oblique waves. Such waves pile water up and initiate longshore currents. If these currents are confined, at every point along the beach the addition of more and more water could only be handled by a continuous acceleration of the current, which does not occur. In this case the head of water is relieved by the formation of rip currents oblique to the shore and to the wind.

Constructive waves are often low frequency waves of swell type. Large swell waves may be present with an absolutely calm sea or even with light offshore winds in the opposite direction. These tend to cause offshore movements of surface water and either a return movement along the bottom or in certain restricted zones. Thus the constructive action of the waves may be helped by the general onshore movement of water.

Tides, surges and currents

The causes of the tides need hardly concern us here, except to state that the interaction of the solar and lunar gravitational effects, the shape of the oscillating basins, i.e. the seas and the oceans, the earth's rotation, and the modification imposed upon tides by coastal configuration produces tides of varying range from virtually nil in many oceanic situations to very high figures at the head of certain bays and estuaries, for example 15 m (50 ft) at the head of the Bay of Fundy and 12 m (40 ft) at Avonmouth in the Bristol Channel. Bird (1968) divides up coasts on the basis of tidal range into microtidal (less than 2 m: 6 ft), mesotidal (2–4 m) and macrotidal (more than 4 m: 12 ft).

Tidal ranges are important in several respects. The inshore limits of wave erosion are related to the heights of high tides, so that the inshore margin of a marine terrace may vary by 6–7·5 m (20–25 ft) (half the tidal range) from areas with no tides to areas with the maximum tidal ranges known. This fact must obviously be taken into account in the interpretation of raised beaches.

Where there is no tidal range wave attack is concentrated at the same level of the coast for twenty-four hours a day, but where the tidal range is great it may be spread over a vast foreshore zone and may only attack the foot of the cliffs for a very short period at each high tide. Furthermore, if large tidal ranges are accompanied by such wide foreshores much of the wave attack and energy will be expended thereon. Thus the optimum conditions for coastal terrace development would seem to be areas with small tidal ranges.

Finally, tidal range is an important factor in the generation of tidal currents (see below) which may locally become of geomorphological importance.

Surges, or storm surges, are accentuations of tidal conditions by suitable meterological situations. The height of tides is increased by onshore winds and by reductions of atmospheric pressure, a fall of 1 millibar raising the ocean level by about 1 cm (0·4 in). Very deep depressions might lead to rises in sea level, therefore, of the order of 0·5 m (1·5 ft) or even more in tropical cyclones. True storm surges lead to even greater rises than this. They may be illustrated by the surge of 1 February 1953 that caused such serious damage round the coasts of the North Sea.

The North Sea is obviously so shaped that very strong northerly winds can cause a piling up of water in the southern part of the sea, either because the escape route through the Straits of Dover is narrow, or because it is quite possible for the winds in the North Sea to be predominantly northerly while the winds in the English Channel are predominantly westerly. Such a situation can easily arise when a depression is situated off southern Norway. It results in a tendency for water level to rise both on the North

Sea and English Channel ends of the Straits of Dover thus preventing the water escaping in either direction.

Such a situation arose on the night of 31 January–1 February 1953 (Robinson 1953). A deep depression with a centre varying between 968 and 978 millibars moved from the Faroes to the mouth of the Elbe, while behind it a ridge of high pressure built up strongly over the Atlantic. The northerly gradient behind the depression was thus exceptionally steep. In fact the gradient wind at 600 m (2 000 ft) was of the order of 250 km/h (150 mile/h) and surface gusts of 180 km/h (over 100 mile/h) were recorded on the east side of Scotland. It was a time of spring tides, which, because of the surge, were some 2 m (6 ft) higher than predicted in eastern England and 3 m (10 ft) higher in estuaries in the Netherlands. On to this exceptionally high tide were piled waves of over 6 m (20 ft) amplitude so that wave attack was absolutely excessive (a series of accounts of the damage is given in *Geography*, 1953, pp. 132–89). It was fortunate that the time of the surge did not coincide perfectly with the high tide, otherwise damage might have been worse. Again, it was bad enough to have a surge associated with high spring tides, but it could have been associated with equinoctial springs seven weeks later with even more disastrous results.

Currents may be set up in various ways (Guilcher, 1958). Wave currents are caused by the fact that a wave is an orbital motion which decreases with depth. Therefore, the forward movement at the top of the orbit is greater than the reverse movement at the bottom. This results in a gentle down-wave drift of water, which can lead to the accumulation of water along the coast and so help to generate rip and longshore currents. These latter have already been discussed and attributed to gravity movements of water accumulating along the shore.

Tidal currents are caused in two main types of situation. If, for some reason, the time of high tide is different at the two ends of a strait, there will obviously develop a strong gradient between the two, giving rise to a tidal current which in narrow straits can become very strong. Even if there is no difference in time a narrow strait can so canalise the tide that a powerful current is developed. The maximum velocity known occurs in the Moluccas and is of the order of 25 km/h (15 mile/h). Velocities of 17–18 km/h (10–11 mile/h) occur between Alderney and the Cotentin peninsula, while even in the Straits of Messina, in the virtually tideless Mediterranean, velocities of 8 km/h (5 mile/h) have been reported.

Discharge currents in estuaries may be very strong because the high tide impounds river water which strongly reinforces the ebb flow.

The attitude towards the role of currents in erosion and transport has varied a lot. At the turn of the century both were attributed mainly to currents. In the interwar period most of the activity was referred to waves. Since then a more balanced view has prevailed in which currents and waves are seen to act together. Currents of the velocity quoted above are obviously capable of severe erosion and considerable transport. Where

currents are feeble their role may lie principally in determining the direction of fall of particles raised into suspension by wave action. Virtually,
then, they determine the direction of saltation.

The effect of the nature of the coast on erosion

Although reasonable surmises may be made concerning the effect of
various properties of the coast on erosion, little is known of the precise
effects which they have. The nature of the rock, its hardness, its reaction to
chemical weathering and the frequency of its faults and joints must be of
considerable importance. The height of the cliffs must be considered, as the
higher the cliff the greater the amount of material which has to be eroded
to cause a given recession of the coast. Finally, the orientation of the coast
with respect to dominant waves, the degree of indentation of the coast and
the nature of the shore profile away from the coast seem to play important
parts in coastal erosion.

The chief lithological effects are probably due to the hardness of the rock
and the degree of its jointing: the harder the rock and the smaller the degree
of jointing the slower the rate of coastal erosion.

Erosion is most rapid in completely uncemented rocks. Much of the east
coast of England is formed of Pleistocene deposits which yield rapidly to
wave action provided that they are not protected by natural constructional
forms or by artificial defences, such as breakwaters and groynes. Similarly
many parts of the coasts of Sussex and Hampshire are formed of soft Eocene,
Oligocene and Pleistocene rocks, which are also readily eroded. The speed
of erosion of these uncemented rocks can be very alarming. Steers (1953)
has quoted examples from the east coast: in Holderness in a thirty-seven-
year period towards the end of the last century the average loss from the
coast was a strip 65 m (215 ft) wide, or an average annual loss of about
1·75 m (6 ft), while the district suffering the most severe erosion lost 82 m
(273 ft) in the same period. At Pakefield, near Lowestoft, one point on the
coast receded 185 m (620 ft) in the period 1883–1947, i.e. an average annual
loss of nearly 3 m (10 ft). A similar average annual loss has been recorded
at Warden Point on the Isle of Sheppey, where the rock is non-resistant
London Clay. Immediately after the 1939–45 war considerable erosion
started on the Sussex coast between Worthing and Littlehampton, due
probably to the decay of the groyne system. For a couple of years the low
cliffs of unconsolidated coombe rock and brickearth were eroded at the
rate of 3·5–6m (12–20 ft) per year before they were boarded in and a new
groyne system established. The extreme example of soft rock is probably
provided by volcanic ash and a recession of almost one mile in the period
1883–1928 of cliffs composed of this material on the island of Krakatoa
between Sumatra and Java has been observed by Umbgrove (Guilcher,
1958): this works out to a loss of over 30 m (100 ft) per year at some points.

For erosion to continue at the rates noted above it is necessary that the material supplied by the cliff should be transported away by the sea: otherwise there may be formed extensive beaches, spits, bars and comparable forms, which dissipate the energy of the waves and prevent, at least temporarily, the base of the cliff being attacked. The eroded material is usually moved away by the action of longshore drift (see below), which, in England, moves material mainly from north to south on the east coast and from west to east on the south coast. Whether erosion is active on any cliff may be established by the nature of the cliff profile: an actively eroded soft cliff shows evidence of scouring at its base, and the cliff above will have signs of active slumping of the unconsolidated material; a dead cliff in incoherent material will quickly become degraded and covered in vegetation (Fig. 8.4). Good examples of dead cliffs are to be found behind Romney Marsh,

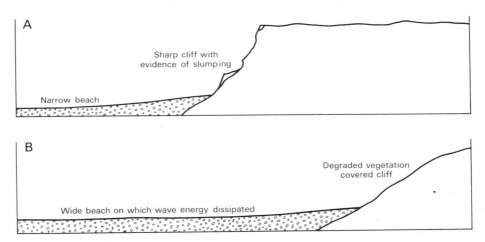

FIG. 8.4. Cliffs in non-resistant rocks
   A. With active erosion; B. With no active erosion

Morfa Dyffryn in Merionethshire, and the dune complex at Winterton Ness in Norfolk. In soft cliffs erosion is very active at first and becomes progressively slower until it halts, unless the debris from the cliff is moved away so that it does not interfere with wave action at the foot of the cliff.

   Erosion of hard rocks is usually very different. Here, the joints and small faults are of very great significance in allowing the ingress of the sea and in affecting the detailed form of the cliff and the rock platform in front of the cliff (Plates 19 and 20). Joints and fissures are probably enlarged partly by the abrasive effect of stones scoured into them by the waves and partly by the pressures generated by the waves when they break against the cliffs. Joints may be enlarged into deep narrow inlets such as the geos developed in the Old Red Sandstone of Caithness. Faults may be enlarged by the sea into caves or even into tunnels through narrow promontories: spectacular examples of the latter are to be observed at Tintagel, north Cornwall, where a fault zone is followed by two through tunnels (Wilson, 1952) (Fig. 8.5), the more important one being the well-known Merlin's Cave beneath the Island. The final result of marine erosion along joints and other planes of weakness is the production of stacks, such as the Old Man of Hoy in the Orkneys and the Needles off the western end of the Isle of Wight.

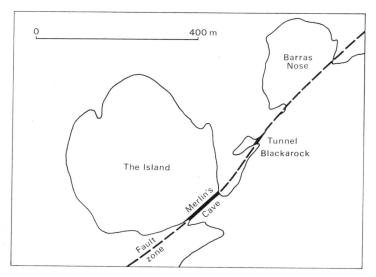

Fig. 8.5. Formation of tunnels along fault zone at Tintagel, Cornwall (*after Wilson*)

The joint pattern is also of considerable importance in controlling the nature of the cliff profile. When the dip of the beds is fairly steep towards the sea, there is a tendency for blocks of rock to break off at the joint planes, usually at right angles to the bedding, so that the cliff profile tends to be dominated by the dip of the beds. Where the beds are vertical, horizontal or dipping inland, joint blocks cannot readily break off and slip down the bedding planes: the result is a tendency for the cliffs to be more nearly vertical.

Interesting forms may result where rocks of differing lithologies are exposed in the cliff face, especially where soft, incompetent rocks are overlain by resistant rocks. This arrangement of beds is conducive to the formation of landslips on a large scale, several examples of which are to be found on the south coast of England. At Folkestone Chalk lies above Gault Clay in the cliffs, and the clay thus tends to become saturated and to have its shear strength greatly reduced. This leads to landslipping. At Seaton in Devon comparable landslips occur where Chalk and Upper Greensand overlie incompetent rocks of Lias and Trias age.

PLATE 19. The coast south of South Stack, Holy Island, Anglesey; see also Plate 20

The possibility of considerable chemical action at water level on limestones and related rocks cannot be ignored. Guilcher in particular (1953 and 1958) has analysed the development of such features in a variety of climates. Even in the British Isles some of the harder limestones are eroded into clintlike surfaces. The difficulty has been to account for the way in which such features develop, as the sea in most regions is saturated with calcium carbonate, and in some places supersaturated.

In a study of Oahu, one of the Hawaiian islands, Wentworth (1939) attributed the solution not so much to the action of the sea but more to the effect of fresh water seeping from the land at sea level. Pools of water, presumably derived from this source, were observed on the upper parts of the beach and were not saturated with calcium carbonate. But although this hypothesis removes one difficulty it creates several more, as indicated by Guilcher (1958). Even in Oahu, where the solution is intense, the coastal limestone regions receive only about 1 000 mm (25 in) of rain a year and this would not appear to be sufficient to account for the amount of solution. Secondly, Panzer has noticed solution forms at the base of several small isolated rocks, which would not have sufficient surface area to allow a large amount of surface water to collect. Finally, important solution forms are to be found in the Red Sea, a region of such aridity that the possibility of fresh water playing a part at the present time can be ignored.

Some other explanation of the solution must obviously be sought and is probably to be found in diurnal variations of the carbon dioxide content of the inshore waters. One suggestion is that, as the solubility of carbon dioxide increases with decreasing temperature, the cooling of the sea at night will lead to an increased acidity and hence to increased solution. A more likely explanation is probably to be found in the diurnal variation of carbon dioxide content caused by the activity of algae. During the day these organisms absorb carbon dioxide from the water for the process of photosynthesis, the decrease leading to the precipitation of finely divided calcium carbonate by the sea. This deposited material is mainly removed by the waves so that at night, when the algae emit carbon dioxide, the ensuing greater acidity of the sea water causes a chemical attack on the rock itself and not merely the redissolving of the finely divided material precipitated during the day.

Even in non-calcareous rocks it is possible that sea water may exert a chemical action on some rock-forming minerals, thus causing a rapid dissolution of the rocks. According to Joly, basalt, obsidian, orthoclase felspar and hornblende weather three to fourteen times more rapidly in salt water than in fresh. There is, of course, some difficulty in ascertaining how much of the chemical weathering of a rock is due to its contact with the sea, as it is quite possible that rocks, subjected to severe chemical

PLATE 20. The coast north of South Stack, Holy Island, Anglesey. The two views (Plates 19 and 20) show a coast of resistant Pre-Cambrian rocks with erosion of numerous planes of weakness forming slitlike caves and small stacks

weathering on the land may, by movements of base level, be brought to sea level. In that case, it would be easy to attribute to the chemical action of the sea the chemical decay already caused by subaerial weathering. It is also possible that some of this weathering might be caused by salt crystallisation effects (see Chapter 3).

Apart from the lithological details of the rock, the height of the cliff must have a considerable effect on the rate at which it weathers back. For any given cut into the cliff by the sea (AB on Fig. 8.6) the amount of debris

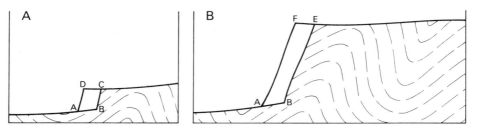

FIG. 8.6. Effect of cliff height on recession

liable to collapse on to the beach will be proportional to the height of the cliff. In Fig. 8.6 the amounts of fallen debris are proportional to the areas ABCD and ABEF: the amount provided by the cliff in Fig. 8.6B will obviously be much greater than that provided by the cliff in Fig. 8.6A. Before the sea can effect any further erosion at the base of the cliff it will have to reduce the cliff debris to such a size that it can be transported away from its protecting position at the base of the cliff. Not until that is done will the sea be able to recommence its erosive activities at the foot of the cliff.

Finally, the exposure of the cliff to damaging waves must be considered. Any cliff or headland which faces the direction from which the greatest waves come will obviously be exposed to severe wave attack. This will be especially true of headlands, due largely to the phenomenon of wave refraction (Fig. 8.7). Any wave approaching the coast from the open sea will usually reach shallower water first off the headlands. As the waves run into shallower water they are retarded, this being noticeable when the depth of the water is reduced to one half of the wave length. The retardation is apparently not due to friction with the bottom, but is caused by changes in the flow pattern enforced on the wave by the shallower water (Russell and Macmillan, 1952). In addition to the retardation this results also finally in an increase in the height of the wave as it is about to break. In Fig. 8.7 the shallower water off the headlands is affecting wave E slightly, causing a slight sinuosity to be formed on the wave front. Waves D, C, B and A show progressively greater degrees of retardation, with resulting greater degrees of refraction until wave A is not only breaking against the point but also against the sides of the headland. The result is naturally a concen-

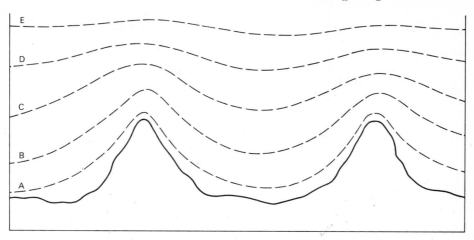

F IG. 8.7. Effect of refraction on approach of waves (for explanation see text)

tration of erosion on the headlands, thus leading ultimately to a smoothing out of the coastline if it is of one rock type. Embayments caused by less resistant rocks will probably not be eliminated, as an equilibrium will be reached between waves, coastline form and rock resistance. Even where the coast does not face directly towards the incoming waves refraction causes the waves to swing round and approach the shore less obliquely than they were when further out to sea (Fig. 8.8). The result is a tendency to increase the amount of erosion on the coast.

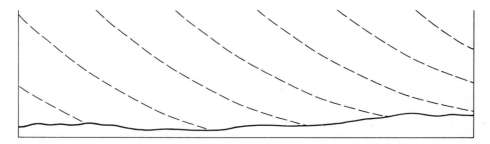

F IG. 8.8. Oblique waves approaching a coast

It is usual today to illustrate the concentration of marine erosion on certain parts of the coast by drawing a series of lines orthogonal to the refracted waves (Fig. 8.9). Where these orthogonals converge erosion will be severe: where they diverge it will be minimal. This provides a graphic illustration of the erosion of headlands and the general lack of erosion in pronounced embayments.

Although it appears to be generally true that erosion is concentrated on the more exposed portions of the coast, one occasionally finds examples of

231

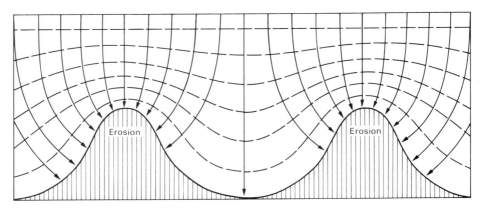

FIG. 8.9. Concentration of marine erosion shown by orthogonals.

marked marine erosion in extremely sheltered positions. In this connection the 7·5 m (25 ft) raised beach, probably of late Pleistocene age, on the west coast of Scotland is very instructive. The beach is found not only in narrow lochs but even in very narrow, sheltered straits between islands and between islands and the mainland. A good example is afforded by the distribution of the beach round the island of Seil. The beach is well developed in the south-west of the island between Easdale and Cuan Sound, where the coast is very exposed to the south-west (Fig. 8.10), i.e. in a position where the beach would certainly be expected. But the beach is also marked on the extreme southern tip of Seil where the exposure to the waves is limited and, even more remarkably, in the narrow strait separating Seil from the mainland (Plate 21). North of Clachan Bridge it has remained as a clear bench on either side of a strait no wider than an average English lowland river. It is possible that the beach on the exposed portions of the coast was originally much wider and that much of it has been destroyed by marine erosion at the present sea level, whereas it has been almost completely preserved in the narrow straits. But even if this were true, it would only alter the degree of the problem, as the rocks on the west coast of Scotland are almost entirely hard and the cutting of beaches in narrow straits must indicate that waves of considerable power have operated there. Many of these straits are noted for strong tidal currents, but although such currents may assist in moving away the debris of erosion, they are generally held to be impotent as a cause of marked marine erosion, at least of rocks as resistant as those which outcrop in this locality.

The nature of the wavecut beach

As the sea cuts into the land, the cliff collapses and the collapsed material is broken down and transported away, there results the slow development of a wavecut platform. In soft rocks a profile drawn normally to the beach

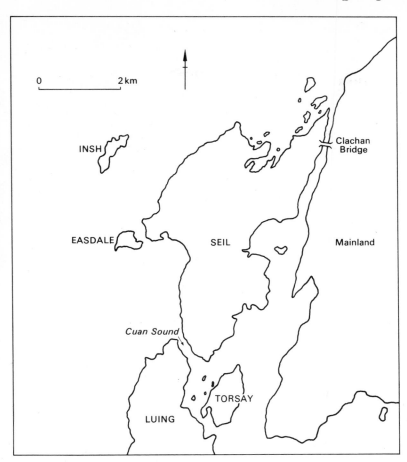

Fig. 8.10. Seil Island and its surroundings

would show a concave form comparable with the long profile of a river. At the foot of the cliff, where debris slipping from the cliff is liable to collect, a steep gradient is needed to enable the sea to comb the material down the beach, but, although this action may assist in the formation of a concave profile, it is probably not the sole cause of the concavity.

It is a matter of common observation that shingle tends to accumulate on the higher parts of the beach while the lower parts are essentially sandy. Many of the beaches of the east and south coasts with their banks of flint shingle in the upper parts and sand lower down show this admirably. There are probably two contributory causes. It has already been noted that the swash often tends to push shingle up the beach, while the back-wash is incapable of moving it down again. A second factor may be found in the nature of particle movement within waves when they run into shallow

233

water. Although in deep water the particle movement is almost completely orbital, this is not so when the waves reach shallow water near coasts. The waves change from regular undulations to sharp crests separated by long flat troughs. As the crest passes there is a strong but short forward movement, while during the passage of the long flat trough little movement takes place. This causes a tendency for material to move abruptly forward during the passage of the crest, while the slight movement associated with the trough is insufficient to affect the material. The operation of these processes is thought to contribute to the shoreward movement of sand and shingle (Russell and Macmillan, 1952). Shingle moved to the upper parts of the beach by either of these causes will pile up until the bank becomes so steep that stones start to roll down the beach, thus maintaining a steeper inshore section to most beaches.

In areas of resistant rocks the wavecut platform (Plate 22), although often showing some degree of concavity, may be considerably flatter than the average sand and shingle beach. Considerable doubt exists as to the precise method of formation of such rock platforms.

One of the major controversies is concerned with the relation between high water level and the abrupt break of slope at the inshore margin of the beach, a feature which has sometimes been called the nip. The relation between this point and high tide level is of considerable importance in the interpretation of raised beaches and old high level marine-cut terraces. Obviously mean sea level has little to do with the height of the nip, which is more closely related to high tide level, but high tide level is in itself a variable level. Further, if splash and spray are effective agents of marine erosion, erosion may take place considerably above high tide level, the exact amount depending on the size of the waves and the way in which they break against the bottom of the cliff.

There seems to be general agreement that the inner edge of the bench may be a metre or so above the level of the highest tides. This is eminently reasonable as effective action by big waves must extend above the level of the highest tide. But when the action of splash and spray is considered, there arises the possibility of the rock bench extending above this, or, if the rocks are suitably stratified, of subsidiary benches at higher elevations. That significant marine action extends to quite high levels is well known. Wentworth (1938) has for instance described an example from the Hawaiian Islands, where a block of tuff weighing about 7 tons has been moved several feet from its natural position at an elevation of 12 m (40 ft) on an exposed cliff. Other examples, often involving the movement of large blocks in structures well above sea level, are also known, while there is the record of the grounding of a 100-ton sloop on the top of the Chesil Bank in 1824. The sloop apparently ran straight on to the top of the bank under sail in spite

PLATE 21. '25-foot' (7·5 m) raised beach in Clachan Sound between Seil Island (left) and the mainland of Argyllshire (right)

of the fact that the top of the bank is 10 m (30 ft) above the level of high spring tides. Jutson (1939) has pointed to a series of generally narrow platforms cut mainly in almost horizontal rocks in New South Wales and extending up to a height of 10 m (30 ft) or more. From the presence of enlarged joints and marine gastropods on these shelves, he was prepared to attribute them to marine action since the sea has stood at its present level. In practice it would be very difficult to distinguish between platforms formed by splash and spray in relation to present sea level and platforms formed by higher sea levels and modified today by splash and spray. Other authors have used the second interpretation in connection with the same platforms. Stepped wavecut platforms may be formed in horizontal rocks (Plate 23), but on these the steps are smaller than on those described by Jutson.

A second major controversy concerns the possible width of the wavecut bench. In this connection thought appears to be divided almost on a national basis, English authors having resorted freely to the assumption that the sea is capable of cutting wide platforms in their interpretation of high level terraces, while American and continental authorities have been much more inclined to doubt whether the sea can cut a very wide bench unless there is progressive drowning of the land or unless the sea is merely trimming up a subaerial surface.

One view of marine erosion would regard it as active only in the early stages of cliff development (A to C in Fig. 8.11), while at later stages the energy of the waves is dissipated on the increasingly wide rock platform so that erosion at the base of the cliff becomes ineffective, a fact attested by the degradation of the cliff by agents of subaerial weathering (D and E in Fig. 8.11). Other authors, e.g. Challinor (1949), have expressed quite opposite views, maintaining that the wavecut bench is cut at such a level as not to impede the action of the waves at the foot of the cliff, so that long-continued marine planation may occur. The truth probably lies between the two extremes. It is difficult to see how the rock platform can entirely dissipate the wave energy, as, if the waves break on it, they must presumably erode it, thus ultimately allowing them to reach the foot of the cliff again. On the other hand, this degradation of the bench must be a slow process, so that the rate of cutting of a marine bench may be expected to become slower as it becomes wider.

These problems are all confused by difficulties of observation, largely the result of many changes of base level in the Pliocene and Pleistocene periods. It has sometimes been argued that marine erosion is not active at present because raised beaches are still preserved on exposed parts of the coast, e.g. on Gower in South Wales and at Brighton. But the fact that a usually narrow portion of beach is preserved is not an argument against modern erosion by the sea unless there is not even a minor cliff at present

PLATE 22. Wavecut platform in Old Red Sandstone, Freswick Bay, Caithness

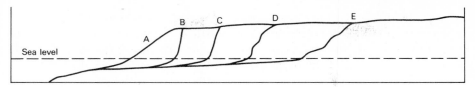

FIG. 8.11. Retardation of cliff recession (for explanation see text)

high tide level, as there is no means of telling how wide a strip of raised beach has been eroded away by the sea. In some areas, for example Arran, the low raised beach shelves gradually beneath the sea and, where this happens without any eroded nip, it seems that there can have been no erosion related to present sea level. The opposite suggestion, that marine planation at the present level is active, has been made from observations of wide platforms truncating hard rocks and apparently related to present sea level. In this case, however, it may be that present sea level coincides with an older sea level to so close a degree that the erosion platform merely represents the trimming up of an earlier one. The possibility that present sea level is approximately the same as an earlier sea level must be taken into account before attributing some of the present high cliffs of the British coasts to recent erosion. It is difficult to imagine that vertical sea cliffs 60 m (200 ft) high, such as those at Hartland Point in Devon, have been cut entirely in the short Post-glacial period since the sea rose to its present level.

In places it may be possible to test whether a wavecut platform is related to present sea level or not. Many British cliffs are degraded, inactive cliffs. This can be seen very clearly on some parts of the North Devon and Welsh coasts, for example round Llangranog in Cardiganshire, where the greater part of the cliff profile consists of a vegetated, degraded cliff with a sharp

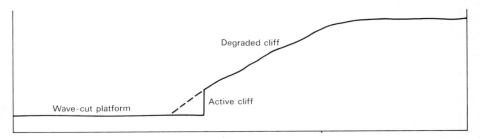

FIG. 8.12. Relation between wavecut platform and active cliff (for explanation see text)

active section at the base (Fig. 8.12). Even some active-looking cliffs, such as those on the Lias of Glamorgan near Southerndown, show on close inspection vegetation growth which suggests, if not complete inaction, at

PLATE 23. Stepped wavecut platform in horizontal Jurassic limestones and clays, Arromanches, Normandy

least intermittent erosion only. In front of the active section there should be a wavecut platform corresponding in width to the amount of recession of the active cliff. If the wavecut platform extends much further seawards than the projection of the degraded cliff it seems very likely that much of the platform was cut in the past at the foot of the cliff which has now become degraded and not in relation to the present active section at the bottom of the cliff, the state of affairs shown in Fig. 8.12.

The belief that the sea acting at one fixed level cannot cut a very wide bench may be derived from the knowledge that sea level relative to the land has changed often in the recent geological past. It appears never to have remained sufficiently constant to enable benches more than a few kilometres wide to form, at least since the beginning of the Pleistocene period.

The constructive action of the sea

Although fine material may be transported far out to sea in suspension, much of the coarser debris eroded from the cliffs or brought to the coast by the rivers accumulates on the beach, where it may be subjected to the constructive action of waves. It may be built up into shore features without undergoing movement along the beach or, more commonly, it is transported along the beach to a point where natural factors allow it to accumulate and to be built up by wave action. The nature of the processes causing this longshore drift is of great importance in understanding many constructional shoreline forms.

Longshore drift

If the wave front approaches parallel to a coastline sand and shingle merely move up and down the beach, but under the more usual conditions of oblique approach the movement is not quite so simple. Particles moved up the beach by the breaking waves do so normally to the direction of approach of the waves, but they roll down in the direction of the steepest beach gradient. The result is that they move along the beach with a type of zigzag motion (Fig. 8.13).

The oblique waves responsible for longshore drift will also set up a longshore current, which will assist in the process of moving material along the beach. Obviously, such a current cannot affect the transport of material in the swash zone, but it may greatly help transport by saltation and suspension in the breaker zone.

For any particular coastline there is a dominant direction of movement depending on the direction of the dominant winds. Thus, along the south coast of England the movement of material is eastwards, while on the east coast, where the dominant winds are north-east, it is southwards, except along part of the north Norfolk coast where it is towards the west. The direction of the drift can be observed from a number of phenomena. On

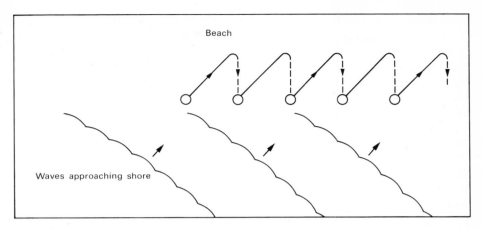

Fɪɢ. 8.13. Longshore drift

the south coast the spits are generally orientated towards the east, as is shown by the spit which previously diverted the mouth of the river Adur east of Shoreham, and by Hurst Castle spit in Hampshire. On the east coast the spits are generally directed towards the south, e.g. Orford Ness, except on the north Norfolk coast, where Blakeney Point is directed westwards. On a smaller scale the piling up of shingle against groynes is always on the side facing the direction from which the drift is coming (Fig. 8.14).

For short periods the dominant direction of the drift may be reversed by abnormal wind conditions. Such reversals, which may last for a period of several years, have been observed on the east coast. At Kessingland in Suffolk at one stage groynes accumulated material on their southern sides,

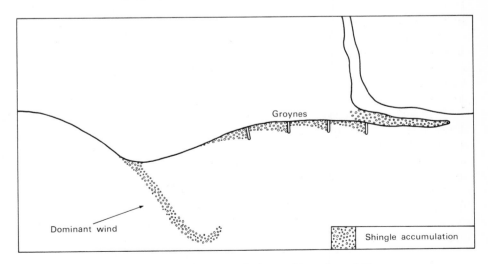

Fɪɢ. 8.14. Relation between shingle accumulation and longshore drift

thus indicating a drift from south to north in a direction contrary to the general direction of drift on the east coast (Steers, 1953).

Although the drift of coarse material on the upper part of the beach must be due to wave action and not to current action, the role of currents in moving material over the sea floor is disputed. One finds direct contradictions such as Lewis's (1931) observation that an 8 km/h (5 mile/h) current off Hurst Castle Spit was ineffective in moving the lightest shell debris except when combined with wave action and Russell and Macmillan's (1952) belief that shingle has been scoured to a depth of nearly 60 m (200 ft) by this same current. They also calculate that sand grains of 0·5 mm (0·02 in) diameter can be moved at a depth of 150 m (500 ft) by a surface current of about 1·7 km/h (1 mile/h). Bird (1968) quotes a number of examples of currents scouring hollows in the sea floor in soft sediments, of scour hollows in narrow straits in which strong tidal currents are known to occur, and of coarse sediments marking the tracks of currents in areas of mainly fine sediment due to the winnowing effect of those currents.

Some of the problems of the relative parts played by waves and currents are in process of being cleared up by the use of traceable sand and shingle. There are three main methods of making such material traceable, but there is one general problem: wide dispersion and burial means that large amounts of material have to be used and that the traced percentage is low.

The simplest method is paint, and paint sufficiently resistant to stand a certain amount of abrasion is now available. Nevertheless, the problem involved in finding painted pebbles among the millions of beach pebbles is on a par with the fabled needle and haystack problem.

Radioactive substances may be added to either sand or shingle. It may be necessary to make artificial materials. Various glasses have been used to make artificial sand. Cement and exotic natural rocks have been used to simulate flint pebbles, because they are more easily drilled for the insertion of radioactive material. When such material is used it must be ensured that it conforms approximately in density and shape with the natural beach material. Fortunately, it has been found that flint pebbles with surface layers of ferric oxide can be rendered radioactive and this eliminates much of the laboriousness of the earlier experiments. Nevertheless, the radioactivity imparted to the pebbles must be weak and transient for obvious health reasons, and this means that radioactivity detectors must be very near the pebbles before they are located, thus increasing the tedium of the experiments. Further, such material cannot be used in places frequented by crowds, while its transient activity means that material which takes time to prepare must be used immediately. The method, therefore, lacks flexibility.

Finally, fluorescent materials, which glow brilliantly in ultraviolet light, may be used. Sand may be coated with a mixture of plastic glue and dye, while artificial concrete pebbles, in which are embedded fragments of fluorescent plastic, may be used to simulate shingle.

On the whole experiments seem to have confirmed that considerable movement of shingle offshore only takes place where pebbles are stirred up by wave action, though currents may contribute to their direction of fall.

Thus it seems likely that in the general process of longshore drift under the action of oblique waves, longshore currents and oblique rip currents may help in the movement of the finer material. The maximum transport occurs in the breaker zone, where wave action is strong and continuous and where the lifting of particles by waves allows the current to operate to fullest effect. The overall volume of transport decreases both up and down the beach from this zone.

Minor constructional features

Although the more generally noticed constructional shoreline features are the large sand and shingle formations, interesting minor features are to be observed on many beaches, among them beach cusps and sand bars, for the formation of which little or no longshore drift is necessary.

BEACH CUSPS

Cusps are a formation found on the upper parts of beaches, especially on shingle beaches. They consist essentially of a series of sharp headlands of shingle separated by sweeping bays (Fig. 8.15). The variation in elevation between the headlands and the troughs is usually less than 1 m (3 ft), but differences up to a maximum of 3·3 m (11 ft) have been reported by Guilcher (1958). The distance between the cusps on shingle beaches is usually of the order of 5 m (15 ft), but on sandy beaches may be much greater ranging up to over 60 m (200 ft) (Russell and Macmillan, 1952).

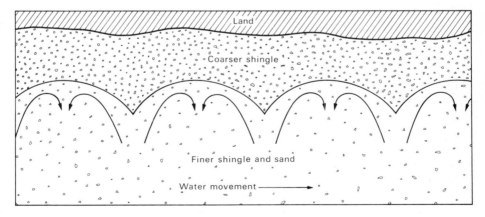

FIG. 8.15. Beach cusps

On beaches with mixed sand and shingle there is a tendency for the coarser material to be gathered into the cusps and for the finer material to form the bays.

The origin of such cusps is obscure, although it is much easier to see why, once formed, they will be maintained. If a beach is ridged at right angles to the land the swash will diverge from the ridges into the furrows, while the backwash will be concentrated into the furrows, thus leading to scouring in them. Such a movement of water on a cusped beach is illustrated by the arrows in Fig. 8.15.

We can say that a given pattern of cusps is associated with given conditions of beach material, beach gradient and wave characteristics. They are rhythmic phenomena and thus analogous to river meanders, or pools and riffles, the basic causes of which remain equally obscure.

SAND BARS

These forms, which have been described by King and Williams (1949), are characteristic of tideless seas, such as the Mediterranean and the Baltic. They are related to the break point of the waves and are sometimes referred to as break point bars. Their formation can be simulated in a glass-sided tank in the laboratory, where observation is very easy (Fig. 8.16).

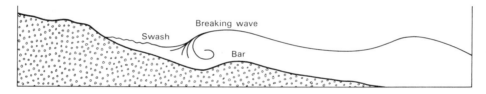

FIG. 8.16. Formation of a break point bar in a laboratory tank

Where the waves are unbroken, i.e. outside the bar, the predominant movement of sand grains on the sea floor is towards the land, while, where a series of broken waves or surf occurs, the movement is mainly seawards. Such processes lead to a concentration of sand and to the building of a bar at the break point. These observations, incidentally, are further evidence against the existence of a general undertow, a point which has already been discussed earlier in this chapter. As the point at which the wave breaks depends on the size of the wave, the position at which the bar will form will also depend on the size of the waves. This may lead to the formation of more than one bar, the outer being related to the break point of storm waves and the inner to the break point of normal waves. In addition, a progressive increase in the size of the waves should lead to the movement seawards of the bar, while a diminution should lead to a landward movement, both being caused fundamentally by the movement of the break point.

244

On flat sandy beaches in tidal seas a somewhat different formation takes place. It can readily be appreciated that with the rise and fall of the tide the break point migrates over a wide zone of the beach so that the formation of a break point bar is not to be expected. Instead, flat beaches, such as that at Blackpool, are often characterised by the formation of a series of ridges and runnels. King and Williams observed in laboratory experiments that, in addition to the break point bar, there is the tendency for the swash to form a second bar at its upper limit. This swash bar may be pushed up above high tide level and is obviously not destroyed by a falling tide as would be a break point bar. The gradient of the swash bar on its seaward side appears to be related to the nature of the waves, short waves building up steeper gradients than long ones. Thus, if the general beach slope is very gentle and the fetch of the waves, and therefore their wave length, limited, the gradient of the swash bar may exceed the beach gradient. This leads to the formation of a ridge behind which a trough or runnel is produced.

Although they admit some doubt as to the exact mode of formation, King and Williams suggest that a swash bar is formed at low tide, that this bar is overrun by the tide, which produces a series of similar swash bars as it advances up the beach. The position and the number of the ridges so formed vary, but there is a tendency for the ridges at high and low neap tide levels to be the most permanent, as would be expected from the fact that the waves have more time to act at these levels than at intermediate levels.

Major constructional features

OFFSHORE BARS

The swash and break point bars described by King and Williams are very small features compared with true offshore bars, which attain lengths of dozens of kilometres on the east coast of the United States. They consist of ridges of shingle and sand projecting above high water level and generally lying a few kilometres offshore. Usually the largest bars are broken through at intervals by tidal inlets especially where powerful rivers reach the coast.

Although it is clear that the bars must be formed by waves breaking some distance offshore, certain details of the origin of the bars are far from clear. To enable waves to break some distance offshore it is necessary that the offshore profile should possess a very gentle gradient. Such profiles are characteristic of shorelines of emergence rather than of shorelines of submergence, a fact which led Johnson (1919) to class offshore bars as features typical of shorelines of emergence. But if the necessary very gentle offshore gradient can be produced on any type of shoreline it is quite likely that offshore bars may form. In the British Isles the best examples are found off the Norfolk coast, as will be described below: although this coast is one of submergence, the offshore profile, composed largely of glacial deposits, is extremely gentle, thus allowing waves to break long before they reach the coast.

The usual explanation of offshore bars is that given by Johnson (1919) and the stages of formation envisaged by him are shown in Fig. 8.17. In the initial stage, Fig. 8.17A, breaking waves excavate material from the sea floor and form a submarine bar, which is slowly built up until it appears above sea level (Fig. 8.17B). The lagoon between the bar and the land is colonised by various types of marsh vegetation and slowly fills up with a mixture of sediment and decaying organic matter (Fig. 8.17C). At the same time the bar is being pushed inland by the sea, which erodes material from the outside of the bar and, in times of storm, flings it over the bar on to the marsh. The result is that the marsh bed appears on the beach in front of the bar (Fig. 8.17D). Finally, after a long time, the bar is pushed so far inland that the marsh is completely eroded and the remains of the bar with associated sand dunes appear on the original coastline (Fig. 8.17E).

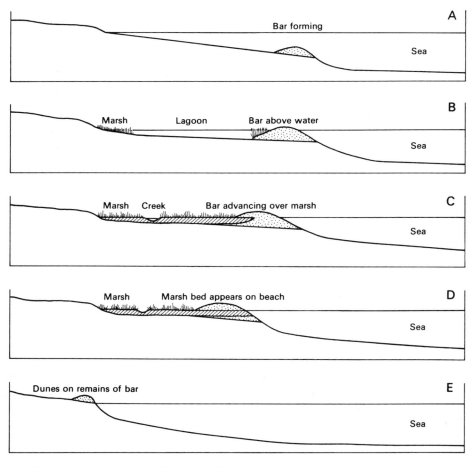

FIG. 8.17. Formation of an offshore bar (for explanation see text) (*after Johnson*)

The origin of the material forming the bar excited some controversy in the early investigation of coastal processes. Some authorities, notably Gilbert, maintained that most of the material was supplied by longshore drift and that the waves merely acted on this material: others, especially de Beaumont, considered that the material contained in the bar was eroded from the sea bed in front of the bar, as is shown in Fig. 8.17. The question was thoroughly investigated by Johnson (1919), who showed that the processes should result in differences in offshore profiles, which ought to be detectable in actual examples. If the material was supplied almost entirely by longshore drift, the profile outside the bar should be continuous with that inside the bar (Fig. 8.18A): if the bar was formed by erosion of the sea floor, the profile outside the bar, if projected back towards the land, should not reach sea level at the coast but some way inland (Fig. 8.18B). The majority of actual examples investigated by Johnson revealed profiles of the second type, so that it may be concluded that much of the material forming the bar is eroded from the sea floor.

One final point which remains uncertain is the manner in which the bars are built up above sea level. Steers (1953) has stressed the importance of vegetation and of dune formation in the history of offshore bars on the north Norfolk coast, which will be discussed below, and it is possible that slight changes of base level may also be important in the initial stages of some offshore bars. Given a slight negative movement of base level, underwater sandbanks may appear above high tide level and could be converted into offshore bars by wave action. Alternatively, a slight positive movement of base level could flood the area behind a line of coastal dunes on a low-lying coast, so that the dunes projected above high water level and formed the basis on which the bar was built.

The most varied and closely investigated series of offshore bars in Great Britain is to be found on the Norfolk coast between Hunstanton and

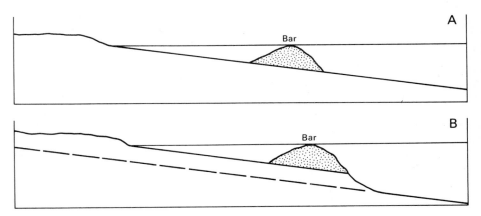

FIG. 8.18. Profiles associated with different methods of bar formation (for explanation see text)

Sheringham. Any material moving southwards along the east coast is stopped by the eastward projection of the Norfolk coast, which has been likened by Steers to a great groyne. In addition, the sea floor, which has a very gentle gradient, is probably composed largely of unconsolidated glacial deposits, with the result that conditions are ideal for the formation of offshore features. Further, there is a general westward drift of material along this section of the coast, so that the shingle formations exhibit a wide variety of features due both to straightforward wave action and to long-shore drift. It must, however, be said that measurements of shingle movement made by Hardy (1964) showed a dominant eastwards movement of material on Blakeney Point, one of the main features of this district. Such a movement, if continued, would presumably quickly deplete the feature and lead to its destruction, unless the bar is being built up by on-shore movement of material at its western end as Hardy suggests. Steers (1964) took up this point and argued that the structure and history, as far as it is known, of Blakeney Point pointed to a westwards movement of material in the long term view. It is well known that local reversals of movement occur and may possibly go on for a number of years.

The earliest stage in the formation of an offshore feature is the deposition of a bank of sand, which may ultimately emerge at high tide. Sand blown from the very wide sand flats at low tide may be trapped on the bank by pebbles or by such debris as barbed wire or any of the usual drift commonly found on beaches. Later, if sufficient material accumulates and the bank is not completely destroyed at high tide, various plants colonise the sand and trap further wind-blown material. Ultimately the bank, with its specialised vegetation and growing sand dunes, appears permanently above the sea. The usual form of such banks on the north Norfolk coast is crescentic, an excellent example being provided by Thornham Western

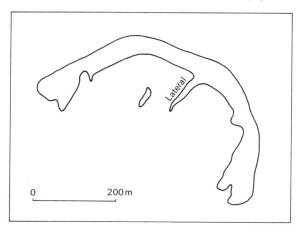

FIG. 8.19. Thornham Western Island in 1939
(*after Steers*)

Island (Fig. 8.19). Once such an island has become permanent, longshore drift takes place along the seaward side, and so leads to modification and complication of the initially simple form.

An idealised development from a simple crescentic bar to a more complex form is shown in Fig. 8.20. The first stage is shown as an arcuate ridge.

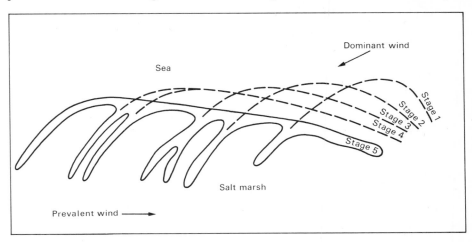

FIG. 8.20. Stages in growth of an offshore bar with recurved laterals

Under the influence of the dominant north-east winds erosion takes place in the east, while longshore drift prolongs the western end until, in stage 2, the bar has a form comparable with Thornham Western Island, the original western part of the island being preserved as a recurved lateral in the middle of the inner side of the island. Later development involves continued erosion in the east and the landward movement of the bar, while the western end is prolonged and further laterals record different stages in its growth (stages 3–5 in Fig. 8.20). It is not entirely clear why the end of such features grows for a period and then is recurved back towards the coast. The prevalent westerly winds on the Norfolk coast are probably of considerable importance in this connection, as once the end of the island starts to recurve, they will tend to build up the laterals roughly at right angles to the main ridge.

Scolt Head Island, the most fully developed offshore bar on the Norfolk coast, has had a history comparable with the above idealised conditions. The present distribution of shingle ridges and their probable earlier continuations are shown in Fig. 8.21. Periods of westward growth have alternated with periods when lateral development was most pronounced. Westward extension was rapid until the formation of the main group of laterals round the House Hills (2–4 on Fig. 8.21). After this phase of pronounced lateral development another period of westward extension carried the island as far as the Long Hills, where another closely spaced group of

249

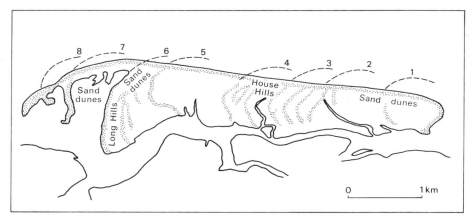

FIG. 8.21. Scolt Head Island. Shingle ridges and probable development (*after Steers*)

laterals suggests a further retardation of the island's growth westwards. Finally, the western point, with short laterals, indicates that the island has recently entered into another period of westward extension, though the point is subject to considerable modification from year to year. The main ridge and the laterals only provide the framework for the island, for, as will be described below, the development of sand dunes and of salt marsh is of the greatest importance in the history of offshore features of this type.

The other major feature of the north Norfolk coast is Blakeney Point (Fig. 8.22), which differs from those already described in being tied to the coast at its eastern end near Sheringham, and is thus really a spit and not an

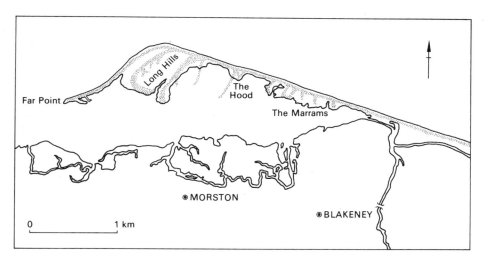

FIG. 8.22. Blakeney Point. Outline at high tide and shingle ridges. (*after Steers*)

offshore bar. The eastern part of the spit, which is not shown on Fig. 8.22, is a long simple shingle bank, but the western part is more complex and gives evidence of periods of westward extension alternating with periods when lateral formation was predominant. The chief groups of laterals are to be found at the Marrams, the Hood, and in the Long Hills. Far Point is subject to considerable change from year to year, as is the western end of Scolt Head Island. The origin of Blakeney Point is open to discussion: it has been suggested that the western end may have been a feature comparable with Scolt Head Island and later joined to the mainland by a simple spit growing westwards from Weybourne: it may have developed entirely as a spit such as Orford Ness or Hurst Castle Spit, which will be described below; or the whole feature may represent an offshore bar driven so far inshore as to become attached to the coast. The equivocal evidence about the direction of movement of material has been discussed above in considering the general direction of movement on this coast.

THE ROLE OF VEGETATION IN THE FORMATION OF COASTAL FEATURES

The shingle ridges forming the skeletons of Scolt Head Island and Blakeney Point constitute only one aspect of the development of such landforms. Sand dunes (Plate 24) and salt marsh, both of which depend in their formation on vegetation, complete the process.

The necessary conditions for the formation of sand dunes are a wide foreshore exposed at low tide so that the wind blowing over it can dry the sand and set it in motion; something to trap the sand and cause it to start accumulating; and plants to colonise the pile of sand and by binding it together to prevent its complete destruction by the wind or high tides. The importance of a wide foreshore in dune formation can be illustrated well from Blakeney Point: at the western end of the point, where the foreshore is very wide, there are well-developed dunes, but at the eastern end, where the narrow shingle ridge faces much deeper water, dunes are absent. Once sand starts to move on the foreshore, which it does by the process of saltation, it will tend to be trapped by shingle patches, by piles of drifted debris, or by tufts of a coarse grass, *Agropyron junceiforme.* The next stage, provided the accumulation is not destroyed almost as fast as it is formed, is the colonisation of the incipient dune by marram grass, *Ammophila arenaria.* Marram grass possesses a deep and branching root system, which effectively binds the dune together, while the plant itself encourages further accumulation of sand. Indeed, marram grass only thrives while fresh sand is being added to the dune. Once accumulation ceases the marram dies out and is replaced by lichens, mosses and other plants: the soil of the dune is enriched by humus from decaying plants and from rabbits and by lime from shell fragments within the sand, so that the nature of the vegetation changes and a more complete cover finally appears on the dune.

The usual reason for the cessation of the sand supply is the formation of a newer line of dunes in front of those previously developed, and these new

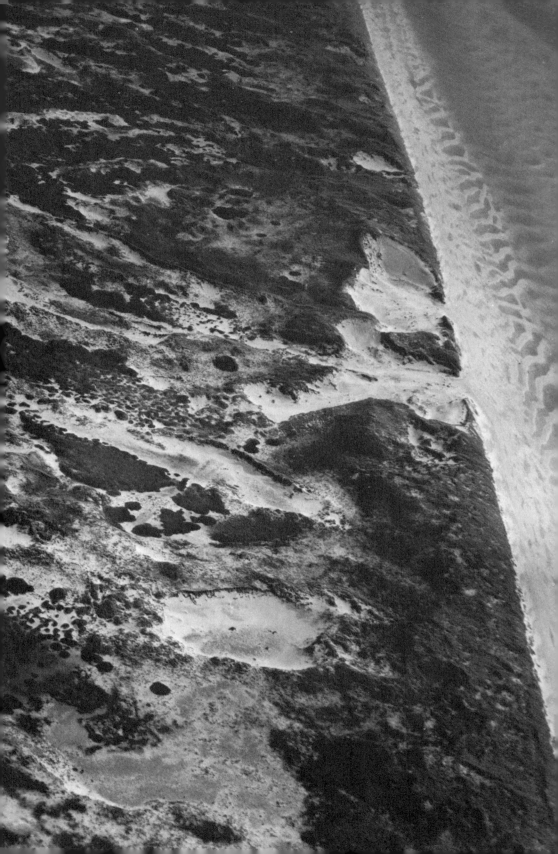

dunes trap the greater part of the sand supply. At this stage, when the marram dies and the dune has only an incomplete cover of vegetation, there is a great danger of a blow out. Either unaided or acting upon a part bared by fire or other forms of human destructiveness, the wind starts to erode the dune. In unfavourable circumstances erosion may be so great as to tear a gap through the dunes, such a feature being described as a blow out. The sand eroded accumulates down wind from the dune, where, being fresh, it is recolonised by marram and stability is thus regained.

Under certain conditions the blow out may not become stabilised, so that the mass of sand migrates slowly in the direction of the resultant of onshore winds with trailing arms of partly vegetated sand on either side of the original blow out. It is then a parabolic or U-dune. In some ways it resembles a barchan (see Chapter 11), but the arms trail, whereas in a barchan the arms or horns lead: the barchan presents its convex face to the wind, the parabolic dune its concave face. If for any reason, such as fire or overgrazing, the arms of the parabolic dune become unstabilised, the whole form may become merely a shapeless mass of sand migrating downwind through the dune belt.

Sand dunes usually accumulate at the top of the beach and on the shingle ridges themselves and materially add to their height, many dunes containing as much as 12–15 m (40–50 ft) of sand (Steers, 1953). Dunes higher than this, for example those on the west side of the Cotentin peninsula near Carteret which reach an elevation of 60 m (200 ft), are piled up over solid rocks.

Salt marsh plays a more important role than sand dunes in extending the area of dry land. The following account of the formation of salt marshes and the plant communities developed in them is necessarily brief, but a much fuller account may be found in Steers (1953) and the papers quoted therein.

Any foreshore of gentle gradient can be seen to possess irregularities in its surface usually in the form of broad ridges and swells and shallow depressions. When such areas are quietly flooded by the tide the material carried in suspension is deposited over the whole foreshore. But, owing to the fact that on the ebb tide water movement is confined to the depressions, the sediment deposited will not accumulate in the depressions to the same extent that it accumulates on the ridges.

The accumulation on the ridges and swells is aided by the presence there of seaweed or flotsam, which tends to hinder the flow of the tide and to cause it to deposit the material it is carrying. When sufficient deposition has occurred the area is colonised by various types of salt-loving, or halophytic, vegetation. In north Norfolk *Salicornia* spp. and *Suaeda maritima* are typical plants found at low levels in the marshes. Such plants, which do not form a very close cover and which in winter are merely bare stalks, do not constitute a very efficient trap for further sediment, except in

PLATE 24. Sand dunes and blowouts at Kenfig Burrows, near Port Talbot, Glamorgan

253

summer when they are best developed. However, they often have seaweed intertwined in their stems and, where this occurs, the dense cover acts as a check on water movement and encourages further deposition. With continuing deposition the nature of the plant cover slowly changes, as various species are adapted to slightly different conditions of flooding. At later stages the marsh cover becomes denser and deposition is encouraged, until finally the marsh is built up to such an extent that it is only covered at high spring tides.

During this process of marsh formation, the original shallow depressions, which were areas of non-deposition in the early stages, are slowly converted to creeks. The ebb and flow of the tide is concentrated in them, so that there is always sufficient scour to prevent their being silted up. The edges of the creeks are often, in East Anglia, colonised by one particular plant, *Halimione portulacoides,* a thick, bushy plant with dense foliage, which, by restricting the flow of the tide, encourages deposition on the margins of the creeks. The banks are often built up in this manner above the level of the surrounding marsh, rather like the levees of a meandering river being built up above the level of the floodplain. Although creeks are thought to be largely areas of non-deposition rather than areas of erosion, the scour of the tide along them may cause some lateral erosion and water draining at times of very high tides from areas behind the zone of creeks may plunge into the heads of the creeks and so cause a certain amount of headward erosion.

Salt marsh development is very significant in filling the areas between offshore bars, such as Scolt Head Island, or spits, such as Blakeney Point, and the mainland. In these areas and between the laterals of the bars and spits, where the movement of the tide is usually unaccompanied by marked wave action, conditions are ideal for quiet deposition. But deposition is not confined to such completely sheltered areas: both on the north Norfolk coast and in the Wash similar salt marshes develop where there are no protecting offshore features.

The rate of accretion varies with the stage in the development of the marsh, as has been shown by careful measurement made by Steers. For the purpose of such measurements a layer of distinctive sand is put down on the marsh and the amount of accretion after a period of years measured by cutting down to the patch of sand with a tool sharp enough not to disturb the section. On the lower parts of the marsh, which are most frequently covered by the tide, accretion at the rate of almost 1 cm per year has been recorded at Scolt Head Island, while at higher levels, where the frequency of flooding by the tide is less, this is reduced to one half or less of the rate at the lowest levels. The rate of accretion may seem slight, but, as it means a gain in height of 0·4–1·0 m ($1\frac{1}{2}$–$3\frac{1}{2}$ ft) per century, it must rank as one of the more rapid geomorphological processes.

When the marsh has built up to an elevation such that it is covered only by the highest tides, it is usually reclaimed for pasture by building a simple embankment around it. Piecemeal reclamations of this type account for the intricacy of the system of banks protecting the low-lying pastures surrounding the Wash. Today, when reclamation is carried out with the use of expensive mechanical equipment, it is not profitable to start until a strip, 200 m or more in width, is ready.

The account of the development of marsh given here applies mainly to East Anglia. On the western coast, where extensive marshes are to be found in Cardigan Bay, the succession of plants differs in some respects from that of the east coast, but the general process of deposition encouraged by plants and leading to a sequence of changing plant communities is similar.

SPITS

Spits differ chiefly from offshore bars in that they spring from the coast and are supplied with material mainly by longshore drift and not from the sea floor. Broadly speaking there are two main types: those which diverge from the coast at a marked angle and those which run approximately parallel to the trend of the coast. In the latter class may be included not only those which deflect a river such as the spit at Shoreham, Sussex, and Orford Ness, but also the bars and spits which often form across bays.

The orientation of coastal features such as spits is largely governed by dominant wind and wave directions. Lewis (1931) formulated the simple rule that spits orientated themselves perpendicular to the direction of dominant waves; in practice they are probably never quite at right angles for this orientation would tend to inhibit the longshore drift necessary for their growth.

Schou (quoted in Guilcher, 1958) used a more refined criterion than the direction of dominant waves. He introduced the idea of a resultant wind obtained by summing vectorially all winds exceeding 20 km/h (13 mile/h) (the lower limit of Beaufort Force 4). The stronger winds are weighted usually by multiplying the frequencies by the cube of the velocities in constructing the vectorial diagram (Fig. 8.23A) in which the lengths are proportional to the calculated vectors. The resultant shown in the diagram is the sort that one might expect in Britain, which is an area of dominant

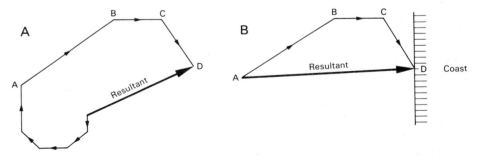

FIG. 8.23. Construction of resultant winds (for explanation see text)

westerlies. The diagram uses only an eight point compass: it is possible and preferable to construct such a diagram with finer divisions.

Of course, the resultant of all winds is probably not the most useful resultant, as on most shores it is only the onshore winds which count in constructional action. Imagine a west coast situation (Fig. 8.23B) with a north–south section of coast. In this case only south-west, west and north-west winds would be onshore, so that it would be better to derive a resultant of onshore winds from section ABCD of Fig. 8.23A.

Most coastal features attached at both ends to the coast orientate themselves at right angles to the resultant provided that the fetch is approximately the same in all directions. If maximum fetch coincides with the resultant the same relationship usually holds. Where they do not coincide the coastal orientation usually lies between the two directions, being nearer to the one which is dominant. Forms which are not attached to the coast or only attached at one end seem to orientate themselves, some at right angles to the resultant like attached forms and some parallel to the resultant for reasons which remain uncertain. Many of the latter are the forms mentioned in the first paragraph of this section as running parallel to the coast.

Schou's theory works best in comparatively enclosed areas where the effective waves are usually caused by fairly local winds, for example Denmark and much of Britain. Where the main constructive waves are swell waves which may have been generated hundreds and even thousands of kilometres away and travelled on as free waves, it is quite likely that the local resultant wind has a far smaller part in the orientation of coastal forms. These conditions hold on open oceanic coasts, for example the Atlantic coasts of Europe and Africa.

An example of the type of spit which leaves the coast at a marked angle, is to be found in Hurst Castle Spit, Hampshire (Lewis, 1931). It is a long shingle spit terminating at its far, or distal, end in a number of recurved ridges (Fig. 8.24). The dominant south-westerly winds cause an eastward longshore drift in the area, and material accumulates off the headland from which Hurst Castle Spit springs. The material is then worked on by the waves and built up into a ridge facing the direction from which the greatest waves come. The tendency for spits to be orientated at right angles to the dominant waves is well illustrated by Hurst Castle Spit which is particularly instructive in this respect for the spit does not continue the direction of the coast west of Milford, which is west-north-west to east-south-east, but orientates itself north-west to south-east, thus demonstrating the connection with wave direction. Once the spit starts to form material is pushed along it by longshore drift, thus leading to its continued growth until it reaches water so deep that wave action is destructive. The sweeping curve to the east on Hurst Castle Spit is due to the fact that offshore banks to the west of the spit cause approaching waves to be retarded and to approach the feature as shown in Fig. 8.24. The eastern extremity of the spit is not greatly affected by waves from the south-west, but comes under the influence of

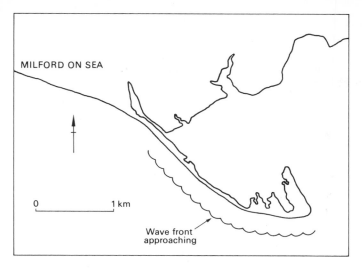

FIG. 8.24. Hurst Castle Spit

north-east winds blowing down the Solent. Waves formed by these north-east winds are responsible for the marked angle at the far point and for the building up of the recurved end so that it faces north-east.

The successive north-east facing ridges at the southern end of the spit probably record phases in the evolution of the spit and are comparable with the laterals on Scolt Head Island and Blakeney Point. The probable evolution of Hurst Castle Spit is shown in Fig. 8.25, from which it can be seen that, with the wearing back of the coast from A to C, the spit will occupy successively the positions AA′, BB′ and CC′, the last being its present position, in which it preserves the recurved ends of former stages.

Orford Ness (Steers, 1953) is a spit of a different type: instead of departing from the coast at a marked angle, it runs roughly parallel to the coast for its entire length (Fig. 8.26). It is remarkable that the river Alde immediately south of Aldeburgh is separated from the sea by only about 50 m of shingle at the narrowest part of the spit, which causes a 17 km (11 miles) deflection of the river.

The spit appears to have started growing from Aldeburgh in a southward direction at a slight angle to the coast with the end recurved towards the coast as in many of the features already discussed. Erosion on the main section facing east and lengthening to the south resulted in the pattern of south-west trending ridges preserved between the lighthouse and Stonyditch Point. The latter represents an important point in the spit's history, as beyond this point it is simpler in form and composed mainly of a few sub-parallel shingle ridges. The importance of Stonyditch Point as the end of the spit for a long period is emphasised by the position of Orford, which was well situated as a port when the Ness ended at Stonyditch Point and

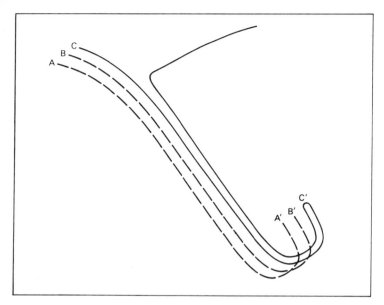

Fɪɢ. 8.25. Evolution of spits of Hurst Castle type

not in its present position some miles to the south. At the present time most of the alterations affecting the spit are confined to the southern end, where it is weakest. The narrow neck just south of Aldeburgh appears to be surprisingly resistant: it was breached, however, by the 1953 floods, but reformed by bulldozers within a few days.

Some spits grow from the mainland towards islands, with the result that the islands become tied to the land by shingle bars, which are known as tombolos. The ideal geological situation for the growth of tombolos is an area of submerged drumlins, where there are many islands and an ample supply of debris for the waves to form the connecting ridges. Spectacular examples of tombolos joining partially submerged drumlins to the mainland are to be found on the coast of Nova Scotia near Halifax (Johnson, 1925).

In the British Isles a fine example of a tombolo is provided by Chesil Beach, a 30 km long (18 miles) ridge connecting the Isle of Portland to the mainland (Fig. 8.27). Although from West Bay to Abbotsbury the main shingle ridge hugs the coast, it departs from the coast south-east of Abbotsbury and leaves between it and the mainland the elongated lagoon known as the Fleet.

A number of problems are posed by Chesil Beach. Compared with most other shingle features in the British Isles the beach is a simple though large ridge, reaching its maximum size at the Portland end, where it is 60 m (200 yd) wide and just over 12 m (40 ft) high. Secondly, there is a progressive grading in the size of the material composing the beach from West Bay

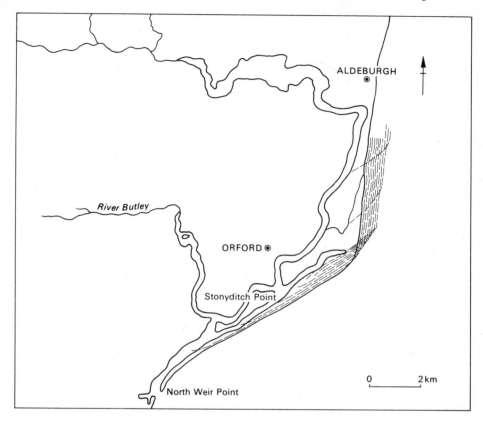

ALDEBURGH

River Butley

ORFORD

Stonyditch Point

North Weir Point

0   2 km

FIG. 8.26. Orford Ness (*after Steers*)

to Portland: at West Bay the shingle is small, about the size of a pea, and it increases progressively to the south-east until it reaches the size of one's fist at the Portland end. Attempts have been made to explain this grading, which is not characteristic of the majority of British spits. It has been suggested that the prevalent west and south-west winds tend to set up a longshore drift towards Portland, so that the majority of coarse material accumulates at that end. When the winds are southerly the drift would be in the reverse direction, but the waves responsible, being of smaller size due to limited fetch and generally lighter winds, would be capable of moving only the finer material back towards West Bay. Although this is a possible explanation, it is not completely convincing, as similar conditions must prevail on many other spits, none of which show the progressive grading of material characteristic of the Chesil Beach. Finally, it is not certain whether the beach originated as a spit and extended south-east from West Bay, or whether it started as an offshore bar which has later been driven shorewards by the waves.

259

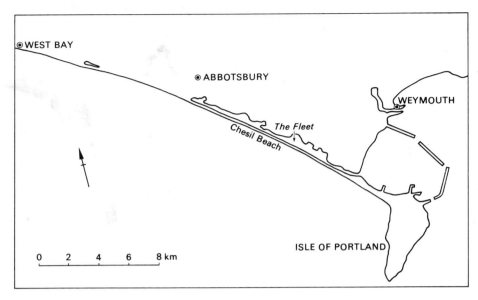

FIG. 8.27. Chesil Beach

A further form of constructional feature is the baymouth bar (Plate 25), which completely seals off the lagoon behind, an example being found at Loe Pool in Cornwall (Fig. 8.28). Of somewhat similar type are those bays, which have spits springing from the headlands on either side and growing towards each other, the distal ends being separated by only a narrow channel. An excellent example of this type is to be found at the entrance to Poole Harbour (Fig. 8.29). Where these double spits are found in positions sheltered from the dominant westerly winds, as at Poole, Christchurch, in Hampshire and Pagham in Sussex, it is possible to regard them as being formed by longshore drift from two directions. In sheltered positions the effect of the westerly winds would be minimised, while that of easterly winds would be accentuated out of normal proportion. On this hypothesis the spits both start from their respective headlands and converge towards the centre of the bay.

An alternative explanation has, however, been advanced by Robinson (1955), who has noted some objections to the foregoing hypothesis. In none of the examples studied is there any evidence of a longshore drift towards the west in the shape of piling up of material on the east-facing sides of the groynes. Further, at Poole Harbour the northern spit, Sandbanks peninsula, appears to be suffering erosion rather than accretion. The danger of a break through the northern end of the spit was so apparent that a sea

PLATE 25. A shingle bar east of Criccieth, Caernarvonshire. The bar blocks the valley containing the marsh of Llyn Istumllyn between the headlands, A and B

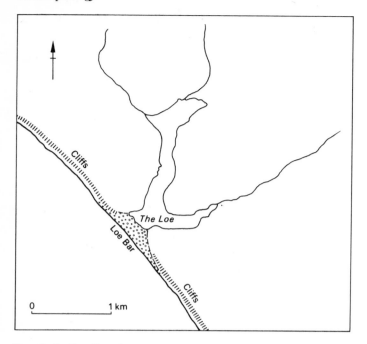

FIG. 8.28. Loe Bar, Cornwall

wall was built along this section in 1890. Such features are not consistent with the idea that the peninsula was built by longshore drift from the east, unless such drift was active only for short periods while during the intervening time the drift was predominantly from the west. In view of this Robinson suggests that such double spit forms were originally baymouth bars which have been breached by waves in times of great storms. On this hypothesis one would expect the eastern spit, Sandbanks peninsula in the case of Poole Harbour, to be in a state of decay due to the cutting off of the supply of material by the breach. Such an idea appears to accord with the known state of affairs at Poole Harbour. Admittedly, there is no evidence of breaching in historical records, but interesting support for the idea is to be found from the history of Pagham Harbour in Sussex, where there are two spits comparable with the Poole ones. In order to counteract erosion it was decided to build a complete shingle ridge across the harbour entrance and this was done in 1876. The ridge survived until a storm in 1910, when it was breached in the middle, thus causing the harbour to revert to its former double spit state. While it is perhaps not completely safe to make conclusions about the evolution of natural features from the fate of manmade structures, the breaching of a former complete bar does, as its author suggests, provide an alternative to the hypothesis of a locally reversed drift.

Although Robinson's suggestions are attractive, Steers (1964) does not

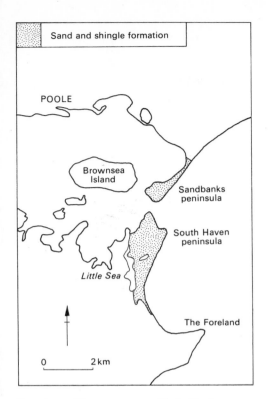

Fɪɢ. 8.29. The entrance to Poole Harbour

believe that all double spits can be explained as breached baymouth bars.
He points out that even at Poole the two spits are not aligned with each
other as one would expect had they been remnants of a former baymouth
bar. Further, there are cases known, such as Bridgwater Bay, where it is a
question of historical fact that two spits have grown towards each other.
Elsewhere other double spits seem to be not explicable in terms of breaching. As usual, it seems that similar phenomena can be produced in different
ways.

Constructional forms already described have been intricate in detail but
simple in general outline, but some natural forms have origins which appear
to be much more complicated. A very good example of a complex form is
provided by the cuspate foreland of Dungeness (Plate 26), which has been
studied in detail by Lewis (1932), whose ideas on its formation are indicated in Fig. 8.30.

In the beginning Dungeness is conceived to have been a simple spit
springing from the coast in the vicinity of Fairlight (A on Fig. 8.30) and extending eastwards across a broad bay now occupied by Romney Marsh.
It is thought that, in the early stages, the sea tended to build up a fan-

263

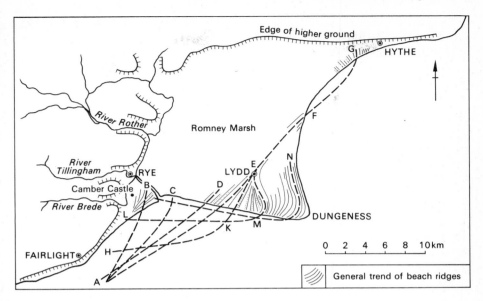

FIG. 8.30. The evolution of Dungeness (for explanation see text) (*after Lewis*)

shaped mass of shingle ridges represented by the phases AB, AC and AD (Fig. 8.30). This development is analogous with the known development of the comparable mass of ridges at Camber Castle. These ridges have been built with considerable rapidity for Camber Castle stands on the inner-most ridge, although, when it was constructed in 1538–9, it stood on the ridge next to the sea. Following the early stages of development came an extension of the spit right across the bay (AEFG): this extension has been suggested to explain the orientation of the ridges near Hythe. Immediately after this stage the ridge was breached, the most important break being near Fairlight. Robbed of its supply of shingle from the west the spit began to be swung round to face the dominant waves, i.e. those from the south-west. The ensuing series of stages, HKEFG, LME and BMN, are consistent with the known orientation of shingle ridges on Dungeness (Fig. 8.30). During this period the spit suffered erosion on its south-west facing portion, the shingle being transported round the Ness itself and built up into new ridges on the east-facing side, where the dominant waves come from an east or north-east direction. The process of erosion on the west and accretion of new ridges on the east has actually been observed by Lewis. Behind the shingle formation Romney Marsh gradually grew up and completely filled the former bay. Although the account above gives, in broad outline, the general evolution of Dungeness, the dating of the various phases is a matter of some difficulty (see Lewis, 1932), while a major prob-

PLATE 26. Dungeness from the south-west, showing the pattern of shingle ridges

lem remains, namely the source of the shingle built into Dungeness. Most of this is flint, which could not have been derived from the Hastings Beds exposed in the cliffs to the west. The Eastbourne area is the nearest chalk district, though much of the flint derived from that area must accumulate at Langney Point and not be available to be transported by longshore drift to Dungeness.

Coral coasts

Coral reefs and atolls present a series of complex problems, involving ecological questions concerning the organisms involved, most of them species of coral, and geomorphological questions of the ultimate origin and present modification of these features. The exact parts played by earth movements and Pleistocene changes of sea level are both involved and imperfectly understood. The problems are treated by Guilcher (1958) and are the subject of a comprehensive review by Stoddart (1969).

Corals themselves require certain conditions for their growth, especially the compound corals that are the important reef-builders. Simple corals are found under somewhat more variable conditions. In the first place they need high temperatures: the mean annual temperature should not fall below 18°C (65°F), and they grow best with temperatures between 25° and 29°C (77 and 84°F). Their resistance to lower and higher temperatures varies from species to species and there are no thresholds at which they suddenly disappear: rather do the reefs become impoverished, probably through interference with reproductive processes; 36°C (97°F) seems to be as high a temperature as most species can stand, but there are exceptions even among allegedly sensitive species. Thus, corals are essentially tropical and virtually confined within 30 degrees of the equator.

The depth to which corals flourish varies from account to account: 25 m (12 fathoms) according to Guilcher (1958), 50–60 m (about 30 fathoms) perhaps less (Kuenen, 1950) or 55 m (30 fathoms) (Steers in Lake, 1949). According to Stoddart most reef corals grow at depths of less than 25 m (12 fathoms) and optimum growth conditions appear at depths of less than 10 m (5 fathoms). The critical point is the effect of depth on illumination, which affects the photosynthetic activity of symbiotic algae. These plants absorb the carbon dioxide released by the corals and so help to keep the water oxygenated. In turbid water the depth to which corals grow will be more limited than in clear water.

The inhibiting effect of sedimentation on corals has been stressed in the past and it has been said that the absence of corals from the mouths of large rivers, for example those of south-east Asia, is due to the amount of fine suspended sediment. However, it is now known that some corals can survive settling sediment and that even reefs may develop in muddy surroundings. Coral species vary in their ability to cope with sediment. Branching corals seem to be less adversely affected than massive corals. Perhaps the general absence of corals from near large river mouths is partly the result of sediment and partly the result of low salinities. Guilcher (1958) quotes salinities of 27–40 parts per thousand as being required for corals, but they have been observed flourishing in the Persian Gulf with salinity as high as 48 per thousand. There is some evidence that low salinities induced by freshwater may result in physiological damage to corals.

Another factor inimical to corals is prolonged emersion, but resistance to this varies with the species. Those which normally live on the inner part of the reef flat (see below) are more resistant to emersion than those normally under vigorous wave action: some species can survive emersion for periods up to three hours.

It used to be argued that turbulence was necessary to ensure the oxygenation of the water and the continuation of the food supply, but it is now thought that the main role of turbulence is in getting rid of excess carbon dioxide.

Coral structures tend to adopt certain recurrent features, both as regards the overall form of reefs and as regards the zonation of reefs. Let us look at

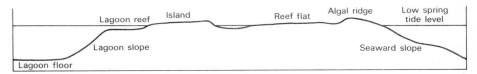

FIG. 8.31. Zonation of a coral reef

the latter first. In simplified form (Fig. 8.31) the zonation of atoll reefs—barrier reefs are very similar—is as follows:

(*a*) *The seaward slope.* This is usually steep, often at an angle of 45 degrees, and sometimes overhanging. It may be interrupted by submarine terraces.
(*b*) *The algal ridge.* This is only well developed on the seaward side of the windward reef of atolls. It is less developed, or even absent, within the lagoon and on the seaward side of the leeward reef. It is characteristic of the trade wind belts of the Pacific, but does not occur in monsoon or Doldrum areas there, and is either very patchy on, or absent from Indian and Atlantic Ocean reefs. The ridge is formed by the algal genus, *Porolithon*, not by *Lithothamnion* after which it used to be named. It extends in places up to a metre or so above the level of the reef flat due to the fact that *Porolithon* can survive in the splash and spray zone and does not need long periods of immersion.

The outer side of the algal ridge is often interrupted by a system of spurs and grooves, the latter running back through the ridge as a series of surge channels, which may even become roofed over in places by the development of *Porolithon* from their sides. Opinions have been divided whether the spur and groove systems are due to erosion or the constructive action of corals. Stoddart quotes a great number of opinions on this subject: it seems that some authorities think that they may have been caused by a fall in sea level which meant that the reef flat became a barrier to water movement, so that surf became channelled down the outer edge of the algal ridge as it returned to the sea; alternatively the spur and groove system may be the most effective form of baffle for dissipating wave energy and is caused by reef-building corals forming the spurs—the grooves, once formed, may of course be accentuated by scouring. This is one of those most probable state explanations, and may be compared with what has been said about meanders and beach cusps.

Inland from the algal ridge there is often a shallow gutter, the origin of which presents a problem. It seems to imply that the algal ridge is not building seawards, unless the gutter is also moving seawards at the same rate.

(*c*) *The reef flat.* This, which is exposed at the very lowest tides, consists usually of a 300–500 m (1 000–1 600 ft) wide surface of dead coral and debris partly cemented with encrusting algae. Live corals are locally present, but the main characteristic is that of a surface-hardened platform of dead coral and debris. Below the surface hardening the material is often unconsolidated.

On the reef flat blocks of lithified material several metres in diameter may be found. In areas subject to tropical cyclones these may be eroded blocks moved under the abnormally large waves accompanying such storms. In other places they seem to be *in situ* and may be the eroded residuals of reefs built in relation to earlier, higher sea levels.

The basic causes of the Indo-Pacific reef flats of the type described above are uncertain. In particular it is not known whether they have been formed in relation to present sea level, in very much the same way as it is not known whether the wavecut platform round Britain was formed entirely in relation to present sea level. They differ from Atlantic reef flats in being considerably higher in relation to tide levels.

(*d*) *The island.* This is probably formed of consolidated, calcareous sandstone, known as beach rock. Accumulations of this material can move downwind through corrosion of the exposed side and the formation of new beach rock on the leeward side.

(*e*) *The lagoon.* Usually there is a narrow reef flat separated from the lagoon floor by a steeper slope. The flat may be partly formed of sand swept from the island and partly of coral. Corals also grow on the lagoon slope and in patches, known as reef knolls, on the lagoon floor.

This zonation scheme is simplified. A fuller division of reefs is given by Stoddart (1969).

It is similarly possible to make complex classifications of the types of reefs (summaries may be found in Daly, 1934; Guilcher, 1958; Stoddart, 1969, and the references quoted therein). As with most classifications of natural phenomena the exact forms of which are governed by the interactions of continuously variable sets of conditions, the classification can be as complex as one likes and the only non-controversial classification is that which puts each reef in a class of its own. Such a procedure, of course, defeats the object of classification.

For present purposes the oldest classification into fringing reefs, barrier reefs and atolls (Fig. 8.32) may be retained as long as it is realised that,

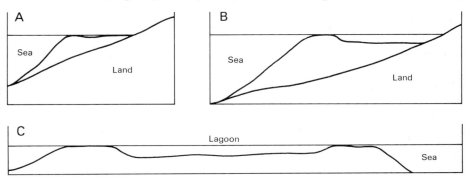

FIG. 8.32. Coral reefs. A. Fringing reef; B. Barrier reef; C. Atoll

although these are commonly occurring forms, there are many others (Guilcher, 1958) which it is very difficult to fit into these simple classes. Barrier reefs and atolls are very like each other in general form and zonation. Barrier reefs may not only lie some distance off major land masses, as shown in Fig. 8.32 and exemplified in the Great Barrier Reef off the Queensland coast which is almost 2 000 km (1 250 miles) long and up to 1 000 m (1 000 yards or more) wide, but may also completely encircle non-coral islands or groups of non-coral islands, one of the best examples of which is the Truk group in the Caroline Islands (Fig. 8.33). In detail barrier reefs consist of whole series of individual reefs and may have small islands on them. Fringing reefs may occur directly exposed to the sea or may form on coasts within barrier reefs: the former often have a zonation similar to barrier reefs and atolls, but the latter have no algal ridge and often very abrupt outer edges.

Some of the general problems concerning the morphology of coral reefs have already been mentioned in discussing their zonation and form. In addition there are distribution problems. The presence or absence of reefs cannot be explained solely in terms of temperature and turbidity, for, although reefs are absent from cold current shorelines, such as the west coast of South America, they do not occur even where the temperatures are high enough for coral in this region. Again, in some areas where conditions throughout appear to be favourable to corals, coral reefs and banks will be found on some submerged banks and not on others, for example in the Laccadive Islands in the Indian Ocean. But the most important fact requiring explanation is the depth at which the bases of the reefs occur: in

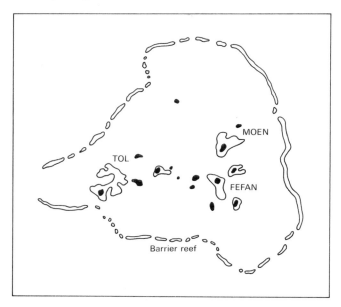

Fɪɢ. 8.33. The barrier reef of the Truk group, Caroline
Islands (*after Daly*)

other words some mechanism is required to provide for the depth at which
coral is found, because this is much lower than the depth at which reef-
building corals flourish. This general tectonic coral atoll problem has been
the one to receive the most attention in the past, though it is now being displaced by morphological, ecological and sedimentation problems.

Although various theories have been proposed, there are two about which
there has been most argument. These are the subsidence theory originally
put forward by Darwin and strongly supported by Davis, and the Glacial
Control theory advocated by Daly. The first requires the subsidence of a
whole series of oceanic islands: the second attempts to use the fluctuating
sea levels of the Pleistocene, which are well evidenced by all sorts of
phenomena, as the explanation of the features of coral reefs. If we believe
in the idea that the most economical hypothesis is the best, then Daly should
win, but it seems at present as though the evidence in favour of the subsidence theory, with its *ad hoc* lowerings of islands, is considerably greater
than that for the Glacial Control theory. Let us consider the latter theory
first and see where it falls short.

Broadly speaking, Daly envisaged a phase of stability in the earth's crust
between the last major earth movements in the Tertiary period. and the
oscillating sea levels of the Pleistocene. This period was sufficiently long
for the widespread planation of oceanic islands, many of them volcanic.
Daly draws attention to the great number of submarine banks and lagoons
of atolls, which occur at depths of up to 90 m (300 ft). Depths less than this

can be explained by the infilling of lagoons and the deposition of some material on banks. The breadth of some of these banks is so great that they cannot be attributed to marine erosion at low sea levels corresponding with glacial periods: the Nazareth bank near the Maldive Islands is about 350 km (220 miles) long and reaches a maximum breadth of about 100 km (60 miles) and lies almost uniformly at a depth of about 60 m (200 ft). The theory thus involves a long phase of stillstand with a sea level considerably lower than the present one in the latter part of the Tertiary period immediately prior to the Ice Age. This supposition conflicts with the conclusions of students of the geomorphology of the lands, who usually regard sea level as being much higher in that period than it is now.

Daly suggests that before the first interglacial period there were no coral reefs in the modern sense, which is disputable. He argues that the low sea level of the first glaciation would enable the sea to act at greater depths and scour much of the mud, which is inimical to corals, from the banks and their edges. In the higher sea level of the first interglacial, any mud left would be at too great depths to be disturbed by waves and hence for the first time the seas would be clear enough for reef-building corals to flourish. The corals were presumably killed off in glacial periods by emersion, by the decrease in sea temperature and by the increased amount of mud stirred up by the lowered sea level. The sequence of events is shown in Fig. 8.34. During the last interglacial reefs occurred (Fig. 8.34A), but with the low sea level of the last glaciation (Fig. 8.34B) these were killed off in most areas. With the Post-glacial increase in sea temperature the bank is liable to be colonised afresh by corals (**xx** on Fig. 8.34B). As sea level rises these will develop and grow upwards, but those on the margins of the bank will be more favoured than those in the area which becomes the lagoon, as they will receive a better supply of food and better oxygenated water than the colonies within the lagoon. Thus the atoll, which had existed in the last interglacial, reformed in the Post-glacial period (Fig. 8.34C).

This theory satisfactorily explains a number of the observed facts, though

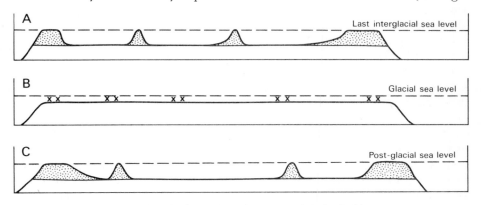

Fig. 8.34. Growth of coral reefs (for explanation see text) (*after Daly*)

271

not all of them. Many of the submerged banks and the bottoms of the lagoons of the atolls seem to be at a remarkably uniform depth, as one would expect if they represent an important phase of planation during a long period of preglacial stability. It must be noted, however, that it is by no means certain in the majority of examples that the lagoons may not contain considerable thicknesses of coral. Daly stressed the narrowness of most reefs and showed, from examples where the rate of reef growth is known, that there had been ample time in the Post-glacial period for the growth of modern reefs. This again is disputable. Where reefs are notably wider than the average it may be assumed that there the corals were not killed off during the Pleistocene. The average width of reefs found by Daly was about 600 m (2 000 ft), while wide reefs in the Banda Sea between Celebes and New Guinea, which are between three and four times the average in width, could be explained on the reasonable grounds that the area is so near the equator that temperatures never fell sufficiently to kill the coral. In the lower glacial sea levels the straits in this region would have been narrower and shallower and the water comparatively warm even for this latitude. Nevertheless, the low sea level would have killed existing corals, which do not survive above low tide level as a rule, so that there ought to be discontinuities in the reefs, and, if those who believe in progressively lower interglacial sea levels are correct, the reefs should exhibit a terraced or stepped form. Davis's (1928) objections on physiographic grounds to Daly's theory do not seem to be valid, and are according to Daly due to a misconception of the Glacial Control theory. The first is that drowned valleys down to about 90 m (300 ft) are explicable as features related to low glacial sea levels and later drowned, but if they are deeper than this subsidence must have taken place. This is true, but even Daly would not exclude preglacial subsidence and it is very difficult to demonstrate the date of subsidence in many examples. Daly himself mentions coral islands which have been elevated and others which have subsided, for his theory does not exclude local subsidence. Davis's other objection, that islands within barrier reefs should show signs of recent cliffing if they had been strongly attacked by marine erosion during low glacial sea levels, is, according to Daly, due to Davis misinterpreting the Glacial Control theory, which does not involve glacial but preglacial planations.

Stoddart (1969) has mentioned several detailed objections to the Glacial Control Theory. If we add these to those included above, we have the following:

1. The theory requires a low preglacial sea level, which is completely at variance with the conclusions of most Pleistocene research workers.
2. The higher interglacial sea levels should, if there had been no subsidence, have produced a series of raised reef terraces corresponding in elevation with the Pleistocene raised beaches and river terraces in other parts of the world.

3. Stoddart uses a rate of rainwater solution of emerged reef limestones of 10–20 mm (0·4–0·8 in) per 1 000 years and suggests that this is not enough to destroy reefs in glacial periods, the last of which he reckons to have lasted 60 000 years. His solution rate would effect 0·6–1·2 m (2–4 ft) surface lowering in that period which is clearly not enough to destroy reefs. Even if we take the general figure of 50 mm (2 in) per 1 000 years for limestone solution used in Chapter 7 it would still only give 3 m (10 ft) of lowering, which is again not enough. Intertidal limestone solution is at a much higher rate, 100 mm (4 in) per 1 000 years, but because of the low sea levels this could not really be expected to destroy reefs. It seems likely that in glacial periods reefs would only have been karstified and not destroyed as envisaged by Daly.

4. Most damaging to Daly's hypothesis is the evidence of coral rock thickness revealed by borings and geophysical work (Guilcher, 1958; Stoddart, 1969). One would not expect coral thickness exceeding 90 m (300 ft) if only glacial control had been involved in the formation of coral islands. A few borings indicating greater thicknesses might be explained away as going diagonally through detritus on the outer flank of the reef, but there are now too many for one to conclude that the results can be explained in this way. The early Funafuti boring reached 340 m (approx. 1 130 ft) while seismic work suggests up to 760 m (approx. 2 530 ft) of coral overlying volcanic rock there. At Bikini shallow water reef limestones have been proved to 780 m (approx. 2 600 ft), while the least depth of the volcanic basement is 1 600 m (approx. 5 300 ft). On Eniwetok olivine basalt was reached in two borings at 1 405 m (approx. 4 680 ft) and 1 283 m (approx. 4 280 ft) respectively. Admittedly, the lowest levels of the overlying limestone date back to the Eocene, but this merely proves long-continued subsidence: horizons of land fauna indicate interruption by periods of emergence. On Kwajalein, the world's largest atoll, seismic work shows a volcanic basement at approximately 1 000–2 000 m (3 300–6 600 ft). In the Tuamotus basalt has been reached at 438 and 415 m (approx. 1 460 and approx. 1 380 ft).

In Bermuda lava has been found at much shallower depths ranging from 20 to 170 m (approx. 70–570 ft) approximately, but in the Bahamas a deep boring ended at nearly 4 500 m (1 500 ft) in Lower Cretaceous dolomite. Even certain elevated atolls seem to have such a thickness of limestone as to indicate subsidence before present uplift.

Thus practically every boring seems to vindicate the subsidence theory and to provide evidence against the Glacial Control theory. Subsidence is common in certain oceanic areas and may be related to the creation and collapse of mid-ocean rises (Stoddart, 1969). Guyots, which are submerged, bevelled volcanic mountains, are known, some having Cretaceous and Eocene faunas on their summits. These presumably sank too fast for compensating reef growth. So, it seems, we must turn to the subsidence theory.

The subsidence theory is simple. One assumes that a subsiding island originally possessed merely a fringing reef. As the island subsided the reef would grow upwards and outwards at the edges, the only places where growth is active for reasons already stated, while the dead part of the reef between the island and the front of the reef would be flooded. Thus, the fringing reef would be changed into a barrier reef and finally into an atoll if the island subsided completely. This theory was advocated in the nineteenth century by Darwin and Dana, and strongly supported by Davis. The evidence from the few areas that have been bored or surveyed geophysically supports this view, but other evidence does not fit so well. If many banks and lagoons are at very similar depths, it would mean a remarkably and almost incomprehensibly uniform subsidence over a very large area, confined, incidentally, largely to tropical and especially tropical oceanic areas. The impression one gets from the various accounts, however, is of a remarkable uniformity of depth from some writers, and of far less uniformity from others. Certainly, some lagoons are abnormally deep and large amounts of subsidence have been recorded from other areas, but Daly's account of uniformity of depth is supported by a considerable number of examples. Kuenen (1950) has also pointed to two features which cannot readily be explained by the subsidence theory. The first is the absence of deep gaps in barrier reefs and atolls, whereas if subsidence has been long continued, there should be deep gaps present, because once the gaps are deeper than the depth at which reef-building corals occur they cannot be filled by coral growth. To account for them by subsidence, it must be assumed that preglacial subsidence was so slow that any breaches could be healed by new coral growth, while the Post-glacial rise of sea level was so rapid that any breaches formed could not be filled. There is a certain amount of evidence to support this contention, which does, however, show the importance of glacial control in the later stages. Secondly, many islands with barrier reefs also have fringing reefs within, whereas if they were subsiding the interior fringing reef should be converted to a barrier reef. In fact islands with multiple barrier reefs are unknown, although a very few double reefs are known. This must mean that subsidence has been steady, for any halts in subsidence should have resulted in the formation of fringing reefs which would later have been converted to barrier reefs by renewed subsidence. The presence of a fringing reef and a barrier reef on an island would seem to be best explained by glacial control, the barrier reef being formed with the Post-glacial rise of sea level and the fringing reef by the more stable modern sea level.

Although there are certain features not readily explicable by subsidence, we cannot revert wholly to the Glacial Control theory. It is incontrovertible that there has been long-continued subsidence on many oceanic atolls. May it not be, however, that others have originated in the way suggested by Daly? Although there are many boreholes and seismic records, there are still a great many more unsurveyed islands.

It seems that the effect of Pleistocene sea levels may, in fact, be evident on a more modest scale. Many of our modern reefs may be interglacial reefs, karstified during low glacial sea levels and now veneered with a capping of modern coral. A discontinuity known as the Thurber Discontinuity occurs in some Pacific areas: above it the corals are younger than 6 000–8 600 years old, depending on the locality, and may represent Post-glacial coral growth when sea level rose above the discontinuity. Some degraded sections of higher level limestone on present reef flats may represent interglacial or Post-glacial higher sea levels. It has even been suggested that some coral reefs, which it will be recalled are mostly dead coral and debris, are being destroyed rather than built up: they could be relics of higher sea levels. So, the Glacial Control theory, while seeming to be mostly invalidated as a basic explanation, may come into its own as the best way to explain many of the details of present reef morphology.

Deltas

It is difficult to know exactly where to include deltas. Many of them are coastal, though others are found in lakes and coastal lagoons, for example the Gippsland Lakes in Victoria. Conditions favouring deltaic accumulation are:

1. A large load of river sediment.
2. Usually a large river. This condition is necessary for marine deltas, otherwise the action of the sea might disperse the sediment. In sheltered lakes deltas may be built by much smaller rivers.
3. Reasonably shallow water offshore, though the necessity for this really depends on the amount of sediment available and the strength of marine erosion. Very deep water must inhibit delta building. For example, the Congo, which virtually debouches into a submarine canyon, has no delta.
4. A coast on which the wave energy is low, though here again how low will depend on other factors such as sediment supply and tidal range.
5. A small tidal range: the Mediterranean, Black Sea and Caspian Sea bear witness to this in the Nile, Rhône, Po, Danube and Volga deltas. However, deltas can be built in areas of larger tidal range provided that conditions 3 and 4 above are met. The Irrawaddy and Ganges deltas are in areas of approximately 5 m (16 ft) tidal range.

Deltas are fundamentally features of river deposition, not marine deposition, though marine sediments may be incorporated in their fronts and intercalated with river deposits if phases of subsidence alternate with phases of delta building. They develop basically because the velocity of a stream diminishes fairly rapidly at its mouth, a condition which must have been accentuated during the Post-glacial rise of sea level: deposition starts, probably aided by flocculation of the finer particles where the effect of salt water is felt, progradation ensues, which further reduces the gradient and so increases deposition by a positive feedback process. Some deltas are virtually continuations of the aggradation type of river floodplain (see Chapter 9).

The subaerial equivalents of deltas are the waste fans that occur at mountain fronts in arid regions and in certain European mountain valleys, especially where there is a lot of coarse debris available, for example in the southern Alps. The reasons are the same; the valleyside torrents are checked when they reach the main valley floor and aggradation takes place, the fan shape usually resulting from the stream shifting its course from time to time.

Although the term, delta, was originally derived from the Greek letter which approximates in shape to the Nile delta, it is generally used with a wider range of meanings today. Deltaic deposition can occur anywhere that a current carrying sediment slows down, especially if the cause of its check in velocity is due to it spreading out on release from a confined channel. Thus submarine tidal deltas can be built where strong tidal currents pass through narrow straits between islands and then fan out. With the reversal of tidal currents deltas can be built up at both ends of the strait. Examples of such double tidal deltas are known from the Dutch Frisian Islands.

Similarly, many estuaries formed by the Post-glacial rise of sea level have infills which are essentially deltaic in character even though they do not emerge at the surface. The mouth of the Loire has a submerged delta of this type, the two distributary channels being near the land on either side and separated by a mass of sediment some of which is exposed at low tide as banks in mid-estuary.

Coastal rivermouth deltas vary in shape and certain forms are alleged to recur, though all sorts of intermediates probably exist. Where sediment is abundant and wave and current action limited, deltas tend to be lobate, the classic example being the Nile. They present convex outlines in plan whether they are simple or compound deltas, composed virtually of a series of sub-deltas. In the extreme case of abundant sediment and weak marine action, the levees on either side of the distributaries may be prolonged seawards to give a digitate delta, for example the Mississippi. Where marine action is relatively stronger deltas tend to be cuspate with concave outlines in plan, for example the Tiber, though one wonders whether these differences in plan might not be due primarily to the frequency and location of distributaries: the Tiber has only one. Delta plans may also be distorted by strong longshore drift.

In very simple small deltas deposition takes place in the embankment form with bottomset beds being overrun by foreset beds, above which the topset beds develop (Fig. 8.35). In practice deposition will be much more complex than this. The distributaries will build levees of the coarser parts of their load, while the areas between will be occasionally flooded, built up with finer sediments, converted into swamps and so become sites of the formation of peats and organic muds. The advancing delta front, especially if there is constructive wave action and longshore drift as on the Rhône

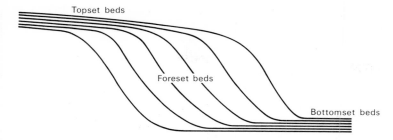

Fig. 8.35. Elements of deltaic bedding

delta, will be fringed with spits and bars, which may cut off coastal lagoons. These will then develop through the usual plant successions into areas of organic sediment rather like the backswamps in the interior of the delta. Very much of the sedimentation will be fine and the classic foreset bed pattern may not be discernible in such sediments.

The whole environment is mobile and in such a state of balance between deposition and erosion that changes are frequent and often important. Imagine a digitate delta prolonging itself rapidly seawards. As the course lengthens the bed of the distributary will be built up and the situation may arise where a levee is breached and a new shorter course to the sea made available. This might well be stabilised as the main course and result in the construction of a new delta segment, until the gradient advantage of the distributaries over it was nullified by the extent of the segment. At this stage the old outlet might be reoccupied or a new breach used.

Such a cycle is shown by the West Bay sub-delta at the mouth of the Mississippi (Fig. 8.36). By 1845 the levee is breached, but no delta exists; by 1875 a well-built lobe fills much of the bay; 1922 shows little further expansion, while by 1958 large areas have reverted to bays and lakes. Some idea of the importance of change may be got from the fact that the Mississippi has built seven major lobes in the last 5 000 years and that the last of these lobes has six sub-deltas, of which the West Bay sub-delta is one. The last four of these sub-deltas were formed by levee breaches in 1839, 1860, 1874 and 1891.

Not every delta will construct new lobes, but some, like the Rhône, have distributaries which vary in importance from time to time.

The formation of new lobes is only one part of the story, for as soon as a distributary abandons a lobe, which is thus deprived of its sediment supply, marine erosion can set in. The sediments of the old lobe will be attacked, destroyed and incorporated in marine sediments. At the same time the inroads made by the sea may provide a potentially steep gradient course to the sea should the levee of the main channel be breached in an appropriate place. One is tempted to see this succession of delta building, nullifying of the gradient advantage as the delta builds out, abandonment of a lobe which is then eroded by the sea, and ensuing recreation of a potential

277

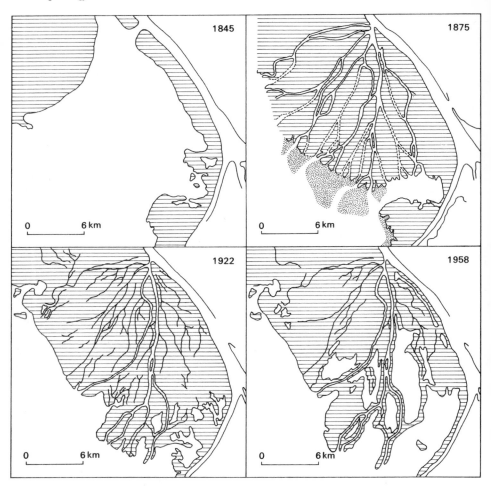

FIG. 8.36. Evolution of a Mississippi sub-delta (*after Coleman*)

steep gradient course, as a possible explanation of cyclic sedimentation without change of relative land and sea level.

In this mobile environment the interpretation of sediment structures is obviously going to be complex. Layers of organic muds of swamp origin will be interspersed with wedge-shaped horizons of marine sediments thinning landwards and representing marine advance, however this may have been caused, levees of coarser sediment will follow the courses of former distributaries, while fanshaped spreads of such material will mark the sites of levee breaches. Old beach ridges may indicate former limits of the delta, while differential compaction, the organic sediments being much more compressible than the mineral silts and sands, will almost certainly further confuse the pattern.

278

Perhaps the most fluid of all deltas are the Arctic deltas, such as the Mackenzie in Canada and the Lena and Kolyma in Siberia. The Lena delta is very complex and is said to have as many as forty-five mouths. Catastrophic summer floods caused by thawing farther south may raise the discharge to the level of the maximum known for the Congo. These floods transport a vast mass of debris including ice rafts and trees into the delta region before the latter has really thawed. The old distributaries may still be frozen and incapable of handling the volume of the river and the load it is carrying, so that there is wholesale bank breaching, flooding and the creation of new distributaries.

It will be realised that the interpretation of deltas is partly a question of geomorphology, involving form and the use of historical records, but the sediment succession is really only explicable by the detailed methods of Pleistocene stratigraphy.

Classification of coasts and shorelines

Classifications of most landforms are in a sense unsatisfactory, because there is usually an infinite variety of forms, some of which fit well into a classification and others only with extreme difficulty. It is doubtful whether many such classifications are of any use at all, for it is more difficult to remember the results of the classification than the individual landforms. Coast and shoreline classifications have usually had one of three bases, descriptive, numerical or genetic.

Purely descriptive classifications may be based on the nature of the coast, whether it is a fjord coast, a ria coast, a low plain coast or any other of a number of types. They may also refer to the relation between structures and the coast, in which case the primary division is usually between longitudinal and transverse coasts. Due to the variety of natural forms there is a tendency to introduce more and more subdivisions until the classification becomes so unwieldy as to be useless.

Numerical classifications were largely a nineteenth-century fashion and principally originated in Germany (Johnson, 1919). They consist usually of a ratio between the length of the coast and some other feature, such as the total land area or a line tangent to the headlands of the coast. As a result they are merely measures of the degree of indentation of the coast, and 'tell little which a good map does not tell much better' (Johnson, 1919, p. 171). It may be that with the present trend towards measurement numerical classifications will come back into vogue.

The most commonly used genetic classification is the simple one formulated by Johnson and includes the following four categories of shorelines, which he uses almost synonymously with coastline here:

(*a*) *Shorelines of submergence.*
(*b*) *Shorelines of emergence.*

(*c*) *Neutral shorelines.* This class includes those shorelines where the exact form is due to neither emergence nor submergence but to a new constructional or tectonic form, e.g. deltas, alluvial plains, outwash plains, volcanoes and faults.

(*d*) *Compound shorelines.* In this class are included all shorelines which have an origin combining at least two of the preceding classes.

In practice, compromise is necessary when assigning actual shorelines to the above classes. This is principally due to the fact that most of the world's shorelines show signs of both emergence and submergence due to oscillations of sea level in the Pleistocene period. The general melting of the ice sheets of the last glacial period has resulted in widespread drowning of coasts, but at earlier stages of the Pleistocene the sea level was higher than at present as is witnessed by remnants of raised beaches. Accordingly, shorelines are said to be submergent or emergent depending upon the most strongly marked characteristics present. Such a practice may easily result in assigning shorelines which are genetically the same to different classes. In Pembroke or south-west Ireland the chief physical features are the drowned valleys or rias, which are so dominant as to cause the shoreline to be classed immediately as one of submergence. West Sussex and east Hampshire have probably had exactly the same sequence of movements of sea level, but the signs of submergence are not so strongly marked. There are shallow inlets and islands in the Portsmouth area, but the main river valleys, although aggraded in their lower reaches, are not rias, presumably because the rivers have brought down so much alluvium that they have filled any rias which may have been present, whereas the rivers of Pembroke and south-west Ireland have been unable to achieve this. Secondly, Sussex shows a good development of raised beach and flat coastal plain, which, together with the smooth shoreline, would tend to make one class it as a compound shoreline.

A classification proposed by Cotton (1952) basically divided coasts into those of tectonically stable and those of tectonically mobile regions. All have been affected by the Post-glacial rise of sea level, but in the second class this may have been accentuated or nullified by local land movements. Thus features of submergence may dominate either class: so too may features of emergence, though these will be inherited in the first class and due to recent uplift in the second. The final categories are:

A. Stable coasts, all basically submergent.
    1. Dominated by features of submergence.
    2. Dominated by inherited features of emergence.
    3. Miscellaneous (fjords, volcanoes etc.)
B. Mobile coasts, affected by tectonic movements in addition to Post-glacial submergence.
    1. Submergent features not offset by uplift.
    2. Dominated by emergent features due to recent uplift.

3. Fold and fault coasts.
4. Miscellaneous.

This classification sorts out Johnson's compound class by splitting it on visually dominating features into his emergent and submergent classes.

Shepard (1963) used a classification which basically divided coasts into those shaped primarily by non-marine agencies and those shaped mainly by marine processes. He then proceeded to subdivide the classes: an outline of the scheme is as follows:

A.   Coasts whose form is primarily due to non-marine agents:
  1. Coasts formed by subaerial denudation and drowned. In detail this can be subdivided e.g. rias and fjords.
  2. Coasts dominated by subaerial deposition. At least two further grades of division are possible e.g. fluvial and glacial exemplify the first of these stages, while deltas represent a second stage division of the former.
  3. Coasts shaped by volcanic activity. Subdivisions are possible.
  4. Coasts dominated by earth movements. Faulting and folding provide two obvious subdivisions.
B.   Coasts dominated by the effects of marine action:
  1. Coasts shaped by marine erosion.
  2. Coasts dominated by marine deposition.
  3. Coasts built up by plants and animals.

Subdivisions are possible in each of these, e.g. coral and mangrove coasts in 3.

Even with a classification as complex as this it is often difficult to place a coast, because the division between A and B is not easy and is very subjective.

An alternative classification has been proposed by Valentin (Cotton, 1954). Here the division is basically into advancing and retreating coasts. The advancing coasts may be due to emergence or to the development of constructional forms either organic or inorganic: they include, thus, shorelines which would have been classed by Johnson as emergent and as neutral. The retreating coasts include those due to submergence and those due to active cliff erosion by wave attack.

There is no limit to the complexity of classification, simply because of all the intermediate forms and combinations of forms that can occur. The formulation of classifications provides, if nothing else, mental exercise for geomorphologists.

9
Movements of base level

The effects of the processes of subaerial and marine denudation described in earlier chapters depend on the influence of base level, which is usually sea level. Any movements of base level set the sea to work at a different level and the processes of subaerial erosion to grade to the new level. Thus, movements of base level constitute interruptions in the cycle of erosion. A major movement of base level must terminate the course of the cycle and initiate a new cycle graded to the new level. Minor movements of less than a hundred metres or so are probably best regarded as interruptions in one main cycle.

In order to avoid the difficult question of whether the sea rises or the land sinks, it is customary to describe any movement involving a rise of the land relative to the sea as a negative movement of base level, or more simply as a rejuvenation since the processes of subaerial erosion are given new activity. Similarly, a sinking of the land relative to the sea is known as a positive movement of base level. The question, whether such movements are caused by worldwide or eustatic movements of sea level or by local movements of the land, is a vexed one.

Movements of base level are of the utmost importance in understanding landforms in most parts of the world for the earth's surface is in an unstable period of its history. The Alpine folding movements culminated no more than 35 million years ago and continued into more recent times in some areas. Many of these Alpine fold ranges, such as the mountains of Japan, the Andes and the arcs through the East Indies are still areas of instability characterised by earthquakes and volcanic activity. In addition, the last two million years have seen one of the greatest glaciations in the earth's history. The waxing and waning of the ice sheets lowered and raised sea level as more or less of the water on the earth's surface was locked up in them. There were at least six cold periods with glaciation somewhere on the earth. These were not of equal intensity, so that the low glacial sea levels may not have been identical. Similarly, the interglacial periods have not been of the same duration and have presumably not resulted in equal melting of the ice. In the short period of Post-glacial time the ice sheets have not completely melted with the result that present sea level is probably lower

282

than it was in some of the interglacial periods. These Pleistocene oscillations of sea level have been superimposed on slightly earlier and usually
downward movements of base level, so that the landscape in detail is
complex and has developed under the influence of many and varied movements of base level.

Positive movements of base level

In their effects on relief positive movements are less spectacular than
negative movements of base level, as the relative rise of the sea has the effect
of hiding parts of the earth's surface beneath the sea or beneath thicknesses
of alluvium.

Along the coast the most obvious signs of positive movement are to be
found in drowned river valleys, usually called rias when they are in areas of
rugged relief and estuaries when they occur in subdued lowlands. Fine
examples of rias are to be found in some of the peninsulas of western
Europe, notably south-west Ireland, Cornwall, Brittany and north-west
Spain. The actual form of the ria depends partly on the relation between
the previous river valley and the rock structures and partly on the relation
between the rock structures and the coastline. Where the river valleys have
followed the weaker beds in the structures, drowning will result in long but
usually simple inlets of the type observed in the extreme south-west of
Ireland (Fig. 9.1). Where the rivers have flowed transverse to the structure
rias formed in them will have a much more irregular plan: the main arm of
the ria may not be straight and it may possess branches developed along
weaker beds followed by the courses of previous subsequent streams. A
good example of this type is provided by Cork Harbour (Fig. 7.14) and
another by Plymouth Sound. If the structure is parallel to the coast drowning will result in a series of lines of islands representing the old ridges and
long narrow arms of the sea along the old valleys, as in the Yugoslav coast
(Fig. 9.2) and the coast of southern Chile. Where the sea breaches former
ridges in this type of coast intricate transverse rias of the type of Cork
Harbour may be formed: an excellent example is to be found north of
Zadar (Zara) on Fig. 9.2.

Whether the rias or estuaries persist depends to a considerable extent
on their depth and on the amount of alluvium being brought into them by
inflowing rivers. In areas of readily eroded rocks the rivers may carry so
much load that one of two things may happen: either the estuary may be
filled soon after it is formed or the aggradation may keep pace with the
rising sea level so that there is not a stage of actual ria or estuary formation.
Most of the rivers of eastern and southern England possess buried channels,
graded to a much lower base level and now completely filled with alluvium.
Excellent examples are provided by the main rivers crossing the South
Downs, the Arun, Adur, Ouse and Cuckmere. The alluvium reaches
thicknesses of approximately 15 m (50 ft) in the Adur and Ouse valleys

Geomorphology

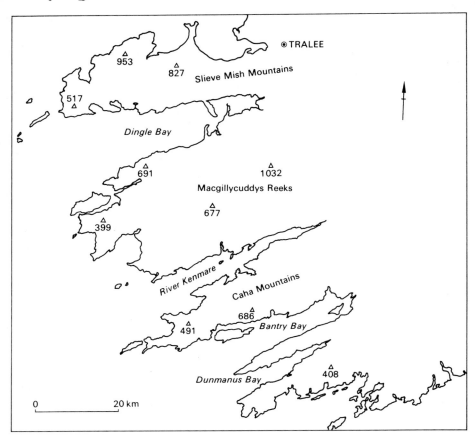

Fig. 9.1. South-west Ireland: rias parallel to structure on a transverse coast

and 30 m (100 ft) in the Arun valley. The aggradation in the lower reaches
of these rivers results in a landform with all the appearance of a true floodplain, a broad tract of easily flooded land over which the river flows in
sweeping meanders, or did so until their courses were artificially contained
in the interests of navigation and flood control. The interpretation of the
sequence of events involved in the aggradation is one in which the geomorphologist has usually insufficient specialised knowledge, as the phases
of marine and river aggradation are represented by different types of
alluvium with differing fossil contents. This subject will be considered a
little more fully below. There is, however, one theoretical criterion by which
rias, which have been later infilled, can be differentiated from lower reaches
aggraded solely by the deposition of river alluvium: it lies in the suggestion
that, if a ria or estuary had existed, the old shores at the side should show
signs of marine cliffing. Unfortunately the rivers, meandering over their
own aggradation, may be expected to produce comparable cliffing where

284

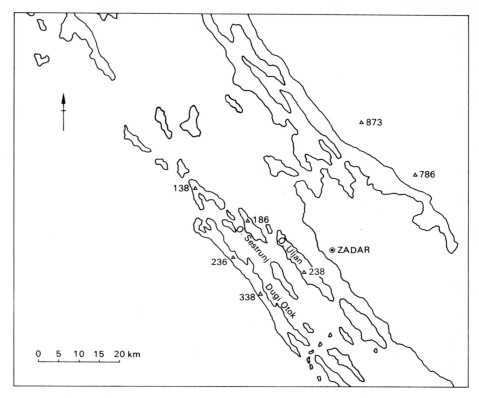

FIG. 9.2. Part of the Yugoslavian coast: drowning of a longitudinal coast

they impinge against the solid rocks of the valley sides, so that, in practice, an interpretation of events based solely on the evidence of landforms is not practicable.

Indeed, it may be true that most of the great floodplains of the world have been formed by aggradation provoked by the Post-glacial rise in sea level and not by the erosional method outlined in the discussion of the Davisian cycle (Chapter 2). Before the days of common subsurface investigation it was very difficult to distinguish between the two. With more money available to satisfy scientific curiosity, with the development of a range of types of boring equipment and with the adaptation of seismic work, the solid profiles beneath alluvium can more often be determined. Usually floodplains are not veneers of alluvium explicable by lateral channel movements, but considerable thicknesses smoothing over more complex relief. The Mississippi is probably the classic case. The English Fens, with their projecting islands of solid rock, are another and will be considered in more detail below. True eroded floodplains may be locally present along rivers, but one needs to use this hypothesis with caution as an explanation

285

of major floodplains. This, incidentally, throws some doubt on the efficacy of the panplanation process (see Chapter 16).

In addition to the aggradation of the lower reaches of river valleys, the Post-glacial rise of sea level caused widespread aggradation of any lowlands which had been eroded to a sufficiently low level. The outstanding example of such aggradation in the British Isles is provided by the Fenland. Here a thick series of alluvial deposits rests in a basin cut largely in the Upper Jurassic clay formations, the Oxford, Ampthill and Kimeridge Clays. Islands of solid rocks, usually Upper Jurassic or Lower Cretaceous, rise through the alluvium especially in the south of the Fens in the vicinity of Ely. The history of the succession of climates and vegetation and the relative changes in the levels of land and sea have been worked out on the basis of stratigraphical and botanical investigations in great detail by Godwin (a simple account is given in Darby, 1938, and a full account of the vegetation zones in Godwin, 1956). As the sea level rose in Post-glacial times two conflicting physical processes took place: a tendency for the sea to invade the area and to leave behind extensive deposits of marine silt and a tendency for a large-scale development of peat in the districts furthest away from the sea. On the seaward side of the Fens in the Wisbech district silt deposition was predominant throughout, while on the landward side in the Cambridge-Ely area peat is the dominant formation. In the intervening regions these deposits interdigitate, periods of marine transgression being indicated by silts and periods of regression by peat beds. In general there were two main phases of peat formation and two marine transgressions, the earlier of the two being much more important. Not only can the physical conditions be interpreted from the stratigraphy, but the actual climate can be estimated from the vegetation preserved in the deposits. On these bases Godwin has distinguished the following phases in the development of the Fens. The dates are based on those tabulated by West (1968).

(*a*) *The pre-Boreal period* (Zone IV). During this period, approximately 8500 to 7600 BC, the whole of the Fenland was a basin of Jurassic and Cretaceous rocks with a varying thickness of glacial deposits covering them. The floor of the North Sea was dry and peat was forming on it, as the Post-glacial rise in sea level had flooded neither the floor of the North Sea nor the Fens.

(*b*) *The Boreal period* (Zones V and VI). From about 7600 to 5000 BC there was a gradual improvement of climate, the predominantly birch forest of the previous period being replaced by pine forest and then by mixed oak forest. Sea level rose and by the end of the period the North Sea had reached approximately its present extent though it did not affect the Fenland. The beginning of peat formation occurred in the deeper river valleys of the area.

(*c*) *The Atlantic period* (Zone VIIa). The period 5000 to 3000 BC was one of increasing dampness and of extensive peat formation. It is generally thought that the climate became more oceanic in character. The growth of peat resulted in the killing of the forest by the exclusion of air from the roots of the trees, and many of these dead trees became entombed in the peat. It was during this period that the main lower peat of the Fens, with its bog oak, was formed.

(*d*) *The sub-Boreal period* (Zone VIIb). An increasing dryness towards the end of the previous period had allowed forest to spread over most of the peat areas, but this in turn was interrupted in the sub-Boreal period, 3000 to 500 BC, by a widespread but shallow marine transgression. True marine conditions probably prevailed only in the north of the Fens but brackish conditions further south are witnessed by some of the fossils. This was the period of the deposition of the Fen or Buttery Clay.

(*e*) *The sub-Atlantic period* (Zone VIII). From 500 BC onwards damper conditions once more prevailed and the recrudescence of peat formation covered much of the Fens, the deposit forming today the main upper peat. A limited marine transgression in the Roman period caused the formation of salt marsh in the northern part of the Fens, while the Fen rivers aggraded their courses and built up levees of silt. The levees caused the areas between to be very subject to flooding and consequently peat formation was further favoured. The levees of former rivers through the Fens are now marked by raised banks, or roddons, which have been favoured settlement sites due to their comparative dryness. The level of the roddons has probably been accentuated by the oxidation and consequent lowering of the level of the peat on either side, but the roddons are thought to have been always higher than the surrounding peat fen (Godwin, 1938). Differential compaction of the peat and silt may have been involved as well.

Such detailed analyses of aggradation deposits as this are not possible by purely geomorphological methods, but require considerations of stratigraphy and especially of fossils. Various fossils may prove useful. The dating of the various phases may be brought about by a study of the archaeology of the district, such methods having proved to be very valuable in the Fens. Beds representing marine invasions may contain marine shells or brackish water shells, thus giving an indication of local conditions. Freshwater beds may be shown to represent either marsh or river conditions by the nature of the freshwater Mollusca they contain. Further discussion of these various methods can be found in West (1968). But by far the most refined technique for the interpretation of aggradation deposits is that of pollen analysis.

The technique of pollen analysis (Faegri and Iversen, 1964; Godwin, 1941) was largely developed in Scandinavia. Under suitable conditions pollen is well preserved in certain deposits especially in peats. Careful sampling of such beds at close intervals by boring tools which reduce the risk of contamination to a minimum, followed by laboratory treatment designed to separate the pollen grains, allows the investigator to build up a picture of the vegetation conditions prevailing during the period of deposition of the bed. For this pollen is the best fossil available as it is the most readily dispersed and therefore the most likely to show correlations over wide areas. The composition of the forest cover and of the smaller plants is usually represented by diagrams showing the relative percentage of the plants present. It has been found that there is a definite sequence of vegetation changes in the Post-glacial period and that these changes can be recognised over wide areas. The actual vegetation may not be identical owing to varying local conditions but there is sufficient affinity to make correlations reasonably certain. By such a refined technique as this it is possible to know what was happening in various aggradation deposits at various periods of Post-glacial history. The method can also be extended to aggradation deposits of interglacial age. Work on these lines has already cleared up many of the mysteries of the Pleistocene and holds more hope of a final elucidation than any other method known.

Finally, the importance of rises of base level in the formation of islands must be stressed. Although islands may be formed by a variety of processes, such as vulcanism, tectonic movements and accumulation, many of the world's present islands must have resulted from the Post-glacial rise of sea level. Reference to the Dalmatian coast (Fig. 9.2) will show how the flooding of valleys and of cols through the intervening ridges leads to the production of many islands.

It is probable that some islands around Britain have been formed by the oscillatory nature of Pleistocene sea level (Fig. 9.3). A comparatively

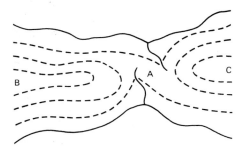

FIG. 9.3. Diagram to illustrate island formation (for explanation see text)

narrow neck of land (A) joining two larger masses of land (B and C) may be imagined. During a glaciation, the streams draining the neck of land will be rejuvenated and will tend to cut into the neck of land, which will be partly flooded when sea level rises in the next interglacial. A series of such oscillations, aided perhaps by marine erosion, is probably an important cause of island formation. The process has been invoked by Everard (1954) to explain the separation of the Isle of Wight. Originally the island was joined to Purbeck by a Chalk ridge the ends of which are still visible in Purbeck and in the Needles at the western end of the Isle of Wight. It was finally breached by the sea flooding through valleys late in the Pleistocene. At the same time the western tip of the Isle of Wight was also almost isolated from the main mass of the island by the flooding of the valley of the western Yar which at present lies well below 15 m (50 ft) OD and extends north–south across the island three miles from its western extremity. Some such process may have been originally responsible for the separation of England and France by the formation of the Straits of Dover, although this would have involved the breaching of a much broader barrier than the Purbeck–Isle of Wight ridge. It should be noted, however, that an alternative hypothesis has been suggested for the formation of the Straits of Dover by Stamp (1927), who thought that they were due to the southwards overflow of a proglacial lake ponded up in the southern North Sea by an ice front stretching from England to Holland. Like the formation of many major landforms, this is a problem where there is room for differences in belief through the absence of any positive evidence.

Negative movements of base level

Negative movements of base level have a much more profound effect on the relief than the positive movements. Landforms developed in relation to the earlier and higher base level are abandoned as erosion starts to work down to the new base level. Some of the landforms, especially if they are depositional, may be quickly destroyed, but forms cut into resistant rocks may be preserved for a very long time. Thus, the record of many negative movements of base level may remain in the landscape to be interpreted by the geomorphologist. The chief landforms resulting are raised beaches and marine terraces, river terraces and breaks in the cross-profile of the valley, breaks in the long profile of a stream usually known as nick points, and incised meanders.

(*a*) RAISED BEACHES AND MARINE TERRACES (Plates 27 and 28)

Provided that the sea has been eroding for a sufficient length of time a negative movement of base level results in a raised beach, consisting as a rule of an eroded platform of solid rocks with or without a covering of beach deposits. Such features are usually called raised beaches if they are at a low level, approximately below 50 m (150 ft) OD, and marine terraces, platforms, benches or erosion surfaces if they are above this level. On the whole the raised beaches are characterised by a platform with overlying beach deposits, while the marine terraces are usually devoid of such deposits. Thus, the terminology corresponds approximately with the nature of the landform, although there are low raised beaches without deposits and a few of the higher marine terraces with deposits.

Low raised beaches are well represented around the south and west coasts of Great Britain, less well on the east coast, and particularly well on the west coast of Scotland. Along the foot of the South Downs north of Chichester there is the well-known Goodwood raised beach at about 30–40 m (100–130 ft) OD, with gravels which are readily recognisable as beach material: a low raised beach a few feet above present high tide level is to be found on the Gower peninsula in South Wales; on the west coast of Scotland a raised beach at approximately 7·5 m (25 ft) OD occurs on many of the islands, such as Arran, and in many of the lochs.

The older and higher marine terraces have been the subject of a very great amount of research in Britain, though there is less general agreement on their interpretation. It has been shown conclusively in the south-east of England that a marked terrace of early Pleistocene age with beach deposits in some localities occurs at about 195 m (650 ft) OD (Wooldridge and Linton, 1955). Therefore, it is clear that there may be expected to be remains of comparable marine terraces at any level between that terrace and present sea level. Most workers who have studied the south and west of Great Britain are convinced that below 195 m (650 ft) OD there is a succession of marine terraces, while some workers have produced evidence to suggest that the succession also extends to considerably higher altitudes. These interpretations have met some opposition on two main counts: very few of the presumed terraces have beach material on them and they are so fragmentary as to rouse the criticism that their interpretation is at times subjective rather than objective. The lack of beach material may be explained in three ways. It is probably fallacious to assume that at every level a pebble beach was formed: if the rocks eroded are weak, such as clay, or both fairly weak and chemically attacked by the sea, such as certain types of limestone, there may well have been no or very little beach material formed. Some present beaches at the foot of chalk and limestone cliffs have only small quantities of pebbles, for example that to the east of Brighton. Secondly, it is possible that the beach gravels have been eroded away since their formation. In Britain much of the area has been glaciated and even the unglaciated part has certainly been subjected to tundra conditions and solifluxion (see Chapter 15) during the Pleistocene. The preservation of beach gravel would then be a question of its accidental occurrence in some specially favourable situation not subject to glacial or periglacial action. Gravels resembling marine gravel do locally occur but they are by no means common, even in areas well away from the ice margin. Finally, it may have been that before the Pleistocene, with its powerful glacial and periglacial denudation, the amount of gravel transferred to the sea by the rivers was much less, so that there was far less material from which the sea could build beaches. The second main objection to the interpretation of

PLATE 27. '25-foot' (7·5 m) raised beach at Imachar, Arran. The beach is cut into Dalradian metamorphic rocks: the old cliffs are to the right and small abandoned stacks in the centre

these fragmentary features as remnants of marine terraces, that they are insufficiently well preserved to be objectively assessed, is a problem common to the interpretation of all relict features undergoing destruction. Between a perfectly preserved marine terrace and one so destroyed as to be unrecognisable there must be a whole series of terrace fragments in various states of preservation. It should really be occasion for surprise that the terraces are as well preserved as they are, not that they are dissected. The real danger lies in a too enthusiastic recording of very doubtful features as marine terraces. Most of the terraces are preserved as flattenings on spurs and it is possible that similar features might result from wasting of a spur due to attack by the headward erosion of valleys on either side. However, careful studies over considerable areas with measurement of heights have shown that the presumed terrace fragments do fall into marked series, so that the basic interpretation seems to be correct, even though features other than terraces may have been included occasionally.

There are other difficulties encountered in the study of the higher marine terraces. In areas of permeable rocks with sub-parallel valleys the terrace fragments on the spurs have breaks of slope both at the back and the front, and, in favourable circumstances, terraces at quite close height intervals can be shown to be distinct from each other. In areas of impermeable rocks, such as the Lower Palaeozoic rocks of Wales, the terraces are often preserved as a series of broad swells reaching approximately the same elevation, but the break of slope between two adjacent terraces is not always observable. It is then possible to argue that the so-called terraces really represent parts of a single warped surface, but a close study of the altitudes usually leads to their interpretation as distinct terraces.

The high sea levels of the Pleistocene, corresponding to the interglacials, are often held to have been responsible for highly degraded marine terraces now found at altitudes up to about 70 m (230 ft) above present sea level. Pleistocene changes of sea level were not simple up and down movements with reversion to a pre-existing level, but oscillations superimposed on a generally downward movement. The descending sequence of marine terraces would then represent successively younger interglacial periods. The correlation of old marine terraces and various phases in the Pleistocene has been a particularly fruitful source of disagreement among glacial geologists and geomorphologists. Much of the earlier work was done on Mediterranean terraces and these, as well as many of the terraces in other areas, cannot be directly related to Pleistocene events. Thus, the presence of solifluxion on a raised beach means that it must precede one glaciation, but this merely fixes its age as pre-last glaciation and gives no indication of how many glaciations have passed since the beach was formed. Although it seems highly probable that sea level changes have been similar over very large areas of the earth's surface, the possibility of local terraces of different

PLATE 28. Raised beach gravel with brickearth above, Selsey, Sussex

293

dates due to other causes cannot be excluded. An example is the terrace around Lake Harrison at a little above 120 m (400 ft) OD (see Chapter 14), which might easily be confused with the well-known British marine terrace at about the same level but probably much earlier in date. The surest way to fix the level of the sea in the interglacial periods would seem to consist in undertaking pollen analyses of interglacial beds containing marine transgressions. This has been done in the case of a number of last (Ipswichian) interglacial deposits in Britain to give a clearer idea of sea level at that period and the amount of East Anglian downwarping since (West and Sparks, 1960; Sparks and West, 1963; Sparks and West, 1970).

The too simple glacio-eustatic view of a descending series of old marine terraces being associated with interglacial high sea levels needs to be guarded against in areas such as East Anglia, where differential warping of the land may well have occurred. This can be illustrated with reference to the Cromerian deposits at West Runton in Norfolk. It has been shown that the Upper Freshwater Bed of the Cromer Forest Bed dates probably from an early interglacial, but there is no sign of a marine transgression in the bed although it covers the middle part of the interglacial, when such a transgression would be expected. The bed lies at about present high tide level, so that there is no indication of marine action at levels as high as at present. Elsewhere the 70 m (230 ft) terrace is sometimes attributed to the first interglacial period. Is it then to be concluded that the Cromer area has been downwarped at least 70 m since the Cromerian interglacial, or, if such an amount of downwarping seems to be excessive, that the sea level of the Cromerian interglacial was not as high as 70 m above the present? The whole subject of the relation of sea levels to Pleistocene events is so confused and the possibility of new information being obtained by the botanical study of interglacial beds so important, that it is better to leave the question open at the moment. It is highly likely that interglacial sea levels are represented by marine terraces or raised beaches, but the levels associated with each interglacial are to some extent in dispute.

The question whether these old marine terraces are due to differential uplift of the land or worldwide lowerings of sea level, is a highly controversial one. The details of the terraces are complex and not of great relevance to such a general study of geomorphology as this, but summaries of various British successions may be found in Wooldridge and Linton (1955) and George (1955).

(*b*) River terraces (Plate 29) and valley cross-profiles

Just as the marine terrace represents the abandoned sea floor, so do river terraces represent valley floors abandoned by the rivers as they start to cut down to the new and lower base level. They are intimately connected with

Plate 29. Small terraces near Keld, Swaledale, Yorkshire. These are more likely to be due to changes in the load-discharge ratio of the stream than to rejuvenation.

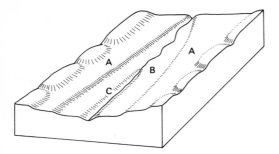

FIG. 9.4. Valley terraces caused by rejuvenation

breaks in the long profile of the streams. In Fig. 9.4 the original floodplain of the river is represented by the terrace A, while, after one rejuvenation, a second terrace B was formed into which the river is again cutting down to form a third terrace C. Each terrace disappears upstream at the point to which the head of rejuvenation has receded: this can be more readily appreciated from a section down the valley (Fig. 9.5).

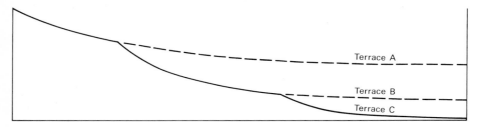

FIG. 9.5. Long-profile of a river showing terraces and nick points

Where rejuvenation intervenes before the river has had sufficient time for lateral erosion to form a flat valley floor, there will not be river terraces at the side of the stream but merely breaks of slope in the valley sides (Fig. 9.6). Such breaks, which occur in deep mountain valleys more often than

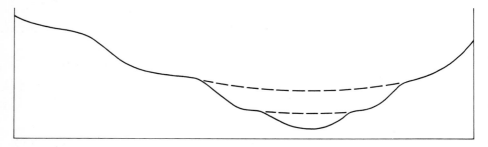

FIG. 9.6. Breaks of slope on valley sides caused by rejuvenation

in lowland valleys, are not always easy to interpret, as similar forms may be caused by lithological differences or by glaciation.

The main characteristic of rejuvenation terraces and breaks of slope in the valley side caused by the same process is that the terraces are essentially paired, so that they occur at the same elevation on both sides of the valley. This enables them to be differentiated from some of the other forms of river terrace, which must be considered in any study of terraces. Generally rejuvenation terraces are cut into the solid rock or consist of a rock bench veneered with a comparatively small thickness of alluvium.

It is possible for a river to reach grade and to start to form a floodplain while it is still transporting coarse material over a comparatively steep gradient. With general reduction of the land through which it flows and a decrease in the calibre of the load, the stream is enabled to flow over decreasing gradients. If this process is combined with lateral shifting of the river the result will be terraces which are not paired (Fig. 9.7) and which converge downstream towards the mouth of the river (Fig. 9.8). Terraces

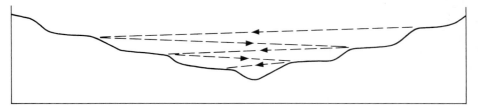

FIG. 9.7. Formation of unpaired terraces: arrows on pecked lines indicate lateral shift of river

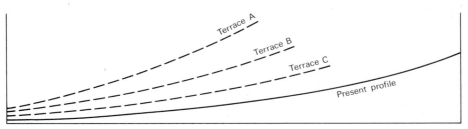

FIG. 9.8. Terraces converging on present profile towards the mouth of a river

of this type have been observed by Lewis (1944) in laboratory models of streams. The formation of unpaired terraces is also likely to happen when a stream proceeds to excavate a mass of glacial material choking its preglacial valley. The resulting terraces will also differ from most rejuvenation terraces in being cut solely into drift and not into solid rock.

Finally, changes of climate may result in yet another type of terrace. Changing climate results in changes in the discharge of the streams and in the nature of the weathering of the valley sides. The latter may well cause the amount and the calibre of the load to alter. In areas near the Pleistocene ice sheets the advent of tundra conditions usually resulted in the rapid weathering of large quantities of material so that the load of the streams was increased in both amount and calibre. The new load could not be transported over the existing river gradients, as these were not sufficiently steep. As a result aggradation took place and the surfaces of such aggradations, now dissected by later erosion, may remain in the form of terraces.

A combination of rejuvenation and climatic change may cause complex terrace forms to be developed. Such a combination occurred, of course, during the glacial periods, when the increased rate of rock weathering caused aggradation in the upper reaches of streams while the lowered sea level caused rejuvenation in the lower reaches. Thus, a glacial terrace profile may be above the interglacial or Post-glacial profile upstream and below it downstream: the terrace profiles, in fact, cross (Fig. 9.9). In this

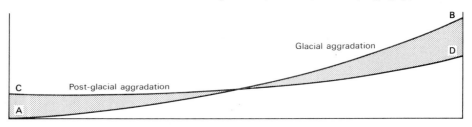

FIG. 9.9. Crossing terraces formed by climate and base level changes in the Pleistocene

figure the normal profile is represented by the curve CD, which is not adapted to glacial conditions as these require aggradation upstream to the level B and erosion downstream to the level A. An excellent example of crossing profiles of this type is provided by the river Durance in southern France (Baulig, 1950). The last glacial period, the Weichsel glaciation, is represented upstream by a terrace which rises about 30 m (100 ft) above the river at Sisteron. Downstream the relative altitude of the terrace decreases and immediately east of the lower Rhône the old course of the Durance plunges beneath the alluvial spread of the Rhône delta to a depth of at least 50 m (160 ft).

(c) Breaks in the long-profiles of rivers (nick points)

As shown in Fig. 9.5 terraces are formed by a head of rejuvenation receding up a river and causing the river to become entrenched in its earlier valley floor. Where the coast has deep water offshore a fall in base level means the formation of a vertical or very steep cliff. One may imagine the river cascading over this in a waterfall or a series of rapids (Fig. 9.10A). Headward erosion is rapid at this point and regrading of the river starts to work back from the mouth. The waterfall is usually held to be slowly flattened out by the erosion of the river so that the new section of the profile occupies successively the positions 1 to 6 (Fig. 9.10A). When the nick point has receded a considerable distance upstream it will have a much gentler form, comparable with A (Fig. 9.10A), which represents the head of an earlier rejuvenation. Conditions such as these are those usually adopted in diagrammatic representations of the process.

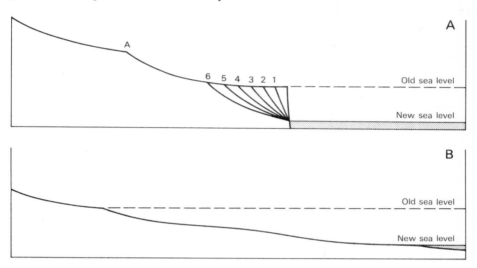

FIG. 9.10. Nature of nick points (for explanation see text)

On very few coasts is there sufficient depth of water offshore to allow for merely vertical displacement of the shoreline following a rejuvenation. Generally speaking the offshore profile is much more gentle, so that a fall in sea level of 30 m (100 ft), for example, will cause a displacement seaward of the shoreline to the extent of anything between a few hundred metres and perhaps 20 km (12 miles). The more rapid the fall in base level the greater will be the migration of the shoreline, as there will be less time for marine erosion to act and push the shoreline back towards the land during the fall. Even if the fall is comparatively slow a lateral displacement of the shoreline must still be expected to occur. Such a movement modifies to a considerable extent the nature of the nick point originally formed (Fig. 9.10B). Instead of being a vertical fall it will be merely a gentle hump, especially if the offshore gradient is very little steeper than the original river gradient. Under these conditions the slight convexity caused by rejuvenation will be quickly rendered even less conspicuous by subsequent river erosion. If the river was not graded before the rejuvenation occurred the nick point may well be indistinguishable from other irregularities present in the stream.

Nick points are often difficult to distinguish from breaks of slope caused by hard bands of rock, especially as nick points will tend to be held up on such hard beds through the difficulty of eroding resistant rock. It is possible for a series of nick points to migrate up a stream and for the whole series to coalesce when they are held up by a resistant bed. In these difficult conditions the question whether nick points related to rejuvenation are present can sometimes be decided by a study of the long profiles of terraces downstream: as many terraces as there are nick points should merge with the present river profile at the hard bed.

(d) INCISED MEANDERS (Plate 30)

If the rejuvenated stream was characterised by meanders it is likely that the meanders will be incised deeply during rejuvenation. It does not follow, however, that incised meanders are necessarily signs of rejuvenation, as if a stream starts to meander before it attains grade it may incise its own meanders to some extent without being rejuvenated.

Incised meanders are sometimes separated into intrenched and ingrown, the chief difference being that the ingrown meanders are more slowly incised, due to less rapid downcutting or to more resistant rocks. The physical differences between the two classes are related to the fact that ingrown meanders are able to perform much more lateral erosion during incision than are intrenched meanders. The meanders themselves tend to develop laterally and the meander belt as a whole tends to migrate downstream. The results can be seen in the form of the spur profiles between the meanders. In the case of intrenched meanders a cross-section of the valley would be more or less symmetrical wherever it was taken (Fig. 9.11A). In addition, the spurs between the meanders preserve the general height of the plateau surface away from the river, except where they are so narrow as to be subject to general lowering by the formation of the slopes on either side. Ingrown meanders, on the other hand, lead to an asymmetric cross-profile of the valley (Fig. 9.11B). The tendency for the meanders to swing laterally is betrayed by the general decrease in elevation of the spurs from the plateau towards the river, while the downstream migration of the meander belt results in the formation of sharp undercut cliffs (positions marked X on Fig. 9.11B) and gentler slipoff slopes (Y on Fig. 9.11B). The differences between the two types, intrenched and ingrown, are not perfect, as there is the possibility of every intervening phase depending on the rate of downcutting.

The floodplain characteristic of meander cutoffs resulting in oxbow lakes is also repeated in incised meanders, especially those of the ingrown type. Continued attack on the upstream sides of spurs and enlargement of the meander leads to a tendency for a break through at the neck of the spur, as it becomes thin (Fig. 9.12A). When the breach finally occurs the old course of the meander is left as a curved through valley, perhaps drained by minor tributary streams, while the remains of the spur form an isolated

PLATE 30. Incised meander of the river Wear at Durham

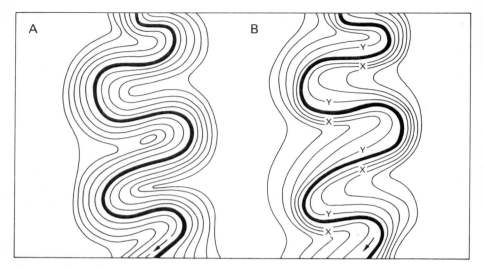

Fig. 9.11. Incised meanders: A, intrenched meanders; B, ingrown meanders

hill, often known as a meander core. The lower course of the river Wye, through the Forest of Dean, illustrates both phases of the process. The most spectacular example of the stage depicted in Fig. 9.12A is provided by Symond's Yat on the northern margin of the plateau, but comparable conditions occur within the gorge itself, at Tintern Parva in the middle of the gorge and near Wintour's Leap a couple of miles north of Chepstow. A clear example of an abandoned incised meander occurs near Redbrook and a less spectacular but probably younger example near St Briavels (Miller, 1935).

Under certain extremely favourable natural conditions the break through the neck of the spur may lead to the formation of a natural arch,

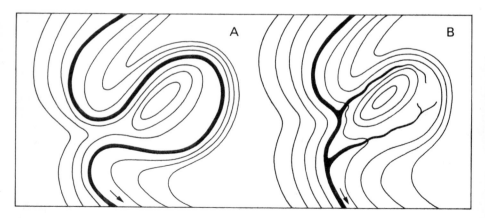

Fig. 9.12. Formation of abandoned incised meander

but this is unlikely in humid climates such as that of Britain, as chemical weathering would probably never allow such a formation. It may happen rarely in arid or semi-arid climates.

If the rocks into which the river is incised are sufficiently weak, ingrown meanders may never develop as the intervening spurs are destroyed by erosion too rapidly by the downstream sweep of the meander belt. Instead, the sweeping of the meander belt would probably lead to a system of un-paired terraces, such as those already described.

There remain certain problems connected with incised meanders (Blache, 1939–40). In areas of rocks of varied resistance incised meanders may be preserved in resistant rocks such as limestones and destroyed in areas of weak rocks such as clay: this is more or less as expected since the downstream sweep of the meander belt would destroy the spurs in clay regions. But there is more to the difficulty than this: the width of the belt of incised meanders in the limestone is often greater than the width of the floodplain in the clay. Unless it is assumed that the original stream had a meander belt of varying width, we must conclude that the meanders have not only been incised but also more fully developed in the more resistant rock. Presumably this might happen due to the greater amount of material to be eroded before a cutoff could occur. However, there does not seem to be a simple connection between resistant rocks and incised meanders, as Blache has pointed to the fact that the lower Loire passing through Palaeozoic rocks does not meander, while the Seine has meanders in the Chalk of a greater size than those found in the Mississippi. In general, areas of tabular relief seem best suited to the preservation, if not to the formation, of incised meanders, e.g. the Mesozoic areas of the Paris Basin and the lower Wye gorge, although why this should be so is not certain.

Methods of investigating features of rejuvenation

A number of cartographical devices and field techniques have been evolved for studying and illustrating terraces and half-formed peneplains. The cartographic techniques are methods of digesting vast amounts of map information and presenting it in a suitable manner: the field techniques are used mainly in plotting data not included in any map.

In any landscape which has been strongly affected by rejuvenation one should expect the uplifted terraces to give rise to an excess of area at certain elevations. This may not be obvious from a visual study of maps, however detailed, because of the amount of dissection since. Thus, the cartographic devices, listed below, are designed to show diagrammatically where the excess lies.

(a) HYPSOGRAPHIC CURVES

These curves, which record the areas at different elevations, are widely used to bring out the major features of the earth's surface, such as the continental shelf, but they may also be used for strictly geomorphological investigation. The data required for the construction of the curve are the areas between successive contour lines. The curve is then plotted by calculating the percentage of the total area lying above each contour beginning with 100 per cent at sea level and gradually decreasing upwards. The areas themselves may be calculated by using squared paper or a planimeter but such methods are laborious. A quicker method is to rule a set of closely spaced lines across the map and to measure the intercepts on these lines between successive contours. Then the sum of the intercepts between each pair of contours is proportional to the area between the contours concerned. One should theoretically use only surveyed contours for the hypsographic curve. On maps of Britain these are usually the 50-foot, the 100-foot, 200-foot, 300-foot and so on up to 1 000 feet OD, but above that only the contours at 250-foot intervals. However, slight inaccuracies in the interpolated contours between the surveyed ones would probably be obscured by errors made in measuring areas, and in practice both surveyed and interpolated contours are used.

The form of the curve for an ideal water eroded landscape not subject to rejuvenation would be concave, as represented by the pecked line on Fig. 9.13 (1). It would vary with the general form of the area studied. Thus, a tilted plane, such as an extensive dipslope, should have a hypsographic curve that is a straight line. Excesses of area at certain elevations are represented on the hypsographic curve by bulges, A and B on Fig. 9.13 (1), and it is at such levels that terraces may be expected.

A more diagrammatic curve of the same type may be constructed by plotting, not the percentage of the total area above each contour, but the percentage of the total area lying between each pair of successive contours. Such a modification leads to a more irregular form with the bulges strongly

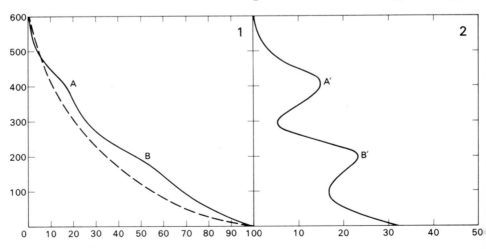

FIG. 9.13. Hypsographic curves: (1) Normal curve, (2) Modified curve

accentuated, as is indicated by Fig. 9.13 (2), which is a replotting of the curve shown in Fig. 9.13 (1), A′ and B′ corresponding to A and B on the first figure.

Hypsographic curves of either type are most suited to the illustration of extensive terraces at rather widely spaced elevations. If the terrace interval is as close as the contour interval the method fails to reveal significant detail. It is unfortunate that the average terrace interval in southern England is as close as the contour interval at least below about 180 m (600 ft) OD, so that the method has limited application, although it will pick out the dominant terraces.

(*b*) CLINOGRAPHIC CURVES

Hypsographic curves may give rise to false impressions of actual slopes, unless it is remembered that the curve does not portray slopes at all. This point has been stressed by Hanson-Lowe (1935), who has emphasised the fact that if a hypsographic curve is drawn for a cone it appears as a concave form, whereas the slopes are really constant. To overcome this drawback he has suggested the use of the same data to construct a curve representing average slopes, a curve which he has called the clinographic curve. For this the irregular areas enclosed by each contour (Fig. 9.14A) may be assumed to be replaced by equal areas enclosed by circles (Fig. 9.14B). Then, as the radii of the circles are known from the measurement of the areas and as the contour interval is known, it is a simple matter to calculate the average slope between any two contours. It is given by the fact that the tangent of the angle of slope is equal to the contour interval divided by the distance between the circles representing the contours (X–Y in Fig. 9.14B).

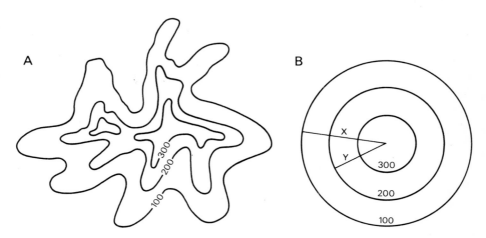

FIG. 9.14. The principle of the clinographic curve (for explanation see text) (*after Hanson-Lowe*)

As most natural gradients are gentle the slopes derived from these calculations may be multiplied by some arbitrary constant for the actual plotting. Breaks of slope will show as convexities on the plotted curve.

Although designed to portray slope, the clinographic curve may give rise to a misleading representation when applied to certain areas. The clinographic curve for a tilted plane would show it to be convex, so that again areas approximating to broad dipslopes are not shown well by this method.

(c) Altimetric frequency curves

Whereas the two foregoing methods make use of the contours of a map, the altimetric frequency curve makes use of the spot heights. The whole of the spot heights shown on a map, except perhaps those in valley bottoms, are listed and grouped into classes. For instance, in the survey of an area of low plateau the number of spot heights falling in each 10 m (25 ft) interval may be counted: then, if the curve shows maximum frequencies in the 50–60 m (175–200 ft) and 80–90 m (250–275 ft) groupings, it would be reasonable to expect terraces at those levels. Another way of constructing the same type of curve is to divide the map into a great number of small squares and to estimate the maximum elevation in each square: a frequency curve is then constructed of the maximum elevations so obtained. As it is not possible to say in advance where the terraces are, the chosen grouping may not bring them out. If a terrace occurred at 48–52 m (170–180 ft) and a 10 m (25 ft) grouping was used, the significant elevations would be divided between two groups and the terrace perhaps not distinguished. To overcome this overlapping groups can be used, e.g. 10–20 m, 15–25 m, 20–30 m and so on.

The success of this method depends on the choice of data recorded on the map to a larger extent than do hypsographic and clinographic curves. It probably works best on maps which show relief by means of hachures and spot heights such as the 1:80 000 map of France and the 1:100 000 map of Germany. These maps usually have a large number of spot heights on summits and terraces. It works less well on British maps, where the depiction of relief is primarily by contours, though interesting results have been derived from British maps by this method (Hollingworth, 1938). Examples of this method applied to north-western France may be examined in Baulig (1935).

(d) Generalised contours

Provided dissection has not gone too far it is possible to attempt to reconstruct the contours of a landscape in an earlier state. Fig. 9.15 shows an area sloping generally towards the south-east and in a youthful stage of dissection, much of the original surface being preserved on the spurs. By ignoring the valleys a reconstruction of the contours can be made as shown by thick heavy lines. They may not be very accurate but they do show a

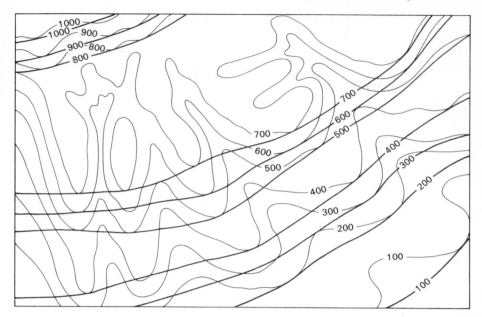

FIG. 9.15. Actual and generalised contours

series of terraces separated by areas of steeper slopes: the most marked terrace is between 700 and 800 m, while a less important one occurs between 500 and 400 m and a third at the lowest elevation shown. In a theoretical example of this type the terraces may be as obvious before the generalised contours were drawn as after, but in some areas, where the ridges are reduced to narrow crests and dissected into outlying hills, such a reconstruction may be of assistance.

(*e*) SUPERIMPOSED AND PROJECTED PROFILES

A similar effect of ignoring the valleys and concentrating on the spurs can be obtained by superimposing a series of sections on one another. Projected profiles consist of drawing the first section completely, while parallel sections behind the first are only drawn in so far as they project above earlier sections (Fig. 9.16). In superimposed profiles the sections are drawn completely so that parts of later profiles below the first are also drawn. The disadvantage of the latter is that it leads to an untidy looking diagram, although it will reveal lower terraces obscured by high ground in the foreground.

FIG. 9.16. Projected profiles of the North Downs near Reigate, Surrey

In drawing such sections it is normal to have them at equal distances at right angles to the guide line which ensures that they are correctly related. However, as it is usually a method designed to illustrate something already known, a certain discretion may be exercised in the placing of the lines of section.

The result is a composite section with some resemblance to a panorama. It differs from the latter in that the viewpoint is not a single point but an artificial one extending the full length of the section. In the right type of country projected profiles may illustrate the relief remarkably well: in Fig. 9.16 it is obvious that there is an area of high ground near the crest of the escarpment separated from a wide terrace on the dipslope by a noticeable break of slope. The terrace, in turn, is terminated to the north by a steeper slope representing that part of the Chalk dipslope exhumed from beneath Tertiary beds.

Field methods

In studies of the results of negative movements of base level in the field, two features are usually surveyed and analysed; the long profiles of rivers and the fragments of marine and river terraces. The methods used for surveying the two differ somewhat, especially in the degree of accuracy involved.

(a) LONG-PROFILES OF RIVERS

As the interval between the surveyed contours is usually 100 feet on British maps and as nick points may be such small features, the long profiles of rivers have to be accurately levelled. This applies particularly to lowland rivers of gentle gradient, but for reconnaissance work in mountain areas, where the gradients are steep and the breaks of slope marked, a less accurate levelling instrument, such as the Abney level, may be used. The actual heights of the nick points are of little significance, because during their recession up various streams they may reach a wide variety of altitudes. But they may be used in reconstructing the former profile of the river and therefore in deriving the sea level to which the river was graded.

As was seen in Chapter 5 the form of a graded river approaches a concave curve as a rule, though it may depart from it under certain circumstances. If a polycyclic river is surveyed and results in a section comparable with that shown in Fig. 9.5, it is possible to extend the segments of the river downstream either by continuing the curve 'freehand' or by finding a mathematical expression to fit each segment and extrapolating the curve on that basis.

Extrapolation based upon continuing the curve by eye is not very accurate and Miller (1939) has stated that it is possible to draw good-looking curves which diverge by 10 m per km (50 ft per mile) but which both provide a smooth extrapolation.

Extrapolation based upon a mathematical formula has the appearance of a much greater degree of precision and has been used by a number of workers in Britain, e.g. Jones (1924) and Brown (1952). The method has however been severely criticised by Miller (1939). The root of the trouble lies in the fact that no segment of a polycyclic river has a perfectly smooth concave profile, so that the formula gives a curve with an approximate fit. This can be dealt with either by fitting the natural curve and the formula curve so that their departures from each other are minimised through the whole length; or by fitting the curve obtained from the formula to the lower parts of the segment of the river on the assumption that grade would be more nearly perfect in this reach; or by refining the formula until it becomes very complicated indeed. Miller has shown that the differences in the derived base level may be very great depending upon which course is adopted. Even when complex formulae are used it has proved possible to find two, which give extremely good fits to the natural curve, but provide base levels differing by about 60 m (almost 200 ft) for a base level at about 120 m (400 ft) OD (Miller, 1939).

The danger of error increases with the length of the extrapolation, which may be many times the length of the segment to which the curve is fitted. Other possible sources of error are to be found in the assumptions that conditions of climate and load-discharge have changed little. But all these factors may have affected the actual form of upper segments of the profile. Further, erosion does not cease above a nick point, so that the form above that point may not be the same as when the river was rejuvenated, especially if climatic change has occurred. Finally the position of the mouth of the river at all stages may not be known so that we may not be certain of which stretch of the former river we are dealing with.

In view of these possible sources of error it is surprising that extrapolated profiles ever yield results of any value. Yet the base levels derived from them are not very different from the base levels derived from the study of marine and river terraces. Two facts probably contribute to this state of affairs: few of the sources of error may be present in the actual rivers surveyed, and most studies involving extrapolated profiles have also considered the terrace evidence so that the old base level was known by other methods. If this is the case extrapolated profiles are best regarded as means to supplement knowledge rather than as a primary method.

(b) Terraces

Although the height of terraces can be estimated by eye by an experienced observer in closely mapped country to a sufficient degree of accuracy, it is safer to determine the heights instrumentally. However, as terraces are usually degraded and as it is usually extremely difficult to determine their precise limits, a degree of accuracy lower than that required for the long profiles of streams is permissible. Such instruments as the Abney level and aneroid barometer (Peel, 1949; Sparks, 1953) can be adapted to the determination of elevations, while the extent of the terraces can be mapped on the 1 : 25 000 or 1 : 10 560 maps by reference to the field boundaries. The advantage of the surveying methods indicated above is that, although they are less accurate than precise levelling, they are much more rapid and allow large areas to be covered. In a consideration of terraces it is probably better to cover large areas with an accuracy of measurement of approximately + or − 1·5–3·0 m (5–10 ft) than to measure a few terrace fragments with great accuracy.

As the aim of investigating terraces is to determine as accurately as possible former base levels, the back of marine terraces (E on Fig. 9.17) should be measured as this approximates to the former high tide level, but the front of river terraces (F on Fig. 9.17) is the closest approximation to

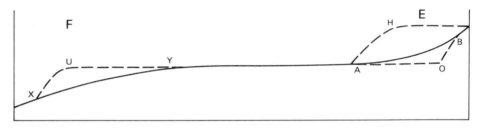

FIG. 9.17. The form of a terrace (for explanation see text)

the level of the centre of the old valley floor which was graded to high tide level. In practice it is difficult to locate these points, as the foot of the old cliff (O on Fig. 9.17) may be obscured by talus or, more often, modified by erosion, in which case it may have been at A and the former cliff be represented by AH. Similarly, if the terrace is a river form, its original sharp front edge (XUY) may have been considerably altered by erosion. One can, then, only obtain approximations to the height of the significant features. Further, the usual smooth convex and concave slopes allow considerable latitude in the decision as to where the back and front of the terrace are located. Lastly, it is often almost impossible to decide whether a terrace is marine or subaerial, for the form itself is not conclusive and there is so often a complete lack of deposits on it. In practice, therefore, it is safest to measure the elevation range of the terrace by determining the altitude at the back and the front, to map the terrace, and to arrange all the terrace fragments on some sort of diagram designed to show the height ranges of the fragments. From the last it can be determined whether the terraces fall into significant altitudinal groups or not, and thus whether they represent intermittent falls of base level punctuated by stillstands or a more or less steady fall of base level with terraces at all levels.

There is, as a result of the sort of difficulties of measurement and sources of confusion mentioned above, a range of possible error in determining the significant elevation on terraces of about 6 m (20 ft) or so. To this we may add the effects of variation of tidal range on the height of the inshore margin of a marine terrace, which may be of similar magnitude. Therefore, when terraces are closely spaced it becomes very difficult to decide whether they are to be correlated from region to region, although within a region surveyed by one observer it is probably fair to examine them by the sort of diagram mentioned in the last paragraph. The haze of uncertainty generated by the possible differences in original elevation, subsequent degrees of dissection and bias introduced by different observers using different maps, different units of measurement, and different surveying and field practices, makes it highly desirable to have criteria of correlation other than absolute height.

Two such criteria suggest themselves. It seems *a priori* that successive stillstands of base level are hardly likely to be of exactly the same length. Therefore they should give rise to marine terraces of varying relative width. Absolute width may mean nothing, because it depends mainly on rock resistance and exposure to wave action. Nevertheless, relative width should be the same from region to region unless exposure to dominant winds had changed or unless the nature of the rock had changed from level to level. The latter seems the more likely, but is capable of being checked from geological maps. In practice one meets sets of terraces which correlate well apart from an awkward member or two, for example the relatively strong development of the 130 m (430 ft) in south-west England compared with south-east England and the opposite position with regard to the 180 m (600 ft).

The other criterion is the degree of complexity. By this one means whether the terrace is one clear unit always found at a constant elevation or whether it seems to consist of a series of minor stages, representing a slowly falling base level punctuated by stillstands. If the simple and complex terraces occur at the same heights in different areas, it seems fair to correlate them especially when the degree of complexity is similar, as it is, for example, in the 100 m and 60 m (330 and 200 ft) terraces of south-east and south-west England. This particular similarity, incidentally, makes the meaning of the different degrees of development of the 130 and 180 m (430 and 600 ft) terraces, mentioned above, even more difficult to assess.

Thus, in the final analysis, field methods leave a residue of doubt: they are, nevertheless, far more real than analyses conducted solely from maps.

10

The importance of changes of climate

Because geomorphology as a science was developed to a large extent by workers in the humid temperate regions of North America and Europe, there has arisen a natural tendency to regard the landforms developed in that type of climate as normal. Any study of a world climatic map will, however, show that this type of climate is no more widespread than several others. Arid and semi-arid conditions stretch over much of Central Asia and the Near East, as well as covering large areas of Australia, Africa and North and South America. Extremely cold climates cover considerable areas in the northern hemisphere and include areas of permanent ice, such as Greenland, and areas of tundra climate, such as the northern parts of the USSR and north Canada. Equatorial rain climates occupy large areas in Africa and South America and comparable monsoon types are found in southern Asia. The extent of these climatic zones, together with others of more intermediate type such as the Mediterranean, makes the attempt to regard our own climate as normal an extremely prejudiced point of view.

Of these other types of climate the most commonly treated by geomorphologists are the arid and polar groups. There are probably several reasons for the concentration on these two.

The polar regions, the high glaciated mountains and the deserts seem to have attracted explorers far more than the equatorial and tropical rain forests. Modern expeditions have usually included scientists with the specific object of gathering specialised information on all aspects of the areas explored, and so landforms have been studied by geographers and geologists. The reports of earlier and less specialised exploration often include references to landforms as they are often the first things to engage the attention of the non-specialist. In these ways a considerable amount of knowledge of the landforms of the most explored regions has slowly accumulated.

Secondly, many of the late nineteenth-century American geologists, who provided much of the basis of systematic geomorphology, were engaged in surveying the semi-arid western parts of the United States. Thus, an early appreciation of these landforms was almost bound to arise. It has been strengthened by the continued attention paid by the Americans to their own arid regions and, in later years, by the general increase of research in Africa.

Thirdly, both arid and glacial climates leave the landforms bare and almost devoid of vegetation and, since the regions as a rule are infertile, natural forms are left unobscured by the activities of man. Such open landforms are in many ways more satisfactory for study than, for example, the cultivated, boulder clay country of East Anglia, where natural forms are obscured by woods, hedges and buildings, or the tangled rain forest of the Amazon Basin.

The concentration on arid and glacial regions was probably also partly instrumental in the development of climatic geomorphology. These two climatic regions are undoubtedly those with the most individual landforms, largely the effects of wind plus intermittent water action, and ice action respectively. Hence, it was natural to become preoccupied with climate and to try to set up a geomorphological scheme in which climate dominated erosion and erosion dominated structure and lithology, so that every climatic zone had its characteristic landforms. Climatic geomorphology has been a very important element in the subject in Europe, especially in Germany and more recently in France. Davis, himself, paved the way for this when he admitted the existence of arid and glacial cycles of erosion. Later attempts to define climatically related morphogenetic regions have been made by a number of people, e.g. Peltier (1950), Büdel (1963) and Tricart and Cailleux (1965).

Climatic differences cause changes in the nature and balance of the processes working on the earth's surface with the result that the ensuing landforms may vary considerably. The actual agents of erosion may be completely different: glaciers do not exist in humid temperate regions except at high altitudes, where the climate is not really humid temperate, while wind, which is of considerable importance as an agent of erosion, transport and deposition in deserts, can almost be ignored except in the formation of coastal dunes. The relative importance may change, as can be seen in the increasing importance of mechanical weathering in cold and dry climates. The speed of action of any agent may be different: the rate of chemical weathering in the equatorial regions appears to be greater than in any other part of the earth. The effective resistance of rocks varies with the changing dominance of the agents of weathering: limestones may be more resistant in deserts than in humid regions as they are less subject to chemical weathering there. Differences in the vegetation cover can affect erosion by direct protection of the ground and by binding the soil mantle together. Some climates have marked seasonal regimes, so that the study of seasonal fluctuations of climate may be necessary to explain the landforms. This can be seen in the example of river action: in climates with rain at all seasons erosion and deposition are almost continuous except in so far as they are varied by floods; in areas marginal to ice sheets the short periods of summer thaw, which result in enormous increases in the discharge of meltwater streams issuing from the ice, are the significant ones; in deserts it may well be that the isolated rainfall, occurring perhaps once every five or ten years, is of greater significance than anything else. These and other differences must all be considered in order to understand the landforms produced by the various types of climate.

Yet it is by no means certain that these climatic variations produce completely distinctive sets of landforms. Details may vary and certain climatic zones may produce more individual features than others, as was suggested above for the glacial and arid climatic zones, but it seems to be doubtful whether fluvially controlled landscapes formed in different climates are as distinctive as is sometimes maintained (Stoddart, 1969). It might well be more prudent to think of climatic influences on forms and erosion rates rather than climatically dominated landforms.

In this connection it must be remembered that climatic regions are not really climatic regions at all. The world can be divided up on a purely climatic basis by using any climatic parameters one likes, so that the number of possible methods of division is very large, but, as most of these are meaningless geographically, it is possible to say that pure climatic regions do not exist, at least from a geographer's point of view. There are very few, if any, abrupt breaks in climate, only steeper and less steep climatic gradients. The former usually coincide on a broad scale with mountain ranges and with areas where ocean currents of contrasting characteristics converge, e.g. the northern Pacific coast of South America. Hence the geographer usually chooses to classify climate on a vegetational basis, assuming that vegetation is mainly a response to climate. In effect, he defines the climatic range of vegetation zones and proceeds to call them climatic zones. It is important to remember this when thinking about climatic geomorphology, because if one assumes that landforms are going to vary significantly with these zones, one assumes that the climatic parameters controlling landforms are the same as those controlling natural vegetation. And that seems to be too large an assumption. In fact, it would be much better to try to determine by objective, and preferably quantitative description whether distinct and separable groups of landforms exist and then to proceed to define any such in climatic terms.

Furthermore, the pure climatic-geomorphological point of view involves two basic assumptions, which one would hesitate to make. The first is that climate swamps structure and lithology, an admission that few geologically trained or biased geomorphologists would make. This particular form of controversy is well illustrated by recent discussions on karst geomorphology, which seem to have resulted in the general attitude that there are certain forms which characterise tropical karst, but that the effect of lithology is as important in tropical limestone landforms as it is in temperate limestone landforms. This attitude seems eminently sane.

The second assumption inherent in climatic geomorphology is that as an attitude it is actualist, i.e. it attributes the landscape to present processes. This may be a reasonable assumption in certain circumstances, e.g. in the denudation of spoil tips or the dissection of incoherent sandstones in the semi-arid western parts of the United States, but it is demonstrably wrong in many areas and likely to be wrong in many others. We know that in most regions there have been frequent changes of climate in the geologically immediate past. Therefore, these Pleistocene changes of climate must be considered as well.

The result of these was that the areas covered by the ice and adjacent to the margins of the ice sheets experienced more or less regular successions of climate, ranging from glacial through periglacial to humid temperate in the glaciated areas and from periglacial to humid temperate in areas near the ice sheets. Each climate is capable of producing different types of landforms. Therefore, it is dangerous to try to explain presentday landforms merely by the operation of present processes of erosion acting over very long periods of time. While some landforms may have been so produced, it is possible that many of them may be relict forms subject to little later modification. The great difficulty is that it is impossible in many cases to determine whether erosion of the type required for the production of the landforms continues, as the period of observation available to any one person is too short. The importance of the processes of glacial and periglacial erosion in the formation of many features is clear, but others have much more debatable origins. The normal forms of glaciated mountains are accepted by all as due to ice erosion and no one would seriously suggest that such features as corries, glacial troughs and overdeepened basins are still being formed. Yet there are a number of other landforms in the formation of which the relative importance of past and present processes is difficult to assess. Among them are such features as the screes of the Lake District, which may be fossil forms due to frost shattering in the closing phases of the Pleistocene; dry valleys in Chalk areas, which have been discussed in Chapter 7; some of the gravel river terraces, which, although composed of coarse gravel, have gradients less than those of present rivers, which appear to be capable of transporting no material coarser than sand and mud.

Pleistocene changes of climate do not apply, however, only to temperate latitudes, but to most of the climatic zones of the earth. The general shift towards the equator of the climatic belts resulted in large parts of the deserts having more temperate and, of greater importance, slightly more humid climate. In the explanation of landforms found in deserts, the possibility of valley formation by streams in the more humid parts of the Pleistocene cannot be ignored, as will be seen in the following chapter. It is generally held that the glaciations were approximately synchronous in both hemispheres so that the climatic belts of both hemispheres tended to move towards the equator simultaneously. The hypothesis raises, however, one difficulty: if there was a shift towards the equator in both hemispheres, it would seem that either a climatic belt was squeezed out or one or more narrowed in latitudinal extent. The second suggestion is probably the more likely, because, in view of the general nature of the earth's atmospheric circulation, it is improbable that the deserts squeezed out the equatorial belt. There was probably a diminution in the width of the deserts due to the encroachment of humid temperate conditions on their high latitude sides and to the stability of the equatorial belt on the other side.

Although Pleistocene climatic changes are by far the most important in the interpretation of present landforms, it may be occasionally necessary to consider the changes of climate in the earlier parts of the Tertiary period. Few areas have been subjected to uninterrupted denudation since then, but, as the earlier Tertiary climates were very different from the present ones, any surviving landforms could present difficulties to the investigator. In Great Britain, for example, parts of the Eocene were characterised by hot, humid climates, as is shown by the resemblance of the contained fossils to forms now occurring in Malaya. The role played by early Tertiary tropical weathering in the production of some European landforms may have been considerable. It could well be that some of the summit surfaces of the Hercynian blocks are Eogene pediplains, while the weathering responsible for the formation of tors may have been tropical weathering of the same date.

However, the significance of climates of the more remote geological past is greatest in connection with their role in the formation of relief now being exhumed from coverings of later rocks. In Britain various landforms have been said to be exhumed from beneath a covering of Triassic sediments. The Trias in this country was a time of desert conditions so that the forms exhumed from beneath its rocks will have to be compared with modern desert landforms. Charnwood Forest, a mass of Pre-Cambrian rocks projecting through the Trias of Leicestershire, was regarded by Watts (1947) as essentially an exhumed Triassic landscape. The present slopes of Charnwood Forest are continuous with those beneath the adjacent Trias, while signs of polishing, grooving and etching by wind action have been found on granite excavated in quarries. Geophysical work (King, 1954) has demonstrated how some major British landforms are more or less coincident with New Red Sandstone features. Thus, the western part of the English Channel coincides approximately with a basin of deposition of this age, although it is unlikely that the detail of the landscape owes much to desert erosion in that period. Generally, the Bristol and English Channels with the high ground of Devon and Cornwall between correspond roughly with the distribution of basin and range in New Red Sandstone times. More questionable, unless there is evidence, is the attribution of uplifted and dissected plateaus to Triassic erosion, burial and later excavation. There are so many periods during which such plateaus could have been formed and dissected that, in the absence of geological evidence in the form of outliers, it is not surprising that theories of origin have varied widely.

Exhumed forms may be of any age, although Triassic forms have probably received more attention than any others in this country. Parts of the Chalk dipslopes of the London Basin are regarded by Wooldridge and Linton (1955) as exhumed Tertiary forms (see Fig. 9.16), while large areas of the Central Plateau of France are considered by Baulig (1928) to be exhumed surfaces of various ages (see Chapter 16). The idea of palaeo-karst landforms seems to have been attracting the attention of European geomorphologists in recent years and possible examples of various ages have been described from different places, e.g. Belgium and Poland (Sparks, 1971).

Exhumed landforms usually only dominate the landscape in the broad view. The majority are erosion surfaces, and in detail the landscape is usually attributable to a combination of forms derived from present and recently past climates.

It is only fair to conclude this short discussion with the statement that there have been and, indeed, are geomorphologists who have fervently opposed the climatic-geomorphological point of view. The most forceful of these at the present time is Lester King, who would have the pediplanation cycle recognised as the standard and who sees no essential morphological differences between arid and humid cycles. The reader is referred to King's works for further details (King, 1967; Chapter 5 and the references quoted in that work).

11
Landforms in arid and semi-arid climates

Although aridity is the dominant characteristic of the deserts of the world, the variety of climate, structure and relief to be found in deserts is large. Excluding the cold tundra, deserts range from 50 degrees North in Central Asia to within a few degrees of the equator in South America, with the result that temperature conditions vary considerably. The intermontane and continental interior deserts have greater extremes of temperature than the west coast deserts, such as the Atacama, where the moderating influence of cold currents is felt. Similarly the amount of rainfall varies considerably. In parts of the Atacama desert practically no rain falls, but in other deserts, especially locally on mountains, the annual rainfall may be as much as 250 mm (10 in). All these factors, together with the susceptibility of the different regions to torrential downpours at irregular intervals, will affect the nature of the weathering and erosion taking place.

There is an even greater variation in relief and structure, and the sand dunes and rock plains, which commonly figure in the popular concept of a desert, form only a fraction of the total area. The Australian desert, for example, is structurally an ancient worn-down plateau of crystalline and ancient sedimentary rocks, the more resistant of which rise to a thousand metres or more (a few thousand feet) above the general surface level. Some of these mountains have the typical inselberg form (see below), for instance Ayers Rock and Mount Olga north of the Musgrave range, while parts of the adjacent Petermann and Musgrave ranges, if clad with more luxuriant vegetation, would be difficult to distinguish in form from mountains developed in temperate climates. Excellent photographs of these different forms are given in Finlayson (1936).

In the desert of Arabia, again, a great variety of scenery is to be found. In the south-eastern corner between the Hadhramaut and the Persian Gulf, lies Rub' al Khali, the Empty Quarter, which consists essentially of a low plateau sloping northwards from about 300 m (1 000 ft) down to sea level and covered to a considerable extent with sand dunes. This area, which approximates to the traditional concept of a desert, may be compared with western Arabia adjacent to the Red Sea rift, where in Yemen rugged mountains rise to well above 2 000 m (7 000 ft) and north of Medina extensive flows of lava are found.

The Sahara, too, has its variety. The almost impenetrable Great Sand Sea of western Egypt and adjacent parts of Libya forms one of the world's greatest dune areas. The contrast between this area and the Tibesti and Ahaggar massifs is enormous. The highest part of the Tibesti consists of an enormous volcanic mass, which seems to have reached the solfatara stage, but has certainly been active in the recent geological past (Gautier, 1928). In form and size it is comparable with Mount Etna and reaches a similar elevation, approximately 3 300 m (11 000 ft). The Ahaggar is also volcanic, but activity ceased earlier than in the Tibesti and it is now more degraded, although the wrecks of the ancient volcanoes still reach to about 3 000 m (10 000 ft).

A different structural type is to be found in the interior basin deserts of the western United States and Central Asia. Precipitation on the surrounding mountains may give rise to streams which quickly disappear where they reach the basin, as in the Taklamakan desert of the Tarim basin (Stein, 1933). The basin often consists of gentle slopes of graded sediments derived from the surrounding mountains leading to a central saline lake or swamp.

In view of the enormous variety of deserts, caused principally by structural and lithological differences but also by climatic differences, it must be emphasised that the features discussed below are not characteristic of all deserts.

Weathering under arid conditions

Although doubts have been cast upon the efficacy of temperature changes in disintegrating rocks by expansion and contraction alone (see Chapter 3), the question deserves a little more attention. It will be remembered that experiments involving large temperature changes and steep temperature gradients between the surface and interior layers of the rock appeared to produce no disintegration until water was present.

Although the diurnal range of air temperature in deserts may be very large, that of the temperature of the surface of rock and dust is often considerably greater. Some instructive figures are given in Hume (1925). Generally speaking, the minima recorded in the air and on the rock surface are the same, but the maxima recorded on rock considerably exceed the air maxima, the actual amount of the excess depending principally on the nature of the surface and the relation of the surface to the prevailing wind. For example, the maximum excess of surface temperature over air temperature varied in Egypt from 4°C (7°F) on slate to 18°C (33°F) on flint, while sands and gravels in sheltered places had temperatures exceeding the air temperature by 20 degrees C (36 degrees F). As, however, the air, even in deserts, is never completely dry, the effects of temperature changes cannot usually be isolated from chemical effects. The humidity may be quite high, a series of observations quoted by Hume showing minimum relative humidities ranging from 16 to 32 per cent and maximum relative humidities ranging from 68 to 96 per cent.

Some of the illustrations of shattered pebbles attributed to thermal expansion and contraction are of flint and quartzites. Hume, for example, shows flints split by straight parallel fractures (Plate IV), while Bosworth (1922, fig. 72) has an admirable illustration of a quartzite pebble from Peru subject to beautifully regular exfoliation. As both flint and quartzite are practically homogeneous silica rocks, not subject to chemical weathering except by strong alkalis, it is difficult to resist the conclusion that in these examples, in spite of the experimental evidence, thermal expansion and contraction alone have effected the disintegration of the pebbles.

Salt crystallisation effects, which have been described in Chapter 3, have recently figured more prominently in explanations of desert weathering. This is really a mechanical effect, although solution is necessary to allow the crystals to form within the fabric of the rock by evaporation. Without any moisture little could happen. Peel (1966) attributes honeycomb weathering in sandstones and taffoni in acid crystalline rocks to salt weathering. The very form demonstrates that they cannot be due to wind blast and the lack of polishing of the rocks as well as the equal development of undercutting of rocks in all directions suggests that a directional agency such as wind is not responsible. Chemical weathering or salt crystallisation or both combined are far more likely to provide the cause.

Whether mechanical disintegration is regarded as acting alone or always or usually in conjunction with some chemical weathering, the fact remains that it appears to play a large part in desert weathering and is responsible for the predominance of angular broken waste material in deserts.

Various surface and subsurface crusts may be developed in deserts, but it is doubtful to what extent they may form under present climatic conditions. Buried yellow and brown crusts seem almost certain to represent past more humid conditions. Calcareous crusts, commonly called caliche by the Americans, are more likely to form by the capillary rise of solutions in semi-arid regions at the present time, though the presence of successions of such indurated horizons in some regions can hardly be explained solely in terms of present climates. Crusts of the most soluble salts in depressed areas may be explained by modern conditions, although they could represent the drying up of bodies of water accumulated in former wetter periods: local evidence might help one to decide in individual cases whether the crust was still developing or being destroyed. Only the less soluble materials, such as calcium carbonate, are likely to influence the development of local micro-relief.

The effects of wind

Sandstorms and sand dunes are spectacular features and indicate significant wind action, but the effects of wind in deserts may easily be overstressed, especially the erosive effects.

The nature of the movement of particles by the wind in deserts has been closely investigated by Bagnold (1941), who distinguished three types of movement, suspension, saltation and surface creep. Only very fine particles, those with diameters less than 0·2 mm, are carried in suspension by the winds normally found at the surface of the earth. Particles larger than this are subject only to saltation and surface creep. Saltation, which has already been discussed with reference to rivers, and the sea, is most significant in understanding the erosive action of sand. In a turbulent flow of air near the surface local upward currents may lift a sand grain. As it falls it is moved horizontally in the general wind direction. Thus the path taken by a moving grain consists of a short, near vertical ascent and a longer sloping fall (Fig. 11.1A). When it strikes the surface it may either rebound, thus repeating its motion, or it may start saltation in another sand grain by impact. The height attained by sand grains undergoing saltation depends on the velocity of the grains, and thus basically upon wind velocity, and on the nature of the surface. On pebble surfaces saltation is noticeably higher than on sand surfaces (Fig. 11.1B), but even on these surfaces grains undergoing saltation rarely reach more than 2 m (6 ft) above the surface.

The energy of the falling grains may not be absorbed in rebounding or in

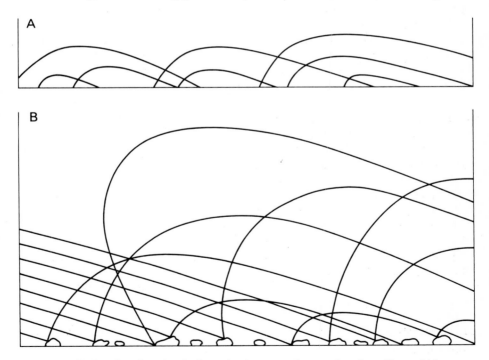

FIG. 11.1. Paths of sand grains during saltation over, A, a sand surface; B, a pebble surface (*after Bagnold*)

starting saltation by impact, but may be used in disturbing a number of grains, which are driven forward for a short distance on the surface. At low wind velocities the jerky movement of these grains may be observed, but as the wind velocity increases this appears to change into a steady forward creep of the surface grains. This surface creep merges into very low saltation.

Thus it may be seen that the only grains actually suspended in the air are the very fine ones. The upper parts of sandstorms, which may rise hundreds or even thousands of metres into the air, are composed solely of dust, and sand movement takes place only very near ground level. When the dust-laden wind meets an obstacle, the dust particles rarely strike the obstacle as they are diverted round it in the general air flow and so exert little or no erosive effect. The sand grains themselves are confined to a layer very near to the surface and thus their erosive effect is very limited in vertical extent, while surface creep can obviously affect only an extremely limited vertical range. The work of wind erosion is therefore very limited.

Its main effect is probably to be seen in undercutting and fluting at the base of upstanding rock masses, while such things as wooden telegraph poles and stone monuments may be undercut at the base and even worn through. Very few features appear to be attributable solely to the action of wind-driven sand and Cotton (1942) mentions only yardangs as likely to be caused solely by wind action. These consist of elongated U-shaped troughs separated by sharp ridges and were first described from the Taklamakan desert. They are usually developed in soft material and the relief rarely, if ever, exceeds 8 m (25 ft) and is usually much less. The undercutting of isolated rocks to form pedestals may be partly due to wind action, but may also be caused by increased chemical action, due to some retention of moisture in the surface layers at the base of the rocks, and subsequent deflation by the wind of the weathered products.

Peel (1966) in his review of the arid landscape generally agrees about the comparatively minor effects of wind erosion, but raises the interesting question whether wind action over very long periods might not exert some effect on hard rocks. He quotes areas of the central Sahara around the Tibesti, where sandstones of normal hardness have been carved into systems of ridges and troughs with a relative relief of the order of 15 m (50 ft) over hundreds of square kilometres. The relief trend conforms very closely to the mostly unidirectional wind flow in this area. If we attribute the relief to the wind, and also allow much time for this relief to have been formed, as it seems we must, it appears that the general wind direction in the central Sahara has been very constant for a long time.

Of greater importance is the process of deflation, or removal by the wind of the fine products of weathering. On the general desert surface the wind removes the finer products in suspension and may shift the sand by saltation and surface creep, thus leaving a surface of gravel or bare rock. These deflation surfaces have been given a variety of names: desert pavement and boulder pavement are self-explanatory terms, while the gravel left after deflation has been termed lag gravel and in the Sahara a stone-strewn surface is known as hamada.

Some desert hollows may also be attributed in considerable part to deflation. An initial hollow, whether scoured by the wind or not, would tend to be damper than the surrounding desert and thus chemical weathering would be greater there. Chemical weathering is of the greatest importance in producing fine material, as clay minerals are formed by this process. If the hollow is subject to drying, there would be a considerable proportion of material fine enough to be removed by the wind in suspension. Phases of weathering, drying and deflation would lead to a steady deepening of the hollow. Whether such a process can operate on a large enough scale to produce major oasis depressions is more doubtful.

The process of deflation of the finer constituents from deserts probably accentuates the dominance of coarse waste, initially brought about by the importance of mechanical weathering. The dust from the deserts is carried considerable distances until it is dropped either by a decrease of the wind velocity or, probably more usually, by being washed down by rain. The deposit, which is characterised by a great dominance of silt-sized particles, is known as loess. The most extensive deposit of this material is found in north China, where it is derived from the deserts of Mongolia and transported to north China by the strong north-westerly winter winds. Deposits of comparable material which extend from the Ukraine westwards along the foot of the Hercynian uplands of Europe as far as the chalk plateaus of Artois and Picardy, were most likely derived by deflation from the glacial deposits of the North European Plain during the Pleistocene period.

Probably of greater importance than the erosion forms produced by winds in deserts are the depositional forms. These again have been intensively studied by Bagnold (1941) who distinguishes between the small-scale sand ripples and ridges and the large-scale dunes, of which there are two main types, the barchan and the seif.

The general distribution of the major sand dune areas in deserts requires some explanation. Peel (1966) estimates that only about 10 per cent of the area of the Sahara is formed of ergs or sand seas and that these remain more or less fixed in position. The source of the sand raises an interesting question, because it is difficult to see how such amounts could have been accumulated by the wind. Opinion is tending to veer to the view that the wind has merely modified pre-existing sand accumulations left by other agencies, including the sea. It seems that the Rub' al Khali desert of south-eastern Arabia occupies a structural offshoot of the Persian Gulf in which marine sands accumulated in the later parts of the Tertiary period and have since been reworked by the wind. Some Saharan ergs may have had a similar origin but others may have occupied basins of centripetal drainage into which sand was transported in periods of wetter climate. If the raw material of most of these ergs has been provided by other agencies, as differences in character between the surface wind-worked material and the sand at depth seem to indicate, then the locations of desert dunes, coastal dunes and periglacial dunes all seem to depend on the concentration of sand provided by some other factor. This hypothesis must not be pushed too far, because one of the main Saharan ergs is the Libyan Sand Sea on the borders between Egypt and Libya in the arid heart of the Sahara.

Reverting to the construction of sand features and taking the small scale features first it may be said that the formation of sand ripples is closely connected with the process of saltation. If one imagines a surface with chance irregularities (Fig. 11.2) on to which grains are being driven by

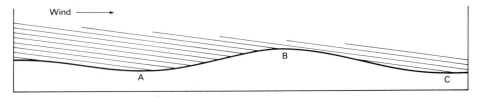

FIG. 11.2. Formation of sand ripples (for explanation see text) (*after Bagnold*)

saltation, it is obvious that the frequency of bombardment by sand grains will be greater on the slight slopes facing the wind (AB, Fig. 11.2) than on slopes facing away from the wind (BC, Fig. 11.2). Thus on AB the rate of arrival of the sand grains and the surface creep will be greater than on the reverse slope, BC. In fact, material will be arriving at B faster than it can be removed. Similarly, in a hollow, such as A, the sand grains are being driven faster towards B than they are arriving, thus accentuating the hollow. As all sand surfaces possess chance irregularities, even though they may be only of the magnitude of a few sand grains, there is a natural tendency for ripples to form.

The wave length of such ripples varies with the strength of the wind, which controls the distance travelled by each grain during saltation. If the surface were perfectly flat saltation would take place theoretically from the whole area and there would be an even shower of grains falling after saltation. But the ejection of grains by bombardment will not be even on a surface such as that shown in Fig. 11.2. Due to the greater bombardment on the slope AB more sand grains will be ejected from this area: these will tend to land again at a distance of one characteristic grain path downwind and thus lead to increased accumulation in that area. Thus, a ripple pattern with a wave length corresponding to one characteristic grain path, which depends ultimately on wind velocity, will be formed.

If this process went on unchecked it should lead to an increasing ripple height, but, as the ripple crests rise, they increasingly interfere with the wind, so that the sand grains on the crests are blown over into the troughs. Thus the maximum height to which ripples can develop is limited.

The formation of sand ridges is by a somewhat different process. In sand containing coarser and finer grains, such as that shown in Fig. 11.3, the fine material may be removed from the area CD, thus exposing the coarser grains. Although these coarser grains cannot be removed directly by the

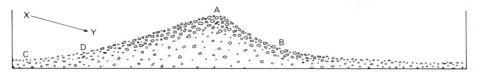

FIG. 11.3. Formation of a sand ridge (for explanation see text) (*after Bagnold*)

wind, the impact derived from saltation is able to move grains six times the size of those forming the saltation. Thus the coarse grains will be driven up the slope DA and will come to rest in the lee of the crest, AB, past which they cannot move because of the complete shelter here. This process leads to the accumulation of coarse grains on the crests of ridges, from which they are not removed, as they are considerably larger than the size the wind is able to move. Ridges formed in this way attain greater sizes than ripples and Bagnold quotes wave lengths exceeding 20 metres (70 ft) and heights of more than 60 cm (2 ft) as common in parts of the Libyan desert.

On an altogether larger scale are the crescentic dunes, or barchans, and the long straight dunes, or seifs. The maximum height of barchans is about 30 metres (100 ft) and their maximum width and length about 400 metres (1 300 ft). Seif dunes are even larger: in the Egyptian Sand Sea they attain a height of 100 metres (330 ft) and dunes twice this height have been reported from southern Iran. The transverse width is about six times the height, and their length may be 80 km (50 miles) or more.

According to Bagnold, the barchan is formed when the wind is nearly unidirectional: it is orientated with the horns downwind. Seif dunes are formed when strong winds blow from a quarter other than that from which the prevalent winds, responsible for the general sand drift, arrive.

Observed examples of these relationships occur sometimes in the lee of cliffs. If a unidirectional wind blows over a cliff face (Fig. 11.4), sand will

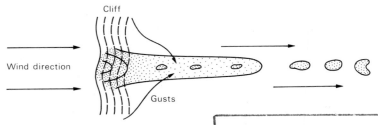

FIG. 11.4. Relation of wind direction to seiflike and barchanlike dunes (*after Bagnold*)

325

accumulate in the lee of the cliff, where it will be sheltered from the prevalent wind but subject to lateral gusts. In this area a seiflike dune forms, but away from the shelter of the cliff the influence of the prevalent wind is again felt and the seiflike dune tends to break down to isolated mounds and finally to barchans.

In addition, Bagnold has observed that barchans may be formed in the troughs between series of seif dunes, in the manner shown in Fig. 11.5. The

Prevailing wind ⟶

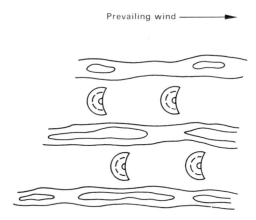

FIG. 11.5. Barchans in troughs between seif dunes

direction of the seif dunes shows the direction of the prevailing wind, which is also responsible for the barchans in the troughs, where the effects of strong winds from another quarter are not felt.

Additional evidence for the association of barchans with unidirectional winds is provided by Kharga Oasis, one of the only desert stations with wind records: at this locality the winds are almost unidirectional and the barchan is the only form of dune present. Finally, in very different conditions very small barchans have been observed to form in very cold ice crystals as long as the wind remained absolutely constant in direction.

A barchan dune will tend to form from a mound of sand, the development envisaged by Bagnold being as follows. In cross-section (Fig. 11.6), the greatest effect of the wind will be felt on the windward slope of a mound of sand, and will lead to a pronounced movement of sand grains to the crest. Just beyond the crest they will tend to accumulate, for the effect of the wind is felt less here. Accumulation at the top of the lee slope leads to a steepening of that slope (Fig. 11.6B) and hence to a decrease of the wind effect on it. The steepening of the lee slope by accumulation at the top goes on until the angle of rest of the material is exceeded (AB on Fig. 11.6C), when shearing takes place along a slightly less steep surface (CD on Fig. 11.6C). Thus, the

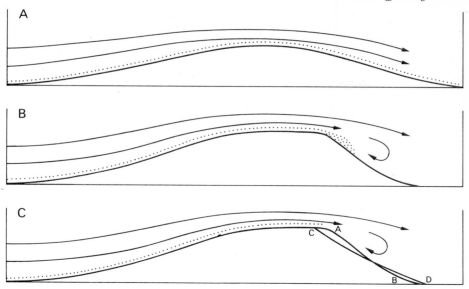

FIG. 11.6. Formation of a slip face on a barchan dune (for explanation see text) (*after Bagnold*)

slip face of the barchan advances by a process of oversteepening and shearing. The flanks of the original mound of sand advance more rapidly than the centre, for the rate of advance is inversely proportional to the height of the slip face. In this way the crescentic barchan form is developed. The wings advance until they are very much in the shelter of the main mass of the dune, where the sand flow is less and their rate of advance retarded.

The formation of the seif dune is less well understood, but transition forms, which can be connected into a series, have been observed by Bagnold. If a barchan formed by the prevailing wind (Fig. 11.7A) is subject to a strong wind from another direction (Y on Fig. 11.7B), it will tend to swing round with the development of an incipient new wing at C and the over-development of wing A. If the prevailing wind, X, again dominates, there will be a tendency for the dune to revert to its original form (Fig. 11.7C). At this stage the wing A is still exaggerated, but would however be reduced provided the prevailing wind, X, was of sufficiently long duration. However, if the strong wind, Y, again intervenes, there will be a further prolongation of the wing A (Fig. 11.7D). At this stage it is so far extended as to receive a supply of sand from the original barchan even during the prevailing wind, X. Thus it will continue to grow both during the prevailing wind, X, and the strong wind, Y, with slip faces developed on the side away from the strong wind. The seif dune differs from the barchan, therefore, in that the slip faces are on the side away from the strong wind and not facing the direction of advance as in the barchan.

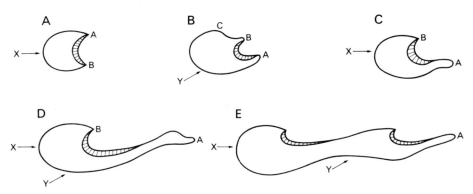

FIG. 11.7. Formation of a seif dune from a barchan (for explanation see text) (*after Bagnold*)

The increase in the availability of aerial photographs and especially of satellite photographs has revealed the quite extraordinary regularity of many areas of seif and longitudinal dunes over vast tracts of country. For example, in the Sahara the dunes are orientated slightly east of south in the north and then veer consistently in direction until they are orientated some-what south of west in the southern Sahara, beyond which similar orienta-tion is betrayed by the old fixed dunes (Fig. 11.8). The same is true of Arabia. This pattern is very consistent with the general pattern of winds swinging round the eastern end of the subtropical high pressure cell: in the north they are the north-west to north, etesian or meltemi winds of the Mediterranean lands, while further south they become the north-east Trades.

It has been postulated by Hanna (1969), adopting a suggestion made earlier by Bagnold, that these dunes are caused by the development of longitudinal helical roll vortices in the general air flow. The idea is illus-trated in Fig. 11.9. It will be seen that, if the maintenance of such longitu-dinal vortices is possible, the presence of cloud streets (long lines of shallow cumulus clouds), which have also been clearly shown by satellite photo-graphs, and seif dunes could be explained by one single hypothesis. Hanna showed that it was mathematically possible for such vortices to form under the right surface, wind and stability conditions and that such vortices would have the right sort of diameter, about 2 or 3 km (1–2 miles), to ex-plain the spacing of the dunes. Once the sand began to be swept to one side to form ridges, heating of the flanks of those ridges would promote the continuation of the vortices and hence ensure a stable dune pattern.

In addition to the barchans and seifs, which are probably the basic forms of dunes, a multitude of other types may be recognised in a morphological classification of dunes. For example, transverse elongated dunes exist in some parts, while great domeshaped dunes seem to dominate in parts of

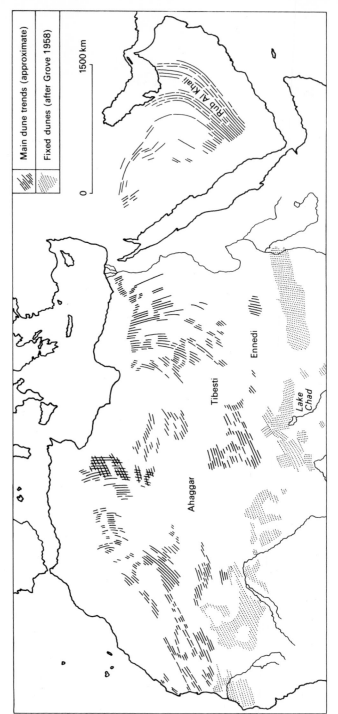

FIG. 11.8. Pattern of dunes in Sahara and Arabian deserts (*after Peel*)

Legend (on map):
Main dune trends (approximate)
Fixed dunes (after Grove 1958)

1500 km

0

Rub Al Khali

Ahaggar

Tibesti

Ennedi

Lake
Chad

329

A

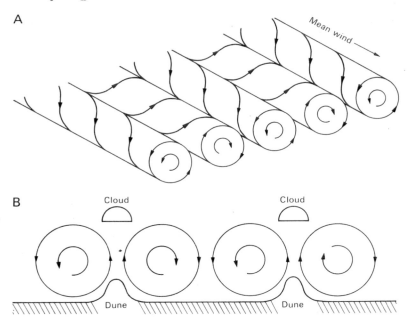

B

FIG. 11.9. Development of dunes by longitudinal roll vortices (*after Hanna*)

Arabia (Peel, 1966). In other areas other shapes occur. One must expect this and while we are trying to understand fully the development of dune shapes and dune systems, it is necessary that researchers should classify dunes on a descriptive basis. It must also be remembered, however, that there is every reason to believe that dunes will represent the interaction between local wind and sand conditions and the resultants of short term and long term changes in these conditions. Hence, it is unlikely that all dune forms will ever be reduced to one or two simple categories, although a few simple categories may describe the basic forms.

The effects of water

Very few parts of the world's deserts are absolutely rainless, though certain sections of the Libyan and Chilean deserts approach complete aridity. Elsewhere, rainfall, although very infrequent, is sometimes heavy, convectional and highly localised. Extreme examples are provided by the Peruvian desert, normally very dry due to the cold northward-flowing Humboldt current along the coast. Occasionally, however, this current is displaced and a warm southward-flowing current prevails for short periods. When this happens, extremely heavy rainfall may result along the desert coast of Peru. For example, Trujillo received 35 mm (1·4 in) between 1918 and 1925, but during March 1925, when the southward-flowing current prevailed, 390 mm (15·5 in) fell, of which 225 mm (8·9 in) was concentrated in the three days, 7–9 March (James, 1959). Although special conditions operate in this desert, very heavy convectional showers occur occasionally in most deserts, such as Egypt (Hume, 1925).

When rainfall of high intensity occurs, the run-off is very great and immediate evaporation not very high, at least in the region of the storm, where the air is saturated. As the water flows away, it suffers heavy losses through percolation into the underlying sand, gravel and rock, and through evaporation outside the storm area. The rate of rise of floodwater in normally dry watercourses is, in many cases, extreme. French legionaries have been drowned while encamped in wadis draining from the southern Atlas of Algeria, due to the rapid advance of floods from storms in the high mountains of which they had heard or seen nothing. King (1942) has quoted the example of the Great Fish River in South-West Africa, which may be dry one day and a surging flood, several hundred metres wide and 9 m (30 ft) deep, the next. The wadis of eastern Egypt are equally liable to this form of flood and many examples are described and illustrated in Hume (1925).

Although these floods, which have been termed streamfloods by Davis, are rare and of short duration, the amount of work of which they are capable is enormous. During the long dry periods between floods debris accumulates in the valleys as a result of weathering and also is probably swept in by wind action, so that the streamfloods have an enormous amount of readily available load. Debris of all sizes up to great boulders may be swept along until the flood subsides. Often it is transported out of the upland valleys and spread out as fans at the upland edges. These floods probably constitute the main agent of erosion in deserts at the present time. Their relative importance is even greater than the floods of humid, temperate regions, because so little erosion is effected between floods.

When streamfloods emerge from the upland front or when heavy rain falls in featureless desert plains, a second type of flow may occur, provided that no main water channels are present. This consists of an intricate, anastomosing pattern of streamlets, often resembling a series of overlapping, deltaic distributaries. Should the rainfall be sufficiently heavy the whole of a flat desert landscape may be completely covered with a sheet of moving water. This type of flow, which is probably rare, has been termed a sheetflood. Its main action appears to be not one of erosion, but merely one of transport: it is the mechanism by which much debris is thought to be moved from the foot of the uplands across the gentle slopes that often border them in deserts.

The traditional picture presented above of the efficacy of the occasional desert flood probably really applies to the semi-arid rather than to the truly arid regions. The significant measure, of course, is the ratio of runoff to rainfall and this seems to vary widely depending on the amount of the rainfall and the nature of the surface on to which it falls. In general, however, there is little evidence to support the idea that many of the occasional desert rainfalls cause catastrophic floods. Peel quotes evidence that the raging streamfloods and sheetfloods mentioned above are the exceptions that have impressed themselves strongly on the mind of chance, lucky observers. In fact, many of the occasional desert rainfalls are very light and incapable of any serious erosion. Therefore, in the true deserts one is thrown back more strongly on to past climatic conditions to explain obviously water-formed features.

In modern semi-deserts the position may be rather different. From first principles it might be deduced that the amount of erosion will increase with increasing precipitation, provided that the surface remains bare, but that this will ultimately be reversed by the development of vegetation which will protect the surface from increasing erosion with increasing rainfall. For certain stations in the United States Langbein and Schumm (1958) have quantified this in terms of sediment yield and shown that this increases to about 300 tons per square km (800 tons per square mile) with a rainfall of about 371 mm (15 in) per year and then decreases as increasing vegetation cover protects the surface.

Although streamfloods and sheetfloods may be effective agents in the formation of semi-arid landforms, many deserts bear the imprint of even more significant water action. It seems very doubtful whether the required amount of water ever falls in present conditions and many authorities have attributed such erosion to a wetter period during the Pleistocene. For example, the valley on the west side of the Nile at Luxor containing the Tombs of the Kings appears to be a water-eroded valley, but the tombs themselves show no signs of destructive water action: this means that water action of any significance probably ceased at least 4 000 years ago (Hume, 1925).

It seems easier to relate landforms in many deserts to earlier wetter periods than to present conditions. Hume gives photograph after photograph of steepsided valleys in limestone and sandstone country in Egypt, which are obviously of watercut form, many of them containing caves, from which springs probably issued at an earlier period. The same author has also many illustrations of closely spaced badlands type of valley pattern on impervious igneous and metamorphic rocks in the same region. Indeed, the chief impression on viewing the Mt Sinai massif from the Gulf of Suez is of the gashed sides of this stark, mountain mass.

Further evidence for a past wetter climate is to be found in the wadi pattern present in many parts of the deserts. These wadis are too long and, in many places, form too integrated a pattern for them to have been formed under present climatic conditions. In Arabia wadis of very great length are known, while some of the patterns in the Sahara have been described by Gautier (1928). The Ahaggar massif was apparently one of the chief watersheds, for enormous wadis radiate from it in all directions. In this region the integration of the drainage pattern is so good that practically every point in the area can be assigned to one or another of the former drainage basins. The comparatively recent date at which water flowed in these wadis is attested by the presence in some of the oases of fish from tropical Africa, while Pleistocene beds in the Atlas have a fauna which has been called a Zambezi fauna. The last degenerate remnant of this fauna, which migrated across the Sahara during a wetter period, was the small Carthaginian elephant. It must not be concluded that the Sahara was well watered in the Pleistocene: it probably had a hot steppe type climate rather than a true desert climate as at present. Nor must it be assumed that the whole of the Sahara shared in this climatic amelioration, for, as Gautier says, there are few traces of wadi patterns in the Libyan desert, which, forming the core of the Sahara, was probably less affected.

Recourse to an earlier wetter phase was had by Peel (1941) to explain the dissection of the Gilf Kebir highlands in the Libyan Desert. The Gilf Kebir consists of a monotonously smooth plateau, the size of Wales, formed of Nubian sandstone and standing as much as 330 m (1 100 ft) above the surrounding flat desert plains. The plateau, which is itself an erosion feature, is deeply dissected by steep-sided wadis, which are irregular in both long and cross-profiles. The pattern of old stream courses can be clearly seen from above on the valley floors, but Peel believes that spring sapping may have been very important in the formation of the wadis. The heads of the wadis are very abrupt and there is no evidence of water having spilled into them from the plateau surface: on the other hand caves within the wadis may well have been formerly occupied by springs during a period of higher rainfall. The Nubian sandstones forming the plateau are permeable and water sinking into the sandstone may have been thrown out at the foot of its escarpments. Supporting evidence for a wetter period is archaeological and consists of evidence of human occupation, especially near the edges of the plateau, of Palaeolithic and 'Neolithic' type.

Evidence for similar water action in the Peruvian desert has been described by Bosworth (1922). In addition to relief forms, which he described as in no way different from those which might have been expected from water action, there are great waste fans at the foot of the Andes in the desert. The principal of these is the Amotape breccia fan which occurs at the foot of the Amotape range, a part of the Andean system rising abruptly 1 500 m (5 000 ft) high above the desert. It is 30–45 m (100–150 ft) thick near the mountains and thins to 1–3 m (5–10 ft) 8 km (5 miles) away. These fan deposits extend into the mountain valleys and have been dissected into terraces by occasional floods emerging from those valleys. Although modern floods are still of importance, much of the Amotape fan dates from the Pleistocene period.

The evidence for water action in deserts is, therefore, strong. Some water action may still occur intermittently, but many authors refer many of the landforms to a former wetter period. Parts of the desert landscape are, therefore, fossil forms subject to little subsequent modification, just as parts of the Highlands of Scotland are fossil glaciated forms, which have undergone no great modification since being glaciated.

How much of this former water erosion can be attributed to pluvial phases in the Pleistocene is very doubtful. Water erosion on bare, unvegetated landscapes was probably as powerful as ice erosion on our own temperate landscapes. Yet the degree of integration of drainage patterns shown in some arid regions seems altogether too high to be attributed merely to Pleistocene pluvial periods. The presence in such deserts as the Sahara of vast lakes such as mega-Chad which overflowed into the Bodele depression south of the Tibesti seems more the sort of feature which would have resulted from increased Pleistocene rainfall. There is plenty of geomorphological and zoological evidence for such features. But temporary lakes, however enormous, and integrated drainage patterns which demanded the erosion of much resistant rock are different kettles of fish. One suspects that many of the dominant erosional relict features in arid regions may be the legacies of even more remote periods in the past.

Pediments, pediplains and inselbergs

Unlike humid temperate regions, where the transition from upland to plain is often a smooth slope, many arid and semi-arid regions are characterised by the juxtaposition of smooth sloping plains, reaching a maximum slope of 7 degrees but often less, and steepsided residual hills. The break of slope between the two is often extremely sharp.

In block-faulted country, such as that of the south-western United States, the gentle slope leading from the ranges into the basins appears to be composed of several elements (Fig. 11.10). The whole slope from the range to the infilled playa lake is usually termed a piedmont. The upper part is often, but not always, a rock-cut surface, normally termed a pediment, while the lower part is an aggradation feature formed of detritus from the ranges and termed a bahada. The bahada deposits form a feather-edge on the outer part of the pediment and it is not always easy to separate the two: the feather-edge of bahada deposits may extend back to the foot of the mountains. Such is the use of the terminology by Cotton (1942) and

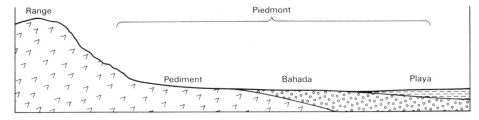

FIG. 11.10. Elements of an arid landscape

Thornbury (1954), but it should be noted that the terms have been used in other senses: for example, King (1942) uses bahada as a synonym for piedmont, while Balchin and Pye (1955) use peripediment for what has been termed a bahada above and bahada for waste fans at the foot of the mountains.

It is obvious that the bahada and playa deposits are formed of material derived from the mountains, the calibre of the sediments becoming finer further away from the mountains, but the origin of the rock pediment has provided one of the most controversial questions in geomorphology and the literature on it is extensive and scattered. Fortunately summary accounts are available in Cotton (1942), Balchin and Pye (1955) and especially Tator (1952).

The sharp break of slope between the pediment and the mountain front seems to point to a change of operative process, but there is no agreement as to the nature of the processes involved. Two major groups of theories have been put forward in explanation of the pediment and the sharp break of slope.

The first is that of lateral planation by streams issuing from the mouths of canyons in the mountain front. One of the principal protagonists of this theory, Johnson, imagined streams being diverted from side to side as they issued from the mountain front and undercutting the foot of the slope there. These migrating streams would form a convex rock fan and in different floods they would take different courses over it. Thus, a pediment would be initiated by a series of coalescing rock fans. On this theory, rock fans should be observable even though many might be obscured by debris, but the number of convincing rock fans reported is small. Further, the mountain front should be highly indented, because all planation stems from canyon mouths, but examples of straight mountain fronts are known. Lastly, it is difficult to envisage the late stages of such a process. As the mountain mass becomes smaller, the efficacy of the streams would presumably decrease and erosion would cease before the mountain mass was finally consumed. Many African inselberg landscapes, which appear to be comparable with the American examples though possibly at a more advanced stage, consist of a series of isolated steepsided inselbergs rising from an almost flat plain. No streams appear to emerge from the inselbergs, yet, to judge from the degraded piles of boulders, which seem to represent the last stages in the destruction of inselbergs, the process of pedimentation goes on to the very end.

The alternative theory regards the recession of the mountain front as a parallel slope retreat (cf. the ideas of Wood, Bakker and Penck discussed in Chapter 4), and the pediment as a graded slope over which the debris from the mountains is transported. The recession of the mountain front has generally been attributed to the action of weathering and the removal of the weathered debris by gravity, but occasionally to rill action. The pediment itself on the other hand is generally held to be formed under the

335

action of sheetfloods or rill wash, forms of unconcentrated flow which are capable of removing debris but incapable of effecting any real erosion. The angle between the mountain front and the pediment thus represents a real break in the processes involved.

For the maintenance of this abrupt angle a delicate balance must be maintained between the rate at which material is delivered to the foot of the mountain front and the rate at which it is removed over the pediment, while the calibre of the material supplied to the pediment should determine the general angle of slope of that feature. The ideal type of rock for maintaining a steep angle between mountain and pediment appears to be granite. With granite there is no gradation of weathering products from the largest boulder to the smallest size of sand, but merely two grades, the joint-controlled boulder and the sand corresponding in size to the individual crystal constituents of the rock. This sand can be transported over relatively gentle slopes; hence the pediment tends to have a low angle and the break of slope between it and the mountain foot becomes very sharp. On the other hand some compact, fine-grained igneous rocks and quartzites produce weathered debris of all calibres, much of which cannot readily be removed over the pediment and hence accumulates at the foot of the mountain front, thus obscuring the sharpness of the angle there.

Landscapes resembling the pedimented areas of the arid south-western parts of the United States are found on an altogether larger scale in Africa, especially in the semi-arid regions. They are the inselberg landscapes, consisting of vast extents of monotonous plain, from which arise steep-sided residual hills. These landscapes are not associated with desert basins, as are the American ones. The vast plains, which have probably been formed by an extreme degree of pedimentation, are termed pediplains, a term adopted by King in his studies of these surfaces.

The pediplanation cycle, as envisaged by King (1967), is in essence simple. The dominating action is the parallel retreat of slopes (see Chapter 4). In the early part of the cycle, river incision and the formation of cliff-like slopes are very important. Once these have reached a stable repose angle, King imagines them retreating parallel to themselves away from the drainage lines leaving continually expanding pediments between the slopes and the drainage lines. The exact angle of slope will be governed mainly by the lithology, but once attained it will be maintained until the residuals on the developing pediplain are consumed by intersecting slopes at the back of converging pediments. The result will be a multiconcave landscape, although creep effects on the upper weathered layer may induce a subsidiary upper convexity. A similar convexity might also develop late in the cycle when soil creep becomes dominant. In the development of a pediplain the drainage density will be of great importance, because the denser the network the faster the destruction of the land surface by encroaching pediments.

It is inherent in this theory that a pediment developing from one river may be shaved down to a lower level by a pediment developing from another, so that the deposits on a pediplain surface may vary in age and, in any case, give one little idea of the date of the opening of the pediplanation cycle. Also included within the theory is the idea that major scarps may retreat across country irrespective of drainage lines. King quotes the case of the Drakensberg escarpment in Natal which he maintains has retreated about 250 km (150 miles) since it was initiated late in the Cretaceous period. Such escarpments are considered to retreat as fast as nick points can retreat up rivers—a very Penckian concept.

The pediplanation theory with its emphasis on backwearing rather than downwearing has as a corollary the idea that old erosion surfaces will be widely preserved in the landscape, and in this sense has the same type of implications as the hypothesis of multiple marine erosion surfaces applied to the British landscape. The history of landscape development in terms of erosion surfaces can be read from the landscape right back to the Mesozoic era in many places. This is the extreme aspect of the belief in the preservation of the past in the landscape, and the direct opposite of the equilibrium ideas gained from studying non-resistant sandstones of the semi-arid American west and waste heaps.

King (1950) has widely applied his concept of pediplanation to the African plateaus and has distinguished a number of cycles of pediplanation in the time between the Jurassic period and the present day. Not only has King (1967) applied these ideas to Africa, but he has enthusiastically correlated these African surfaces with surfaces in areas, presumed to have formed part of Gondwanaland, i.e. South America, India, Australia and Antarctica, and even with surfaces in the northern continents. Such extended correlations, intended as a basis for stimulating discussion, have provoked considerable controversy (see discussion after King, 1950).

The residuals on these pediplains will vary with the rock type, but the formation of bornhardts, or granite-gneiss inselbergs (Plate 31) has been discussed by King (1948). The first requisite is a coarse-grained granitic or gneissic rock, not greatly effected by chemical weathering in the climates under consideration. Other coarse igneous and metamorphic rocks of a more basic character may also form inselbergs though the form is usually not as perfect as that developed in granite (Plate 32). Secondly, the rock should possess a widely spaced system of rectangular joints, to control the drainage pattern, and also the curved exfoliation type of jointing mentioned in Chapter 3.

Typical bornhardts seem to be associated with the second cycle of erosion in rejuvenated landscapes. Presumably at the end of the first cycle the drainage is well adapted to the major joint system as a rectangular pattern of subsequent streams. Upon rejuvenation these streams are incised along the joints (Fig. 11.11A). From the valleys pediments are developed and extend back into the rectangular interfluves from all sides (Fig. 11.11B), ultimately reducing the interfluves to steepsided inselbergs (Fig. 11.11C). King indi-

cates spalling and chemical erosion as factors in scarp retreat, the fine products of chemical erosion being blown away and the coarser falling to the base of the scarps. Occasionally the break of slope at the foot of the scarp is free from debris, but this does not seem to be normal in the African examples. The detail of the form of the inselbergs formed by these processes is, to a considerable extent, controlled by exfoliation-type jointing and curving sides are common on African inselbergs. If the general rectangular system of joints is very close, true bornhardts may not be developed, but will be replaced by piles of roughly rounded boulders known as castle koppies. Castle koppies may also, but more rarely, represent the last stages in the degradation of bornhardts. King's conclusions have been to a considerable extent supported by Mabbutt (1952, 1955) in South-West Africa. He has stressed the relationships between the detail of granitic landforms and the nature of the joint pattern present, and strongly supports King's idea of the association of inselbergs with the second cycle of erosion.

A slightly different sequence of erosion has been suggested by Peel (1941) for the Gilf Kebir, which was discussed above in connection with desert valleys. The climate in this section of the Libyan desert is, of course, more arid than that normally occurring in African inselberg landscapes, but the general assemblage of a mass of plateau, flanked by inselbergs rising from a vast plain, is not dissimilar. But the flat plains surrounding the Gilf Kebir appear to possess no traces of pediments. The author favours a process of spring sapping, which extends the wadis into the plateau, coupled with some lateral corrasion when the ephemeral streams are capable of cutting down no further. In this area, the inselbergs, which retain their characteristic steepness to the end, may be finally demolished through the assistance of wind erosion in undercutting their bases, as many of the inselbergs in the open desert show marked signs of undercutting on their windward sides.

These suggestions obviously run somewhat counter to the standard pediplanation hypothesis. Peel later (1966) introduced other suggestions to account for some aspects of these features of really arid parts of the world. He reiterated the difficulty of accepting the break of slope at the foot of residuals as due to a change of process, because this 'explanation' begs the question of which causes what, i.e. the change of form might control the present change of process. Further, the average pediplain is alleged to be a concave, water-formed surface reaching angles of 5 degrees and occasionally more near the feet of residuals. Yet, in the southern Tibesti and Tassili areas further west, theodolite observations showed angles as low as 10 minutes. Now a British lowland river will flow easily on this sort of gradient, which is about 3 m per km (15 ft per mile), but it would seem that a sheetflood would be so inhibited as to be evaporated almost before it started. Peel suggests that the foot of residuals is kept sharp by the rapid

PLATE 31. Acid gneiss inselberg, Shai Hills, Accra Plains, Ghana

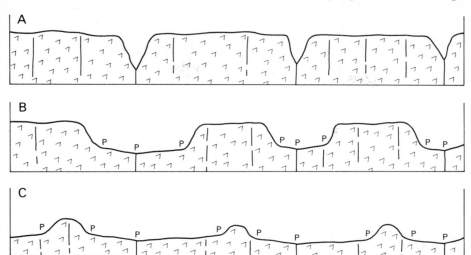

FIG. 11.11. Development of inselberg landscape in jointed crystalline rock by extension of pediments (PP)

shedding of any precipitation by the hill and its conservation near the foot which is hence more weathered. Supporting this is the observation that some of these residuals have annular depressions at their feet, a feature very difficult to explain by the standard pediplanation hypothesis. The near flat desert surface can hardly be called a pediment, and Peel notes that the French are tending to call such features by the noncommittal term, *glacis d'érosion*, pending further explanation.

These comments by Peel lead one to the necessity of keeping an open mind on the question of whether there is one universal hypothesis applicable to all forms of arid and semi-arid planation. In this respect his ideas run counter to those of King and accord more with those of the school of climatic geomorphology. There is obviously considerable scope for more observation and thought on the causes of vast areas of near flatness in arid and semi-arid regions. It may even be that the Davisian cycle of semi-arid erosion which was designed to explain the block-faulted deserts of the United States will prove to have local applicability along with many other ideas.

PLATE 32. Basic gneiss inselberg, Shai Hills, Accra Plains, Ghana

12
Landforms in the humid tropics

It is not possible to make a rigid distinction between the landforms of the arid and semi-arid areas, described in the last chapter, and those of the more humid regions to be considered in this. The climatic boundary is really a zone of transition, and the position is further confused by the possibility of more humid forms occurring as relics in the more arid parts and vice versa. Within the humid tropics two main climate vegetation zones are involved: the seasonal-rainfall savanna and monsoon regions, and the all-seasons-rain equatorial regions.

Birot (1960) makes the following general remarks about the distinctiveness of denudation processes in the tropical forests. The weathering of rocks, and it is predominantly chemical weathering, is very rapid, much more rapid than the movement of debris on slopes. In turn, the movement of material on slopes is more effective than river erosion. This state of affairs is almost the reverse of that found in most other regions. It may mean that the balance of processes is unique in tropical forest regions, but it could also mean that this is the normal state of affairs, other regions having had much of their waste stripped off by more violent erosion in the Pleistocene, so that the present apparent dominance of stream erosion over slope erosion and slope erosion over weathering are abnormal.

Recently the question of the intensity of tropical forest denudation being more apparent than real has been discussed by Douglas (1969), who has pointed out that it is easy to confuse the effects of intensity with those of time, and that, except where the natural ecosystem has been upset by man or volcanic activity, the evidence for much more rapid denudation in the humid tropics is by no means clear.

General weathering processes

One of the features of the landscapes of the humid tropics mentioned by Tricart (1965) is the comparative scarcity of bare rock outcrops, except in two situations. The first is where groups of inselbergs occur, whether developed in crystalline rocks or as tropical karst (see below). The second is the fairly common occurrence of rock outcrops in the beds of rivers. The scarcity of bare rock greatly reduces the role of physical weathering, although this may play a minor part in assisting chemical weathering.

It is well known that the thickness of rotten rock in tropical forest areas can be very great. Alteration to depths of up to 100 m (300 ft or more) is known in crystalline rocks, and, according to Birot, soil thicknesses of 30–40 cm (12–18 in) occur on slopes as steep as 60 degrees. But it must not be thought that the whole thickness of altered rock is waste in movement. In the wetter parts of Brazil the upper part of the mantle to a thickness of about 4 m (12–15 ft) consists of a yellowish sandy layer at the base of which are found pebbles derived from older superficial deposits and fragments of quartz veins. Below this lies rotted rock, deeply red in colour and rich in clay minerals. Downwards this passes into mottled material and then into solid rock. Only the uppermost sandy layer appears to be in movement, for in the rotted layers below the texture of the rock is often preserved, for example quartz veins in their original positions, the banding of metamorphic rocks and traces of fractures.

Although the texture may be so preserved, the rock is rotten and may crumble at a touch. Even quartz crystals are attacked and sound crystals are confined to the lower parts, where they have been weathered for a shorter length of time. The transition from this type of material to sound rock varies in abruptness. Where the rock is very closely jointed weathering penetrates over such a large surface area that core stones are small and rapidly destroyed, so that the transition zone from rotten to sound rock is narrow. Where the rocks have a wide joint spacing, of the order of 2 m (6 ft) or more, weathering may penetrate deeply before the large core stones are completely destroyed, so that transition layers from completely rotten to completely solid rock of 10–20 m (30–70 ft) thickness may occur (Tricart 1965). These conditions are those favouring the formation of tors by subsurface weathering (see Chapter 3). Where there are no joints or where they are tightly closed there may again be a rapid transition from rotten to sound rock.

This great thickness of rotted rock has several consequences. On the economic side, roads can often be cut with the use of machinery and without explosives, but this advantage must be set against the difficulty of finding sound rock for building such constructions as harbours and breakwaters in some areas. On the physical side, rotted rock is much more liable to shear failure than sound crystalline rock so that mass movements on slopes are probably more common and more devastating than in other climatic zones.

As suggested above the great thickness of weathered rock can be explained either by a very rapid rate or by a very long period of weathering, the latter implying unvaried climate. Evidence in support of each can be cited.

Rapid weathering is suggested by observations such as the following. White mica is reported to have been altered to clays in ten years or so in

343

quarries in Madagascar (Birot, 1960), although this is usually one of the more resistant minerals. In about thirty years the augen gneiss used in the building of the Department of Mines in Rio de Janeiro showed rusty alteration of its biotite mica, while in 300 years similar rock used in foundations showed the beginning of kaolinisation of the felspars. In addition, crystalline rocks in humid tropical conditions often show solution pits on near-horizontal surfaces and Karren on slopes: their surfaces can in fact become virtually a crystalline rock karst, thus suggesting a high rate of chemical weathering.

Yet, on the other hand, Tricart (1965) quotes the amount of weathering on certain Pleistocene terraces to substantiate the opinion that an unvarying Pleistocene climate at least must have been required to produce the observed thicknesses of weathering. In Brazil the amount of alteration on Middle and Late Pleistocene deposits is relatively small, but kaolinised profiles are known on older Pleistocene deposits near Salvador and also on ancient Andean piedmont terraces. In one example near Salvador a 20–30 m (70–100 ft) thick crystalline gravel deposits of Early Pleistocene age had been rotted to the base, while up to 15 m (50 ft) of the gneiss below had also been altered.

Douglas's (1969) recent analyses of suspended load and silica load in solution fail to show any very great increase in the tropical rain forest areas. He concluded that variations in runoff were chiefly instrumental, so that the root cause of any greater transport of such materials in tropical rivers was precipitation, which is high in such regions though 1 500–2 000 mm (60–80 in) per annum is probably the normal range in the lowland tropics.

One can advance theoretical reasons why chemical weathering rates should be high in tropical forest regions. Because of the high temperatures all the year round, plant growth is at a very high rate and the annual production of vegetation, measured in terms of weight per unit area, is several times that of the temperate forest. Birot quotes 2 or 3 as the factor, but it no doubt varies with the precise nature of the tropical and temperate areas which are being compared. The whole process of humification of the litter is also speeded up, so that, as the vegetation decomposes, the supply of carbon dioxide and organic acids is probably two or three times that available in temperate forests. The presence of acid crystalline rocks and sandstones over wide areas of ancient shield lands in the tropical forest and savanna zones does not encourage calcicolous plants, so that the resulting humus tends to be of acid type.

The part played by mechanical action is probably very small, because the mechanisms involved, temperature change and crystallisation, have very limited fields of action. Beneath the forest cover in areas which inherently have small temperature ranges a nearly constant temperature and humidity must prevail for much of the year. Where inselbergs occur, however, the bare rock favours a larger role for mechanical weathering.

Temperature changes of the order of 30 degrees C (54 degrees F) are possible when bare rock surfaces are cooled by rain, while salt crystallisation effects, such as taffoni, can occur where alternate wetting and drying occur. If weathering products get into cracks in rocks, expansion and contraction effects may also help. Yet it seems that the rate of mechanical weathering here is very much less than the rate of chemical weathering in vegetated areas, unless it is assumed that inselbergs are in process of being destroyed, an assumption which few tropical geomorphologists seem to have made.

Another mechanical help to weathering which may occur in the semi-humid tropics and possibly in the early stages of decay in the humid tropics is the expansion and contraction of the clay mineral, montmorillonite. This clay mineral is only stable in the presence of bases which are freed from crystalline rocks in the early stages of weathering. As soon as acid leaching starts, as it rapidly does under tropical forest, the clay mineral formed is kaolinite. Thus, montmorillonite may be expected on rock hills in the forests and more generally in the savanna regions, where acidification of the soil profile is less pronounced. Montmorillonite has one of the greatest expansions of all clay minerals on wetting: hence, it may be important in opening up large-scale fractures and minor intergranular fractures by a mechanical expansion and contraction effect. The necessary wetting and drying is much more likely in the savanna than in the forest.

From time to time recourse has been made to other acids in tropical weathering. The most notable of these is nitric acid, which is produced by lightning discharges and which might add considerably to the general weathering attack, for thunderstorms are characteristic tropical phenomena. But measurements of the acidity of rain water have so far not substantiated at all the theory that nitric acid might play an important role.

Silica weathering

A feature of tropical weathering is the apparently much greater mobility of silica. Quartz is generally thought to be extremely resistant to chemical attack in temperature regions, but corrosion of quartz crystals and veins seems much more widespread in the tropics. The ashes of certain savanna grasses have silica contents which may be as high as 5 per cent and occasionally 7 per cent. It is very unlikely that this silica is derived by direct chemical attack on quartz, for colloidal silica is an intermediate product in the weathering of many other minerals and it is more likely to be absorbed by the plants from that stage.

The understanding of the weathering of silica presents problems. Silica is soluble in alkalis, not in acids, but unless the alkalis are strong, for example sodium or potassium hydroxide, the rates of solution are slow. In the humid tropics, with all the acids released by humus decay, silica weathering is most difficult to understand. In the savannas, with bases being less leached and even returned by capillary action during the dry season, the formation of an alkaline environment favourable to an attack on silica is more plausible. But whether the pH value would be high enough seems to be a debatable question. Yet the field facts show that silica is attacked: as stated above quartz veins are attacked; superficial films of silica may be deposited on rock surfaces making them very resistant to erosion—there must obviously have been solution before deposition; silica crusts (silcretes) may form in some areas nearer the desert margins while some ferricretes contain mixtures of iron and silica.

While it is true that silica solubility does rise with temperature, the figures for amorphous silica are 60 ppm at 5°C (41°F) and 120–140 ppm at 25°C (77°F) and quartz is less than one-tenth as soluble as this, the increase is so small that probably increased runoff is more important than increased temperature in increasing the dissolved silica load (Douglas, 1969). Only above 35°C (95°F) does the increase become important. Such temperatures may be found in shallow pools on exposed rock surfaces, where solution and deposition of surface films on evaporation may take place.

The etchplain or weathering front

In either tropical forest or savanna regions it is possible to envisage a double denudation surface: the surface of the ground is subject to erosion but is already in such an advanced stage of chemical weathering that little further alteration is possible; at the base of the rotted zone is another surface, the weathering front, where weathering but not erosion is active. The latter surface, whether still protected beneath the overlying rotted rock or stripped of that cover by process changes engendered by changes of climate or base level, is sometimes called an etchplain. The originator of the term, Wayland, described cases where the erosion surface above was essentially an erosion plain so that the etchplain was virtually a reproduction of it at a lower level. Successions of etchplains could be caused by uplift phases alternating with weathering and stripping phases. It is possible, however, if river erosion is slow and weathering very rapid, for interfluves to be weathered deeply below the surface and then lowered by the stripping of the rotted mantle. It is also possible in a landscape where the weathering surface has been partly stripped, but much of the generating surface remains, for an apparently two-cycle landscape of juxtaposed flattish surfaces to be formed.

The term, etchplain, is perhaps a little unfortunate, because etching is an engraving process in which most of the plate is protected by wax or some such substance, while the corrosive agent, hydrofluoric acid in the case of glass, acts only where it is cut through by the pattern. Although subsurface weathering is deepest along the joints, there is not the complete contrast between protected surface and etched lines as there is in the engraving process. In addition, the general expression 'etched into relief' has been applied on both a regional scale and on the minor scale of the surface of a rock to denote the stage when the original surface is preserved on the harder parts while the softer parts have been eroded. This use seems nearer to the true meaning of the word. On the other hand 'weathering front' seems to express the idea much more clearly.

Surface crusts or duricrusts

The formation of surface or near-surface, indurated crusts in tropical regions presents problems vexed by the proliferation of terms and the fact that some of these are used in different ways by different authors. There is also the usual problem of knowing what to do with terms that have been defined genetically. The oldest term in use is laterite, introduced in 1807 by Buchanan for a reddish, cellular, ferruginous material, much of which hardened on exposure to air and which was used for building—hence the name, which was derived from the Latin for a brick. With many authorities, laterites and laterisation became synonymous with tropical soils and their formation. This is unfortunate, for we need purely descriptive terms un-complicated by genetic overtones. In English, duricrust (i.e. a hardened crust) or its French equivalent, *cuirasse*, seem reasonable terms. As the nature of the material may be important both in understanding the distri-bution patterns of crusts and considering their resistance to denudation, the words ferricrete (iron crust), silcrete (silica crust) and calcrete (calcareous crust) seem appropriate.

The origin of such features, if indeed there are single origins, seems to present complex problems of natural chemistry. Accumulations of iron and aluminium hydroxides in the form of crusts could be effected in situ if the silica were leached out by tropical weathering. They could also occur if laterisation was similar to podsolisation with the accumulation of the hydroxides in the B horizon, which was later brought to the surface by the stripping of the horizons above, a hypothesis favoured by some authorities including Prescott and Pendleton (1952): complete stripping over vast areas is, however, a process difficult to envisage. They could also be preci-pitated by ferrous bicarbonate rising with the water table and being con-verted in contact with the air into ferric hydroxide. Hypotheses such as these have one idea in common, that ferruginous crusts are a product of local soil profile development.

There are, however, cases where the ferruginous cement seems to have been derived from elsewhere (Lamotte and Rougerie, 1962; Tricart, 1965) by percolating solutions, so that the duricrust is not strictly the result of the processes of soil development. This is most clear where quartz gravels, which are very poor in iron, are cemented by ferruginous material. Such a deposit, although a duricrust, cannot have been formed by local weathering which could have provided neither the quartz pebbles nor the cement.

347

Iron minerals, perhaps the remains of iron-fixing bacteria, are transported in the usual way by streams. They will tend to be concentrated where water is spilled overbank and evaporation occurs. They will also tend to be concentrated where drainage is impeded and an iron-saturated ground water rises and falls near the surface. Thus, in natural hollows, on old lake floors, on pediplains and any other surfaces of low relief, impeded drainage and iron-rich ground water may result in the development of ferruginous crusts. Another natural locality for deposition is at the foot of uplands where iron-rich runoff is supplied to an adjacent pediment: such runoff may cement either soils or alluvial material of a variety of types.

It seems that clays exert a strong attraction on iron hydroxide particles, so that they become fixed on the surface of clay minerals, the iron, therefore, being dispersed in the clay and colouring it. Real ferruginous crusts are rare in clays and this perhaps explains their absence from the wet forests where clay minerals are the main products of weathering and where the upper part of the layer of rotted, kaolinised rock is stained a deep red. Iron crusts accumulate more readily in sands and gravels, which are much more likely to occur in the savannas than in the tropical rain forest. Incipient iron and manganese crusts can sometimes be seen in gravels in this country at about the level of the water table.

On the whole ferricretes occur under the wettest conditions conducive to the formation of crusts. With increasing aridity and the retention of bases near the surface, silica is more soluble and may be deposited as silcrete. In even more arid regions calcrete may be formed by the deposit of calcium carbonate. The climatic transition from arid to wet sees the transition from calcrete through silcrete to ferricrete, a distribution noted in many areas, for example Australia (Langford-Smith and Dury, 1965). In addition it is possible to get changes in crust type in the vertical section at any one locality if climatic conditions have changed.

The origins of these crusts pose difficult problems on the margins between pedology, chemistry and geomorphology, which do not particularly enlighten us about landform development. The concept that duricrusts are really cemented sedimentary rocks rather than local soil profile features prepares us for the general consensus of opinion that many of the thicker beds are of great age, dating probably in different areas from various stages of the Tertiary, perhaps occasionally from the later parts of the Mesozoic. In many places they remain very nearly in their original quasi-horizontal and low-angle attitudes. Therefore, given some form of rejuvenation, they act as the hard beds in the landscape, forming the scarped protected edge of plateaus and small scale cuestas. The ferruginous crusts are most apt for this purpose, but silcretes form similar features in Australia (Langford-Smith and Dury, 1965).

A special phase of development is cited by Tricart (1965). An iron crust commonly develops on the pedimentlike slopes surrounding a residual upland. Cementation is least effective at the foot of the upland slope, because there will be greater runoff there than on the pediment and hence less percolation to build up a thick ferruginous crust. The generally larger size of detrital fragments there will also make cementation more difficult because of the greater volume of voids. The result is a potential zone of weakness at the junction between ferricrete-capped pediment slopes and highland or inselberg slopes. Given uplift this zone is often eroded to form a gutter or trough separating the original hills from inward-facing scarps developed at the edge of the ferricrete on the outward-sloping pediments. In an extreme case with a change of climate it is possible for the original hills to be completely eroded away and for the ferricrete escarpments to overlook a central depression.

Because the ferruginous crusts develop mainly at the low points of the landscape and because they are resistant they cause a general tendency for the inversion of relief, most marked in the type of example discussed in the last paragraph.

Slope development

In the humid tropics the removal of material from slopes is slower than the rate of production of weathered, rotted rock. Yet it does not follow that slope processes are acting more slowly than they are in humid temperate regions. According to Birot there are three main reasons why this should be so.

In the first place the rainfall itself is heavier. Vast areas of tropical forest have between 1 500 and 2 000 mm (60–80 in) of rain per year, whereas the corresponding normal rainfall for temperate forest would be 500– 1 250 mm (20–50 in) per year. Not only is rainfall greater, but it is more intense, with, additionally, larger drops and hence a greater splash effect. The frequency of rain ensures a reduced infiltration capacity and hence more runoff.

This rainfall drops on to a usually poorly vegetated forest floor. Many of the leaves of tropical forest trees have drip-tips, which seem to be adapted for the rapid drainage of leaf surfaces as they are of common occurrence and not confined to related species. These, as it were, channel the rain on to the forest floor.

Not only is the forest floor not protected by a herbaceous layer, which would impede runoff, but the humus layer is thinner than in many temperate forests because of its more rapid oxidation. Furthermore, the weathering products under tropical forests are usually clays, which increase impermeability.

All in all, this means that there is probably more surface wash over tropical forest floors than over temperate forest floors. The wash is probably not capable of transporting anything except fine sand and silt, except where it can be concentrated. Such concentration occurs at the foot of trees where threads of water flowing down the trunk may excavate material on the downslope side and expose the roots of the tree. Birot quotes examples of thirty-year-old trees which are associated with erosion to a depth of 15 cm (6 in). Slope wash of this sort and downhill movements of material caused by termites, falling trees (though these are often held in position by lianas) contribute to the general downhill movement of surface material in the tropics. In short the forest is far less protective than has sometimes been asserted (Ruxton, 1967).

The other very important process is the landslide. One might have expected this from the combination of rotted rock and heavy rainfalls, especially as much of the weathering product is clay. Kaolinite may be less susceptible to shrinkage on drying than other clay minerals, and hence less liable to become saturated through its fracture pattern, but even so it is far more liable to landslips than solid rock. The thickness of rotted rock is so great that the root systems of the vegetation cannot stabilise the whole layer. In fact the tropical forest tree root network is very shallow: the cycle of bases passing from vegetation into humus and back into vegetation is effected within the first few metres of waste. Below that the waste mantle is deficient in bases practically down to the contact with solid rock, and, as already stated, is not bound by any stabilising agent. Saturation of this layer by rainwater, coupled in some areas by vibrations set up by earthquakes (Simonett, 1967), precipitates landslides. The vibration may act through the thixotropic character of the clay, i.e. the state when vibration converts a substance apparently solid into a liquid—a crude illustration can be obtained by banging a wet beach with a toy spade. The vast majority of landslips affect slopes of 42–48 degrees, as was shown by Wentworth in Hawaii. On slopes steeper than this waste mantles are usually too thin for mass movement. On slopes gentler than this the susceptibility to landslipping is less. The landslides themselves may be rotational slips of masses of rotted rock, or, where the rotted layer is thinner, slides of material over the surface of the solid rock beneath.

Freise, who studied mainly the landforms of the Brazilian tropics, went so far as to envisage a cycle involving forest impoverishment and slope failure (Bryan, 1940). Virtually, the cycle begins with the development of soil and forest on steep (35–50 degree), bare surfaces of crystalline rock. When the weathered mantle reaches a thickness of 10–11 m (30–35 ft) subsurface earth flows occur. These are followed by deteriorating conditions, with the development of surface crusts, forest deterioration and drying out of the soil. Grass invades the forest and is insufficient protection for the soil which is exposed to rain wash and earth flow and is finally stripped by these processes and by landsliding to expose bare rock again. The cycle is thought to take 300 000–400 000 years. It is probably doubtful whether anything as cyclic as this can take place without some climatic change, for it implies that weathering greatly exceeds slope action early on and the reverse later on, without there even being a significant gradient change for it is essentially a form of parallel slope retreat. It does emphasise, however, the dominance of mass movements and parallel slope retreat in the tropics.

These mass movements in conjunction with rain wash up to the crest of the divides (Ruxton, 1967) result in a strong tendency for knife-edge interfluve development where the area is forested and chemical rotting of the rocks predominates.

Stream activity

By contrast, stream erosion seems to be inhibited in the wet, forested tropics. Rock outcrops are common in stream beds, which, therefore, appear to be ungraded. In areas of crystalline rocks rapids are common, while rivers plunge over the edge of resistant sedimentary beds with few signs of incision. In fact, river channels are often broader at falls than in the reaches between, whereas the reverse usually holds in temperate latitudes with gorges downstream of falls showing how the latter have retreated. Birot argues that these features are endemic to the wet tropics, for they are so common that they cannot all be attributed to very recent rejuvenation or to tectonic activity.

The reasons underlying this apparent general ineffectiveness of linear stream erosion are obscure. It is true that most of the weathering products are clays, so that the amount of clastic debris available for erosion is small. Hence, above waterfalls the accelerated flow has no tools to use, and, having no tools, it cannot abrade fragments from the falls to cause pot-holing and undercutting below. Crystalline rocks tend to provide either very large boulders or fine debris: the former cannot be moved, while the second can perform little erosion. Even quartzites are rotted in the waste mantle and it is a matter of observation that they are worn down in a distance of a few kilometres in the bed of a stream. It has even been suggested that surface hardening of rocks may protect the river beds: Tricart suggested that alternate wetting and drying might lead to surface films which would protect the bed; on the other hand, Birot stated that such a process should lead to a case-hardened major bed, with a permanent minor bed incised into it, but there appears to be little sign of such an arrangement.

If one gets down to fundamentals, it is obvious that deep valleys have been incised in the humid tropics, so that much erosion must have taken place. It is also obvious that linear erosion cannot be as slow as it seems to be, otherwise interfluves would be quickly destroyed, though the lack of recession on many waterfalls remains an unexplained phenomenon. Hence, in spite of their ungraded appearance, rivers must be active agents of removal. Perhaps the answer, or at least part of it, is to be found along the following lines. The lack of erosion by the rivers on the rocks in their beds may imply tendencies to shift laterally and carve away the rotted rock at the foot of the side slopes. The rock outcrops abandoned by the river would then be exposed to the full force of chemical rotting, so that when the river migrated back it would skim off the uppermost layers of rotted rock until it was retarded by catching up with the weathering front. The process envisaged need not necessarily involve great lateral movements of the river to allow thousands of years of weathering of its former course. Smaller and more frequent lateral movements in comparatively narrow valleys might well be effective in lowering valley floors by frequent skimming-off of thin layers of rotten rock. On this view the exposed rocks of the stream beds represent the crests of tor or ridge forms produced where massive jointing impeded the downward migration of the weathering front. The river cannot incise itself in them. Having exposed them, it will migrate laterally leaving them to be destroyed by chemical attack before it swings back to remove the rotted debris. The detailed micro-relief of the rocky parts of the bed (see the plates in Tricart, 1965) could well be weathering effects slightly modified by hydraulic action. Much of this is conjecture, but an awkward set of facts needs some form of reconciliation.

Lithological effects

In the humid tropics, just as in other climatic zones, rock types affect the relief. Many of the effects are similar to those occurring in temperate climates. Any rocks producing permeable debris tend to have convex slopes and low drainage densities, for example sandstones and, under certain conditions, granites. The *meio laranja* (half-orange) hills of tropical Brazil are convex and associated with granites producing permeable debris. On the contrary rocks producing clay weathering products, for example schists and slates, tend to have concave slopes because of the greater runoff associated with impermeability. Rocks with thick waste mantles of impermeable material will be much more prone to landslides than those with permeable mantles. Rocks which resist chemical denudation, but may not be very resistant to mechanical erosion, should have more regularised river profiles than those in which an irregular weathering front develops. All these effects tend to produce nuances in the relief just as they do in other climates.

Limestones possess highly individual relief features here as elsewhere. Although there are certain special types difficult to match in temperate areas, there is a whole range of tropical limestone landforms depending on limestone lithology just as there is elsewhere (for a summary account and references see Sparks, 1971).

There has been considerable argument about the exact role of climate in karst development. Corbel has argued from the amounts of dissolved calcium carbonate in drainage water that karstification is slow in the tropics and attributed this to the decreased solubility of carbon dioxide. If this is so, the obviously intense surface karstification of some tropical limestones has then to be attributed to the length of time that climate has remained unchanged, perhaps through the whole of the Tertiary. Yet this idea cannot be maintained everywhere because intensively karstified limestones of younger Tertiary age are known. It may be that the lower solubility of carbon dioxide is offset by its greater total availability than in the less richly vegetated cold regions. It may also be that the increased rapidity of certain phases of the chemical weathering reactions increases the surface karstification at the expense of underground karstification, and it is, after all, the surface karst which is so spectacular in some parts of the tropics.

Two karst types peculiar to the tropics seem to occur. In the usual terminology they are respectively Kegelkarst or cockpit karst and Turmkarst (Plate 33) or tower karst, though the terms are not used consistently by all authors (see Sparks, 1971). Cockpit karst has been studied intensively by Sweeting in Jamaica, where the hills are conical and the cockpits tend to be shaped like inverted cones. The general order of relative relief is 100–150 m (300–500 ft), the diameter of the hills usually some 300 m (1 000 ft), and the hill slopes about 30–40 degrees. Tower karst consists of very steepsided hills, sometimes called mogotes, with side slopes of 60–90 degrees, separated by flatfloored depressions. The relative relief and magnitude of the features is very similar to that of cockpit karst at least in Jamaica, where both occur. It is especially well developed in south-east Asia, Malaya, Indo-China and southern China. Sweeting has suggested that tower karst develops from cockpit karst when the cockpits reach the water table, so that lateral solution in the inundated hollows extends the floors and steepens the intervening hills by undercutting.

However spectacular these forms are, it must not be forgotten that they are not the only tropical limestone forms. The chief difference between these and temperate forms is probably the dominance of holes: they might be described as holes with hills whereas much temperate karst is plateau with holes. Yet these forms of karst seem to demand special lithological conditions even in the tropics: in Jamaica hard, crystalline limestone seems to be a prerequisite with probably also a strong vertical jointing system for the development of tower karst. In areas of marly, impure limestones a doline karst, not unlike that of southern Europe, develops. Similar lithological controls have been reported from Puerto Rico.

Savanna plains and inselbergs

King would apply his pediplain cycle, described briefly in the last chapter to both semi-arid and the more humid savanna regions, because he believes in the essential uniformity of the erosion cycle irrespective of the climate. On the other hand, other geomorphologists have applied comparable but not identical cycles to other climatic zones. An example of this is the savanna cycle of Büdel (Cotton, 1961), which relies on a wetter climate and much greater chemical rotting than does the pediplain cycle. Such a cycle may also exist in tropical rain forest regions.

353

Büdel visualised deep chemical rotting and the formation of the two surfaces described above: the upper surface of rotted rock subject to stream
action and the buried, irregular, weathering front. On the upper surface
(Fig. 12.1A) streams in practically imperceptible valleys shift over the

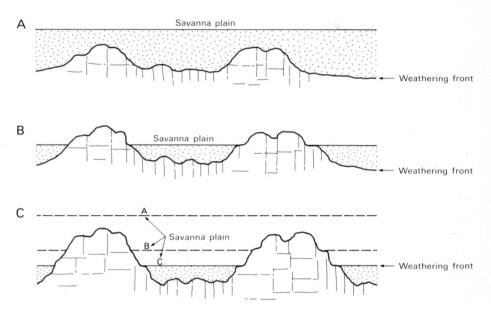

FIG. 12.1. Double erosion surface in savanna regions

surface rather like the distributaries over a waste fan and, during the course
of time, plane down the surface of rotted rock. They are engaged in the
erosion and transport of finely weathered debris and so can attain very
gentle gradients. They rarely encounter solid rock. In fact they act rather
in the way suggested above for streams in the tropical forests, except that
they may shift laterally much more freely. Such rock rotting and subsequent
planation extends into the valleys of adjacent highlands, just as pediments
do, and as a result masses of rock are isolated. Other inselbergs may be
formed when, as a result of rejuvenation, the weathering front appears at
the surface because temporarily stream planation exceeds the rate of
chemical rotting (Fig. 12.1B). Once inselbergs have emerged they may
remain owing to their immunity from chemical weathering and the fact
that physical weathering on them is slower than chemical weathering elsewhere. In fact, it may be that very high inselbergs are due to successive
cycles of weathering and rejuvenation (Fig. 12.1C).

PLATE 33. Tower karst in south-east China

A development of savanna plains on these lines seems to accord better with the association of duricrusts with such features. The general form is the same as that produced by a pediplain cycle, although the agents may be different. It may even be that the pediplains and inselbergs of the semi-arid regions are the products of periods, probably in the Tertiary, when savanna climates extended far more widely than they do now. Rotting and stripping is a hypothesis which can also be extended to account for the formation of torlike features in Europe from ancient (Tertiary) deep weathering. Many of the upland surfaces of Europe with their remains of weathered mantles may be explained in the same way, e.g. the Eogene surface of the Central Plateau of France with its sidérolithique capping (see Chapter 16). Inselberg development also takes place within the humid tropics.

An unprofitable argument has often arisen as to whether inselbergs are due to greater rock resistance or are merely residual in their positions. To some extent one is hampered here by the circularity of reasoning which argues that inselbergs must result from rock differences and, hence, could betray differences in rock resistance imperceptible to the geomorphologist. Birot (1960) has pointed out that, if inselbergs are not due to lithological differences, they should generally occupy divides and hence be the last parts of the landscape to be destroyed. If, on the other hand, they are not related in position to the network of divides and particularly if they extend in linear belts across country, they are probably determined in their positions by lithology.

Most authors on inselbergs have attributed a large part of their formation to variations in the frequency of jointing, good inselbergs like good tors developing where joints are more widely spaced and the joint blocks more massive as a result. This seems to apply in all climatic zones. Birot states that the sugarloaf mountains of Brazil are on the whole developed on massive augen gneiss, whereas the weathered zones are on biotite gneiss. But there are areas of augen gneiss which do not form sugarloaf hills and the distinction seems to be one of joint frequency. Hence joint frequency seems to be a more basic cause than the differences between augen gneiss and biotite gneiss.

Certain rock types such as aplites and rhyolites, which are not very susceptible to weathering because of their mineralogical composition, their impermeability and smooth surfaces, which do not allow water to accumulate and so promote weathering, usually form true lithological residuals (Birot, 1960).

Apart from such occurrences, the form of inselbergs, including sugarloaf mountains, seems to depend on jointing plans to a large extent. The more massive the jointing the more massive will be the inselberg. Finer jointing gives castle koppies (see Chapter 11). The rounding of jointed blocks would seem to require more humid climates to promote spheroidal weathering either now or in the past. Large-scale, domed forms, for example the sugarloaf mountains, either follow a pre-existing joint pattern in the rocks or promote a pressure relief joint system in conformity with their shape. Most geomorphologists today would probably accept the latter explanation. Birot quotes an interesting objection to this idea by Cailleux, who argued that the expansion due to pressure relief would be counteracted by the contraction due to lower temperatures near the surface, but it is probably very unlikely that the two would so exactly balance as to avoid the setting up of stress. Curved pressure relief joints are much more likely to dominate the form of inselbergs where other jointing is widely spaced.

It will be apparent that knowledge of the humid tropics is limited and that at the present time there is a lot of general theory, which might be described in its general style as Davisian. It is after all so difficult to generalise about landforms over such a large and not readily accessible area of the earth's surface.

13
Landforms in glaciated highlands

Introduction

The general cause of the climatic changes responsible for ice ages is little understood, but has been the subject of numerous theories, which have been discussed by Flint (1957), to whom reference should be made. It is generally agreed, though some would dispute the view, that during the Pleistocene period there was multiple glaciation, some of the glaciations being divisible into minor advances and retreats in certain areas on geological evidence. Before the Pleistocene, ice ages occurred in the Permian period, though not in Britain, and also in the Pre-Cambrian. Any general theory to account for ice ages must obviously account primarily for this irregular distribution of the phenomenon.

Whatever the cause, the effect of the changes was to cause a general lowering of temperature in midlatitude regions. But it is not necessary to envisage a catastrophic fall in temperature to levels now reached in winter in Antarctica and the interior of Siberia. It has been calculated that the summer temperatures in Britain may have been lowered by about 11 degrees C (20 degrees F), which would put the July mean in the south of England to about 7°C (45°F) and in Scotland to a couple of degrees above freezing point (Manley, 1952). The winter temperatures may have been approximately 11 degrees C (20 degrees F) below the July means.

Provided that the precipitation remained roughly the same as at the present day, these lower temperatures would ensure that a far higher proportion fell in the form of snow. With the much lower summer temperatures, much less of the snow would melt in the summer. Even at the present time, snow usually lingers through the summer in two shaded gullies, one on the north-east side of Ben Nevis and the other in a similar position on Braeriach in the Cairngorms at altitudes of about 1 125 m (3 750 ft) (Manley, 1952). Obviously, if the snow total was much higher and the summer melting much less, snow would not only linger but accumulate on the highlands and lead to the development of ice caps.

The amount of the snowfall rather than the lowness of the temperature is the important factor, as can be clearly seen from a study of the pattern of the Pleistocene glaciations. In the British Isles the centres of dispersal

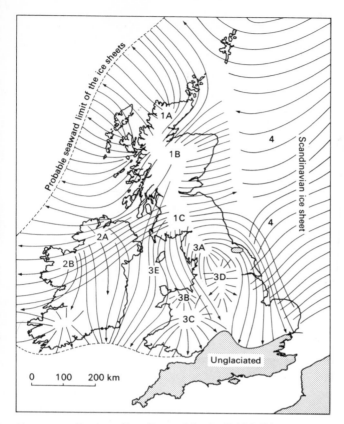

FIG. 13.1. Centres of ice dispersal in the British Isles.
The arrows show the direction of movement of the ice sheets.
The local ice caps are: 1A, North-West Highlands;
1B, Grampians; 1C, Southern Uplands; 2A and 2B,
Northern Ireland; 3A, Lake District; 3B, North Wales;
3C, Central Wales; 3D, Southern Pennines; 3E, Irish Sea
(*after Stamp and Beaver*)

of the ice were mainly in the west (Fig. 13.1), especially the north-west
Highlands and Grampians in Scotland, the Lake District and Wales. This
accumulation on the west implies that the source of the snow was the movement of air from the Atlantic towards Britain. A vivid picture of probable
meteorological conditions in this period has been given by Manley (1952),
who visualises the main Polar Front extending across Britain with vigorous
and stormy depressions passing across southern England and northern
France. The warm air in the depressions would tend to rise over the cold
air north of the almost stationary Polar Front and precipitation falling
through the cold air out of the warm air aloft would naturally fall as snow.
Even if the Polar Front was displaced northwards and warm air reached

the central and northern parts of Britain, the precipitation would still probably be snow due to the intense cooling effected by the land and its ice caps.

The east side of Britain may have been colder, but as the North Sea was dry land at the time of maximum glaciation, any winds blowing from the east were dry winds. Unlike present conditions when outblowing cold air from Europe picks up enough moisture from the North Sea to cause heavy snowfalls in the eastern part of Britain, Pleistocene conditions ensured that the maximum snow was in the west. Roughly comparable conditions can be found in northern Europe, where the main ice cap was in north Scandinavia, while Russia and Siberia to the east, although probably much colder, had much lower snowfall and consequently thinner ice sheets. Indeed, the glaciation of Siberia seems to have developed from a series of minor ice caps associated with the mountain blocks (Flint, 1957).

The excess of accumulation of snow over ablation (sublimation and melting) is slowly converted into ice. This is achieved by melting and refreezing in summer, leading to the formation of a loose mass of snow and ice particles. Two terms are applied to this material, and their meanings are not clearly differentiated. Firn literally means snow which fell last year, while névé applies to a loose mass of compacted granular snow and ice. Where the temperatures are too low for summer melting and refreezing, the snow may be converted to névé and finally to ice by the pressure of the overlying snow.

The rate at which snow is converted into ice depends to a very great extent on the temperature. This is readily observable in many winters in Britain. Heavy wet snow falling with temperature near the freezing point is converted to a very icy mixture on the ground. The process concerned is that the pressure between grains induces thawing followed by refreezing when the water migrates into the voids between the grains. Dry powdery snow at low temperatures is converted to ice much more slowly, because the weight is less and the temperature lower, so that there is less chance of thawing and refreezing at grain margins. Embleton and King (1968) state that in the Alps snow with a specific gravity of about 0·08 may be converted to ice with a specific gravity of about 0·9 in 25–40 years, while in Greenland the corresponding period is 150–200 years. The rate of conversion to ice is one of the characteristics which serve to distinguish temperate from cold glaciers.

When an appreciable mass of ice is present it will start to move outwards, its velocity depending principally on its surface gradient. Away from the centres of accumulation of ice, usually on lower ground, the fate of the moving sheet of ice depends mainly on the balance maintained between the supply from the centre of dispersal and the rate of ablation in the marginal regions of the glacier or ice sheet.

The nature of the glacial action varies with the position of the area concerned in relation to the ice sheet. Generally speaking, the highlands from which the ice is dispersed are primarily regions of erosion (Plate 34). If the ice sheet covered the whole area there may have been a general grinding down of the surface. If part of the highlands stood above the level of the ice, all the signs of sharpening of relief caused by frost shattering may be present. In the lowlands nearest to the highlands, there is usually a mixture of erosional and depositional effects, many of the latter being of the type formed when ice movement is still pronounced, especially drumlins. In the marginal regions of the ice sheet the effects are primarily depositional, the signs of erosion being few and slight. In these marginal regions, there must be a constant struggle between the tendency of the ice sheet to advance as it is slowly supplied with ice from the highlands, and the tendency for ablation to cause the ice front to recede.

In all areas there may be glacial diversions of drainage. The drainage during glaciation is controlled by the relief of the period, and that relief is formed of the surface of the rocks plus the surface of the ice. When the latter melts, the drainage may be so far entrenched into its course that it is not readily diverted. Thus, many glacial diversions of drainage may almost be regarded as special cases of superimposed drainage.

Outside the areas actually glaciated, there seems to have been a region of tundra climate, usually known as the periglacial zone, in which processes of erosion, although not glacial, were so different from present ones that their effects on relief were marked. The chief actions in periglacial conditions are solifluxion processes, depending upon the alternation of winter freeze and summer thaw.

In the British Isles these various zones may be roughly defined. The glacially scoured highlands are typified by the Highlands of Scotland. The effects of frost shattering above the ice are clearly seen in such places as the Cuillin Hills of Skye, while uplands scoured by the ice are probably represented by the coastal regions around Arisaig, Mallaig and the seaward end of Loch Shiel. In lowlands adjacent to the centres of ice dispersal drumlins are common, for example in the area to the south of the western part of the Southern Uplands of Scotland and in the Eden valley. Much of the Midlands and East Anglia provide examples of the predominantly depositional effects of ice sheets. Finally south of the Thames–Bristol Channel line the effects were periglacial. In any of these areas, except the last, diversions of drainage may be found.

The divisions made above are not absolute ones, but serve as a convenient basis for the description of landforms in this and the next two chapters.

Isostatic movements in glaciated areas

The weight of a large ice cap causes such a load to be placed on the crust of the earth that it is depressed during glaciation and slowly recovers when the

ice melts. The actual amount of the depression will depend on the thickness of the ice, assuming the density of ice to be constant, on the density of the material beneath the ice and on the rate at which that material yields to the load placed on it. There seems to be a limiting size for an ice cap below which it will not cause any depression of the crust. This limit cannot be defined exactly, but very large ice caps such as those covering the Laurentian Shield in Canada and Scandinavia during the Pleistocene period certainly caused marked and easily demonstrable depressions. Similarly, Greenland beneath its present ice cap has been shown to possess a basin-shaped form and this has been attributed to a similar process. The Pleistocene ice cap of Scotland, though a lot smaller than those mentioned, also caused a depression of that country, but this was less marked due to the smaller size of the ice cap. Smaller ice caps, such as those occupying Wales and the Lake District during parts of the Pleistocene, apparently caused little or no depression of the crust. Daly has suggested that the critical size of ice caps at which depression occurs is a diameter of 500 km (about 300 miles) and a thickness of 1 km (about 3 250 ft).

A large amount of information about the recovery of the crust from its isostatic depression can be obtained from the study of shorelines of seas and lakes abutting on the land mass during its recovery. These shorelines generally show a domeshaped uplift to have taken place since the ice melted, the line of no uplift corresponding approximately with the limit of the former ice sheet. In general the newer shorelines can be traced much nearer to the centre of the former ice cap than the older shorelines, which were limited in extent by the considerable area of ice still in existence during their formation.

The recovery of a land mass from depression is a slow process and does not seem to go on at an even pace. In the marginal regions, where depression was least, recovery is early and a balance fairly quickly restored. Nearer the centre, partly because the ice sheet lasted longest there and partly because the depression was greater, recovery is much slower and is not always complete even at the present day. Apart from this, there are lines along which there is a change either from no uplift on one side to perceptible uplift on the other or across which the rate of uplift increases markedly. Such lines are known as hinge lines, a self-explanatory term. In the Great Lakes region of North America the hinge line, separating horizontal from warped shorelines, seems to have migrated northwards during the recovery of the area from its ice loading, thus indicating that balance was recovered earliest in the marginal areas, as suggested above. In Scandinavia, on the other hand, the hinge lines seem to have remained in essentially the same places. Thus, it can be seen that the recovery of an area from depression beneath an ice cap is a complex process.

PLATE 34. Glaciated granite highland from north Goat Fell, Arran. To the left Cir Mhòr and in the centre Caisteal Abhail with a corrie between (see Plate 36): right of Caisteal Abhail is the eroded dyke of Ceum na Caillich (see Plate 13)

An earlier explanation of warped shorelines was that they were due to the attraction of sea levels and lake levels towards the ice sheets, and did not necessarily prove any isostatic depression and upwarping. Any water level slopes upwards towards an outstanding mass on the earth's surface, and it would seem to be true that the ice caps must have caused some deformation of water surfaces. But the gradients of the surfaces can be calculated for ice caps of given size and shape, and it has been shown that, even assuming ice caps of impossibly great dimensions, the gradient of shorelines attributable to this cause is far less than that actually observed in Scandinavia and North America.

The best known and probably the most complex example of recovery is provided by Scandinavia. As seen above, the upwarping is not regular and continuous, so that at times it was overtaken by the Post-glacial rise of sea level and at times it exceeded that rise. This resulted in a series of alternating salt-water and fresh-water bodies occupying the site of the present Baltic Sea. The number of detailed stages and substages which may be recognised is very great, but the essentials of the history can be grasped by reference to a few major stages (Wright, 1937).

The earliest major phase was that of the Baltic Ice lake, when ice still kept the rising North Sea out of the low area occupied by the lakes of Central Sweden, Wener and Wetter. This early ice lake had an outlet in the region of the present Sound and Belts between Denmark and Sweden, and probably another outlet at one stage through Lake Onega to the White Sea. With the retreat of the ice north of the lakes depression of central Sweden an opportunity was provided for the drainage of the Baltic Ice lake westwards and the influx of salt water into the Baltic. At this stage, which is termed the Yoldia Sea from the name of a fossil common in its deposits, the exit via the Sound and Belts was closed, because the low area across the Swedish lakes was at that time appreciably lower. The Yoldia Sea left marked shorelines, which can be used to reconstruct a map of contours, or isobases, showing the amount of upwarping of Scandinavia since that time (Fig. 13.2). The area where the higher isobases are missing shows the region still occupied at that time by the ice sheet. The maximum uplift of over 275 metres (nearly 1 000 ft) is a staggering amount, especially when it is realised that this cannot be the total recovery since the last ice age. At the maximum of the last glaciation the ice sheet extended north–south through Denmark and then swung west–east over Germany some distance south of the present coast. Thus it had thinned and retreated considerably long before the Yoldia Sea was formed, so that the maximum uplift of the shoreline of that sea cannot represent the maximum Post-glacial uplift of Scandinavia. Although it may be difficult to conceive how a faulted mass of crystalline rocks could behave in the apparently elastic way suggested above, the magnitude of the gradients involved must be remembered. The total depression suggested for the Scandinavian ice cap is 730 m (nearly 2 500 ft), which would involve a gradient of 600 mm

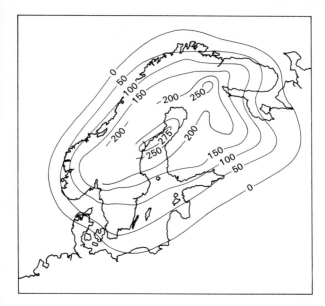

FIG. 13.2. Isobases (in metres) of the Yoldia Sea
(*after Wright*)

per km (approximately 4 in per mile) or 1 in 1 650 approximately. In other words it would be imperceptible to the naked eye.

But the isostatic recovery of the lakes region of central Sweden was rapid and soon the connection with the North Sea through that area was restricted. The Yoldia Sea became the Ancylus Lake, so named from a small freshwater snail characteristic of its deposits. At first this lake overflowed through the lakes region, but later the outlet was transferred to the Sound district between Denmark and Sweden. This latter area was not, however, rising as rapidly as sea level, so that the Baltic region was once more invaded by the sea through this area and the Littorina Sea formed. The Littorina Sea is the ancestor of the modern Baltic, and the entrance to that sea has remained between Denmark and Sweden ever since. There were, however, considerable changes within the Baltic itself. As sea level was rising much faster than the land in the south of the region, the deposits of the Littorina Sea are transgressive in the south, but they are not in the north, where, on the contrary, the land was rising faster than the sea. Even now, the isostatic recovery of Scandinavia is not complete, but appears to be slowing down. At the head of the Gulf of Bothnia it is still of the order of a metre per century.

Obviously in such a region as Scandinavia the pattern of beaches and terraces due to the interaction of a rising sea level and a land mass recovering from its depression by an ice cap is extremely complex; much more complex than might be inferred from the summary account given above. More detailed accounts will be found in Daly (1934), Wright (1937) and Flint (1957) and in the detailed papers quoted by those authors.

The classic views on the upwarping of Scotland were summarised by Wright (1937). They involved a 100 ft (30 m) beach, which was considered to be contemporaneous with the early stages of deglaciation for a number of reasons. Cold conditions during its formation were evidenced by Arctic fauna, and erratics deposited in clays as though they had been derived from ice rafts. The beach is missing from the upper parts of valleys, where, it was suggested, the ice still lingered so that no beach could form. This beach disappeared below sea level well within the limits of the later 25 ft (7·5 m) beach (Fig. 13.3). A series of retreat stages from the level of the 100 ft beach

FIG. 13.3. Limits of late Pleistocene 100-ft (30 m) and 25-ft (7·5 m) raised beaches in the British Isles (*after Wright*)

culminated in deposits below present sea level. The succeeding deposits of the 25 ft beach therefore represented a rise in sea level which occurred in the warmer parts of Post-glacial time, in fact the Neolithic.

In recent years an intensive re-survey of the evidence around the Firths of Forth and Tay has been undertaken by Sissons, Smith and Cullingford, who have concentrated on geological and geomorphological features, depositional raised beaches merging into outwash, peats and estuarine or carse clays. This re-analysis has involved a tremendous amount of precise levelling, boring, and the working-over of existing borehole data, for in some places the raised beaches are below the level of deposits of peat and carse clay. Nothing but precise levelling will serve for areas like East Fife where the height range, 14–22 m (50–70 ft), embraces six shorelines, the closest pair being about 0·6 m (2 ft) apart. The cruder surveying methods used for much older, higher, marine terraces usually have possible errors of about 2 m (6 ft), so that they are not adapted to this sort of problem.

The period represented by the beaches is the Late-Weichselian and the Post-glacial, generally a period of deglaciation but punctuated by the Perth and Monteith (Loch Lomond) re-advances in Zones I and III of the Late-Weichselian respectively. The general positions of land and sea levels appear to have been as follows.

(*a*) In the Late-Weichselian the land seems to have recovered isostatically faster than sea level rose in the inner parts, but probably less fast in the outer parts. This means that the oldest Late-Weichselian shorelines diverge from the younger ones towards the centre of the former ice cap in the southern Grampians. The plot of recognised East Fife shorelines shows the declining gradients (Fig. 13.4) on the lower shorelines (Cullingford

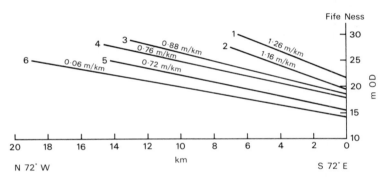

FIG. 13.4. Attitude of Late-Weichselian shorelines in eastern Fife (*after Cullingford and Smith*)

and Smith, 1966). If one extrapolates the shorelines eastwards their order is reversed some 25 km (15 miles) offshore except that 4 and 5 have not crossed. Thus, in these outer areas sea level rose faster than the land. The difficulties about 4 and 5 could be due to the moving position of the hinge line or to the fact that straight extrapolation provides too simple and idealised a view of the situation.

(*b*) The fact that shoreline 6 has only about half the gradient of shoreline 1 shows the quick isostatic recovery of the area during a period when the ice retreated only 25 km (15 miles) (Sissons, 1967). It can be calculated from an inferred position of sea level at this time, an assumption that the centre of isostatic recovery was in the south-west Grampians, and a further assumption that the shoreline had a constant gradient that a total of about 225 m (750 ft) of isostatic recovery has taken place since East Fife shoreline 1. There was probably some recovery before this, but, even if the latter is discounted, it implies a gradient practically double that of warping in Scandinavia (see above).

(*c*) The simple pattern of less steeply inclined shorelines as isostatic recovery occurred may have been interrupted by conditions during glacial re-advances. Sissons argued that such re-advances are synchronous and must have implied a general fall of sea level. Therefore, the resulting shoreline should have been low, but in Scotland the Perth re-advance shoreline is high. This can be explained if we assume that the re-advance caused isostatic depression of the land followed by rapid recovery, an idea which accords with the rapidity of isostatic reaction suggested in (*b*) above.

(*d*) Later on the rate of rise of sea level overtook that of the recovery of the land to give a series of Post-glacial shorelines which cross the Late-Weichselian set. These form the 25 ft beach of the older terminology. An idealised representation of these relationships is shown in Fig. 13.5, in which the number of shorelines does not represent those recognised in the field.

This detailed attempt at reconstruction based mainly on landforms is so far only applicable to the area studied. The relations between these features and the rock wavecut platforms of western Scotland remain obscure. Sissons suggests that the latter may be interglacial features. If they are it is difficult to understand how they apparently slope downwards out from

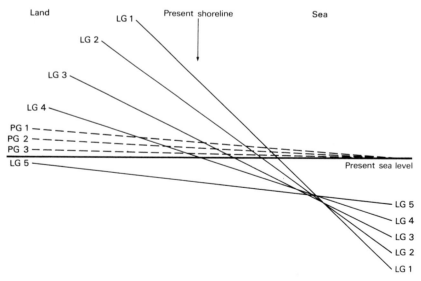

Fɪɢ. 13.5. Idealised representation of Late- and Post-glacial shorelines in south-eastern Scotland

a centre in the Loch Linnhe area (McCann, 1966). With interglacial features complete isostatic recovery should mean that they are horizontal: incomplete recovery should mean that they are lower in their central part: more than complete recovery is presumably not possible. Whatever the answer, the western 7·5 m (25 ft) terrace is so marked an erosion feature in hard rocks that there seems little chance of attributing it to an ephemeral Post-glacial episode, especially when the sea at its present level has hardly started to cut an erosion terrace in most areas.

A more general and conservative approach to the question of the isostatic recovery of Scotland has recently been taken by Donner (1970). Much of the geomorphological and geological evidence used by Sissons is not dated. Donner includes only evidence of marine transgression and regression which can either be zoned by pollen or directly dated by radiocarbon methods. As the Post-glacial pollen zones are dated by radiocarbon elsewhere, pollen zonation alone allows a fairly reliable estimate of absolute age. By plotting the height of the evidence of marine activity against time a curve of relative land and sea levels may be obtained. Donner's work is less detailed than that of Sissons for two reasons: it covers the whole of Scotland and the acceptance of evidence is limited. The graphs produced by Donner for relative land and sea level positions vary from area to area because of the differing relations of those areas to the centre of isostatic recovery, but they all show one general trend (Fig. 13.6). The Late-Weichselian was a period of rapid isostatic uplift outpacing the rising sea level. About 6500 BC this slowed down and the sea then rose faster than the land to give a maximum transgression at about 5000 BC, since when the declining rate of isostatic recovery of the land has led to the present relative positions of land and sea.

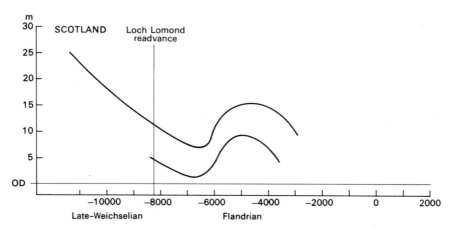

FIG. 13.6 Relative positions of land and sea in the Post-glacial in Scotland (*after Donner*)

In essence the schemes of Donner and Sissons are similar and not fundamentally different from that of Wright. They include a Late-Weichselian group of beaches (divided by Sissons but not by Donner) and a Postglacial group of beaches (again, divided by Sissons but not by Donner), which represent the peak of the Flandrian rise of sea level. These two represent the 100 ft and 25 ft beaches of earlier authors, but it is better to scrap these labels because of the great variability of height. The chief difference lies in Sissons's idea of rapid isostatic depression in periods of ice re-advance, a conclusion which cannot be escaped if it is assumed that ice re-advances were accompanied by a general fall of sea level. Perhaps these falls will be revealed as more carbon dating and pollen zoning become available, so that both types of evidence will converge. Whatever happens, it will be interesting to see what emerges from this active field of research. The general pattern of changes in land and sea levels implicit in all these ideas (Fig. 13.6) is very similar to that derived from western Norway.

North America provides another example of the same phenomenon on the scale of the Scandinavian upwarping. Again the zero isobase coincides closely with the limit of the ice, and once again clearly indicates the fact that such warpings must be the effect of the loading caused by the ice. The amount of recovery is similar to that of Scandinavia (Fig. 13.7), but there is not the complex interaction of seas and lakes, although the detailed history of the Great Lakes is due to a considerable extent to isostatic upwarping (Daly, 1934).

Corries and arêtes

The higher parts of a highland mass may project above the general level of the ice, either in the early stages of glaciation or in the late stages, or, in certain areas of high relief, throughout the glaciation. In such a position the surfaces of the rocks are subjected to very great diurnal temperature changes, as the rock is bare. Lower down, where the ice cover is complete, the diurnal temperature changes beneath a glacier are much smaller, so that shattering due to alternating freezing and thawing is less important. There can be no doubt that rock is shattered above the snowline, for screes and piles of broken debris are very common. Whether temperature change alone can effect this is doubtful, but this question has already been discussed in Chapters 3 and 11. Much of the work is probably performed by water percolating into cracks and freezing, with the consequent heaving-off of pieces of rock. For water to be effective in this way, it is probably necessary for the rock to be well jointed, but the hard rocks of most of the world's highlands are well provided with joints and fractures, thus paving the way for their disintegration by freeze-thaw action.

At about the regional snowline, which can only be defined to within a hundred metres or so as it varies with exposure and aspect, being higher on leeward and south-facing slopes, the characteristic form of mountain

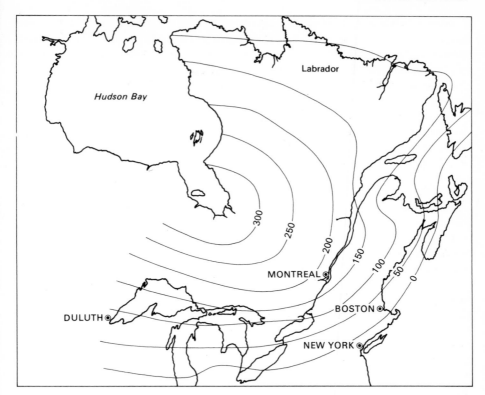

FIG. 13.7. Isobases (in metres) in north-eastern North America of Post-glacial marine limit (*after Daly*)

glaciation is the corrie (also termed cirque and cwm) (Plates 35 and 36). Such features are common in most glaciated areas. The typical form is a deep armchair-shaped hollow often containing a lake, either in a true rock basin or in a basin blocked by moraine. The origin of corries is one of the main problems of glacial geomorphology.

Snow will accumulate in any preglacial hollow, especially on the lee side of mountains. Although the snowfall may be heavier on the windward side, it is often swept off by high winds, which can hardly affect it on the leeward side. Further, corries are most common on north- and east-facing slopes, due to the smaller summer melting which can take place there.

The snow-patch formed in the preglacial hollow is subject to diurnal and seasonal thawing. The meltwater so formed percolates into crevices and, by freezing, tends to disintegrate the rock. The broken waste may be removed by processes of solifluxion (see Chapter 15) and the hollow deepened. As long as only snow accumulates in the hollow, there is probably no snow or ice movement to transport material away or to erode the hollow. This process of hollow-deepening by freeze-thaw and solifluxion is usually known as nivation. It is unlikely to produce more than a small landform and certainly not true corries, which have back walls often a few hundred metres and occasionally up to 1 000 m high. Nevertheless, nivation hollows may provide the sites for corrie development and it has been suggested that there is a progressive evolution from nivation hollows to true corries. Examples of seemingly intermediate forms can be found, but one is faced here with a common geomorphological problem: whether landforms existing together in a landscape represent different stages of the development of a single type of landform or whether they are the final stages of development of a number of different types of landform.

A number of theories have been advanced in explanation of corries, but all meet some objections. The two main features requiring adequate explanation are the high back walls and the overdeepened basins which are sometimes, but not always, present in the floors of corries.

Willard Johnson, after the exploration of bergschrunds (the major crevasses which occur near the backs of most corrie glaciers), put forward a hypothesis involving freeze-thaw action at the bottom of the bergschrund. Material shattered by freeze-thaw at the point where the bergschrund touched the back wall of the corrie (Fig. 13.8) would be transported in the base of the glacier and would erode the basin of the corrie proper. There are two main objections to this hypothesis. It has been shown in a number of present-day bergschrunds (Battle and Lewis, 1951) that the variations in temperature are too small, a range of little more than one degree, from 0 to − 1 °C (2 degrees F, from 30 to 32°F), being recorded by Battle. Further, the exploration of bergschrunds in Greenland by Battle left him with the impression that freeze-thaw was not of great effect in them. The second difficulty lies in the height of the back wall of the corrie. The bergschrund explored by Johnson is said to have been 45 m (150 ft) deep, but modern opinion is inclined to the view that crevasses in ice are unlikely to remain open below about 30 m (100 ft), as they are closed by the plastic yielding of the ice below this (1954 discussion). Even if this figure were doubled and bergschrunds 60 m (200 ft) deep admitted, the theory could only explain corries the back wall of which did not exceed 60 m in height, whereas, as stated above, many corrie back walls are much higher.

A variation of this hypothesis was suggested by Lewis (1938, 1949), who thought that meltwater streaming down the back wall of the corrie would extend to greater depths than the bergschrund, and that the freezing and thawing of this meltwater would tend to disintegrate the whole of the back wall of the corrie and not merely that section of the wall where it was met by the bergschrund. The way in which the necessary freezing is effected well down in the glacier, where the temperature remains very near 0°C (32°F), is again a difficulty.

If the whole of the back wall of the corrie, including that part above the

PLATE 35. A corrie lake, Llyn Idwal, Caernarvonshire. Behind are the Glyders

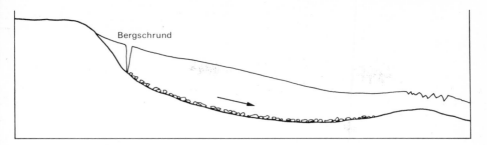

Fɪɢ. 13.8. Corrie glacier with bergschrund.

ice, is attacked by frost weathering, the great height of some back walls may be due not to vertical deepening of the corries but to the headward extension of the back walls into the adjacent mountains, but some mechanism is still needed to explain the erosion of the walls, if freeze-thaw is not an effective agent.

The broken rock of the back wall appears to be removed by being pulled away by the ice, as it moves down the corrie, and remains frozen to the rock. If the rock is not shattered, such a process cannot take place for the tensile strength of ice is far too small for it to pull lumps out of rock of much greater tensile strength. Fractures would occur either within the ice, leaving lumps adhering to the rock, or at the ice–rock contact. Freeze-thaw may play a part, but sufficiently great temperature changes only appear to occur near the glacier surface. Another possible fracturing mechanism has recently been stressed by Lewis (1954, 1960): it is jointing parallel to the surface caused by the relief of load on rocks crystallised at great pressures and temperatures, and is usually spoken of as pressure release jointing. This process has been mentioned in Chapter 3 and again in connection with granite and gneiss inselbergs in Chapter 11. Many observations show a tendency for joint systems to develop parallel to the present surfaces of the rock, and such jointing has been observed around corries. If the rock is so weakened, then it is possible for the plucking action of the ice to pull material away from the back wall without the shattering action of freeze-thaw. In addition, very few rocks are entirely unjointed, so that, even where there appear to be few signs of pressure release joints, sufficient weaknesses may occur for the rock to be plucked by the ice even without freeze-thaw.

It has also been suggested, mainly by Lewis, that corrie glaciers are subject to rotational slip. The dirt bands between the annual accumulation layers have been observed, especially in some Norwegian corrie glaciers, to become progressively steeper down the glacier, although they tend to flatten again near its end (Fig. 13.9). On the face of it this appears to represent a rotational movement, though an alternative explanation relating the

Pʟᴀᴛᴇ 36. The back of a corrie: between Cir Mhòr (foreground) and Caisteal Abhail, Arran

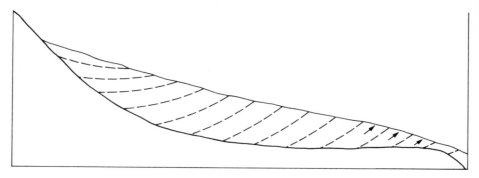

FIG. 13.9. Ideal arrangement of dirt bands in a corrie glacier

attitude of the dirt bands to the velocity distribution in a glacier has been put forward by McCall (1952). In his early application of the rotational slip hypothesis, Lewis imagined multiple upward thrusts near the snout of the glacier along these dirt bands (arrows in Fig. 13.9), but this view has not been substantiated and is now withdrawn. McCall's measurements on the Skauthoe glacier, Jotunheim, Norway, showed that the greater part of the glacier movement consisted of slipping over its bed and that there was no clear evidence for discrete shear planes, although there was an upward component in the ice movement at the snout of the glacier.

The basic mechanism behind the rotational movement of a glacier as a whole is the imbalance resulting from melting at the snout and accumulation leading to ice formation in the upper parts. Thus, the gradient is steepened until slipping occurs.

However corries are formed, they seem to extend backwards into the mountain mass in the valley heads of which they start. This leads to a reduction of the area of comparatively smooth pre-glacial relief between them and to its final replacement by frost-shattered, knife-edge ridges, known as arêtes, while the original mountain mass is reduced to a pyramidal peak, or horn (Plate 37), formed from the intersection of arêtes (Fig. 13.10). The classic example is the Matterhorn, while a smaller British example is Cir Mhòr in Arran.

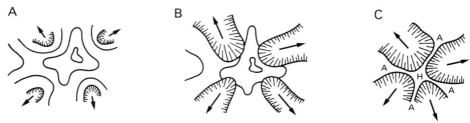

FIG. 13.10. Formation of arêtes, A, and horn, H, by backward development of corries

PLATE 37. A pyramidal peak, Cir Mhòr, Arran

At the maxima of glaciations the corrie landscape may be completely submerged beneath the ice and the general grinding action of the ice sheet may smooth out the sharpness of the frost-shattered forms, except for isolated summits projecting through the ice and known as nunataks. Whether the pattern of corries is entirely destroyed seems doubtful, as some are so large and well developed that one suspects that, although they may have been submerged at times, they suffered repeatedly from phases of corrie glaciation in the Pleistocene.

Glaciated valleys and associated features

Corrie glaciers originating in valley heads moved downvalley, merged and became valley glaciers. In these valleys they left characteristic features. The valleys are usually broad and flat-bottomed (Plate 38), the characteristic section often being likened to the letter U. Compared with river valleys they are usually straighter, with the valley side spurs truncated. Tributary valleys often hang markedly above the main valleys (Plate 39) and there is frequently a pronounced shoulder on the sides of the main valleys. The floors of glacial valleys may possess enclosed basins, steps and roches moutonnées, the last being landforms which are smooth on the upstream side and present a rough, quarried appearance downstream. Some of these features are not confined to glaciated valleys. U-shaped valleys may be found in chalk regions which have never been glaciated; steps in valley floors are a common feature of rejuvenation; hanging valleys may be formed where a main stream has a rate of downcutting greatly superior to that of its tributaries. But the general assembly is unmistakable and points clearly to the effects of glacial erosion. It should be mentioned here that a theory, which assigned a relatively protective role to ice, was elaborated by Garwood (1910) early in this century but that the theory has few adherents today. It is discussed fully by Cotton (1942).

Eventually the formation of these features will have to be understood in terms of ice movement and erosion, but our knowledge of this is far from complete at present. The measurement of rates of movement is very difficult at depth and laborious and slow on the surface. The interpretation of the observed movements in terms of the physics of ice, a problem mostly tackled by physicists naturally enough, is not always easy. The interpretation of past erosion forms—at the present time too few are being formed fast enough for theories to be based on them—involves deeper entry into the field of speculation.

The movement of the surface of a glacier can be obtained by measuring the displacement of a row of stakes driven in line across a glacier and surveyed accurately from stations which must be located on solid rock. The surveying has to be of a high standard of accuracy, especially when the

PLATE 38. The Nant Ffrancon valley, Caernarvonshire: a glacial trough

A

B

Glen Rosa

Garbh Allt

time interval between observations is short. When the time interval is long it is difficult to ensure that the stakes are firmly enough in position not to be lost during winter under snow or thawed out in summer. Furthermore, very steep temperature gradients over the surface of the ice can lead to awkward refraction problems in surveying. This method is most readily adapted to corrie glaciers and to valley glaciers where firm surveying stations are readily set up. It is obviously less useful for ice sheets. Further, most surveying will be in summer when movement is probably at a maximum (see below).

It is usually found that the velocities of glaciers are highest in the central parts and taper off towards the sides as shown in Fig. 13.11A. Occasionally, they remain steady almost to the edges and then taper sharply as shown in Fig. 13.11B. Such movement is called Blockschollen or plug flow. Its im-

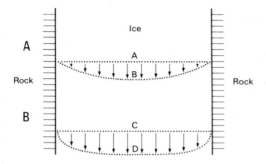

Fig. 13.11. Types of glacier flow. For explanation see text

portance is that it initiates strong shear near the valley walls which may effectively increase erosion there.

Measurements of the velocity profile in depth are much more difficult. In corrie glaciers tunnels have been driven in horizontally and their deformation measured, but such a method cannot be used in a thick valley glacier. In valley glaciers tubes have to be inserted by drilling a hole with heat, and then their deformation measured with an inclinometer. The greatest depth reached by a tube is about 500 m (1 650 ft). All the measurements that have been made indicate a velocity profile of the type shown in Fig. 13.12A. It is a little uncertain what happens at the very bottom of the profile: either there is sliding over the bed, as was suggested for corrie glaciers, or there is very strong shear within the lowest layers of the ice, which are firmly frozen to the bed.

PLATE 39. The hanging valley of the Garbh Allt, Arran. The valley heading in a compound corrie left of the picture swings left at the edge of the granite and falls steeply into the glacial trough of Glen Rosa. In the distance are Holy Island (A), a composite sill reaching a height of 310 m (1 030 ft) and the north-facing escarpment of the Clauchland Hills (B), also a sill

The possible difference of movement in the lowest layers is important in the differentiation of what are termed cold and temperate glaciers. Cold glaciers are essentially those of the Arctic and Antarctic. There is very little information available about them, but it is suspected that they remain firmly frozen to their beds, as meltwater to lubricate them is not available at any season. Their general rate of movement is slow compared with temperate glaciers. Temperate glaciers may be lubricated either by meltwater or by summer rainfall. This may facilitate sliding over the bed. Increased rates of movement have been observed in summer and after heavy rainfall. Thus there is considerable variation in glacier velocity. Normally rates of movement are measured in cm per day and, according to Embleton and King (1968), a rate of 100 cm (4 in) per day is rarely exceeded. Very high velocities have been recorded at the edge of the Greenland ice cap in summer: conditions here are ideal, a great reservoir of ice being funneled down valleys with a steep gradient to the coast. Under these conditions movements of the order of 20–30 m (60–100 ft) per day have been observed. From all this it seems most likely that temperate glaciers in summer are likely to be the most effective agents of erosion especially if they are sliding over their beds and undergoing plug flow.

So far as erosion is concerned it is what happens between the glacier and its bed which is all important rather than what happens within the ice of the glacier. Ice itself behaves in an approximately plastic manner: it resists deformation to a certain threshold and then deforms at a constant rate with time. This is accomplished by reorientation of crystals, thawing on crystals faces subject to pressure and freezing on those subject to least pressure, and movement along the cleavage planes of ice crystals. It behaves effectively like a metamorphic rock. Because of its nature ice can transmit pressure on to its bed equal to that derived from a layer of ice 22 m (about 75 ft) thick. When it is subject to higher pressures than this its own strength is exceeded and it yields. Therefore maximum erosive effects probably increase with ice thickness to about this limit and then remain constant. It is possible to calculate the likely magnitude of damage to the bed using values for the strength of ice and rock (McCall in Lewis, 1960: summarised in Sparks, 1971).

One of the major problems is represented by over-deepened basins in glacial valleys. They also exist in corries (see above) and may there be due to rotational movements of the corrie glaciers. Several possible ideas have been put forward to explain these.

One possible explanation is the theory of extrusion flow, advanced principally by Demorest in America and Streiff-Becker in Switzerland. They imagine that the upper layers of ice are rigid and brittle, while below a certain depth (about 30 m (100 ft)) the pressure of the overlying ice would cause the lower ice to flow more rapidly, in fact to be squeezed out (Fig. 13.12B). Thus where the ice is thick the maximum velocity would be near the base of the glacier, whereas where it is thin the maximum velocity would

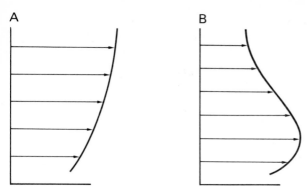

FIG. 13.12. Velocity profiles in glaciers: A, as measured;
B, on theory of extrusion flow

be at the surface (Fig. 13.12A). Applying this hypothesis to a stepped valley (Fig. 13.13), one would have extrusion flow in the basins, C and D, and normal flow on the steps, A and B. Thus, most erosion would take place in the basins and lead to overdeepening. Extrusion flow has never been directly observed and one of the chief arguments for it is the indirect one that the measured surface velocities of a glacier seem to be insufficient to account for the removal of all the firn accumulation in the upper parts of a glacier, unless one assumes a greater flow at depth. Measurements made of the deformation of tubes in various glaciers however, have all shown velocity profiles of the type shown in Fig. 13.12A. In addition, if extrusion flow takes place, it is difficult to see why the surface layers are not carried forward at equal velocity by the underlying layers, unless an extremely narrow valley causes such great friction as to retard the surface. At present, the hypothesis of extrusion flow is not regarded with favour by those physicists engaged on research into ice movement in this country.

Another form of movement is envisaged by Nye (1952) in his theory of glacier flow. He distinguishes between extending flow and compressing flow. The former occurs where the velocity increases down glacier, for example over a step (B on Fig. 13.14): the latter occurs where the velocity

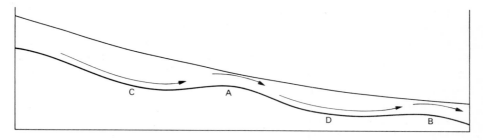

FIG. 13.13. Location of maximum velocities according to theory of extrusion flow

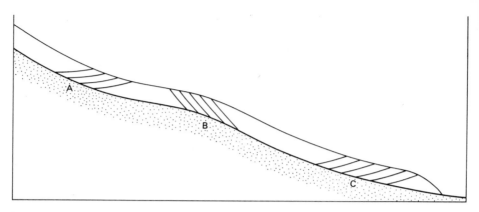

FIG. 13.14. Possible slip surfaces in a glacier (*after Nye*)

decreases down glacier, for example in a basin (A) and near the snout of the glacier (C). Nye has calculated the slipline field (i.e. the surfaces along which slipping is likely to take place) in such situations and the sets of slip-lines which may undergo movement in natural glaciers are shown in Fig. 13.14. Where the flow is compressing, at A and C, there is a tendency for movement up from the bottom of the glacier towards the surface and faults, if they occur, will be reversed or thrust faults. Where the flow is extending, at B, there will be a downward movement of the surface and any faults will be normal ones. It must be noted, however, that the two slipfields do not meet, and that Nye does not envisage the rotation of a large section of a glacier as a unit, although the result will be somewhat similar.

If a rotation occurs, as suggested by Lewis, or if alternations of extending and compressive flow occurs, as suggested by Nye, and if debris in the base of the glacier scours the bed rock, an explanation of glacial basins, including corrie basins, is obviously possible. The common occurrence of glacial striae and of smoothed and polished rock surfaces clearly indicates glacial erosion, but it is not a simple process to understand. The rate of movement of a glacier over its bed is very small, so that debris in the ice has ample chance to be forced back into the ice rather than to erode strongly the rock it presses against. Presumably a large boulder with a small area of contact with the bed rock and a large area of contact with the ice is necessary for severe erosion to take place.

There are other aspects of ice movement, but their connection with land-forms is obscure. One of these is the development, when a glacier apparently gets out of balance, of a wave of high velocity, a kinematic wave, which runs down the glacier with a velocity of anything up to six times the normal. Kinematic waves are usually accompanied by rapid but temporary advances of the glacier snout. Velocities in excess of 100 m (330 ft) per day have been observed in such waves, but the reliability of such observations has been questioned.

The straight valley with its truncated spurs, the hanging valleys and the valleyside shoulders are features explicable by either vertical or lateral corrasion by the ice or by the two actions combined. The truncation of spurs, which is most likely due to their being overridden by the glacier and ground away, seems to point clearly to strong vertical erosion by the valley glacier, but the hanging valleys and valleyside shoulders could be caused by either vertical or lateral erosion (Fig. 13.15). Theoretically, if vertical

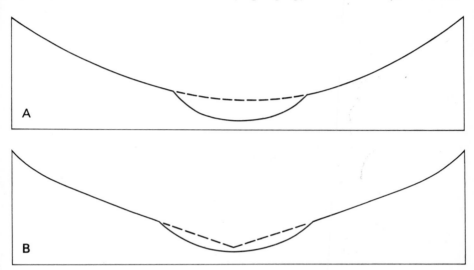

FIG. 13.15. Hanging valleys: A, due mainly to vertical erosion; B, to lateral erosion

erosion had dominated, the profiles of the hanging valleys, when projected downstream, should be well above the level of the valley floor (Fig. 13.15A), whereas, if the erosion had been mainly lateral, they should accord roughly with the valley floor (Fig. 13.15B). Most probably both types of erosion occur. The vertical erosion is mainly the result of glacial scour, while the valley widening may be effected by freeze-thaw and pressure-release jointing above the level of the glacier occupying the valley. At the time of maximum glaciation, the surfaces of the main and tributary glaciers probably met at an accordant junction, the differences in depth of erosion in the valleys being due to the different sizes of the glaciers occupying them. It is also possible that some hanging valleys might be due to two phases of glaciation, separated by a return of stream erosion in an interglacial period. In this case, the broad upper valley would represent the work of the first glaciation, into the floor of which an interglacial stream incised itself. Later, the incised stream valley was converted to a U-shape during the second glaciation.

Steps in the valley floor and roches moutonnées (Plate 40) have certain features in common: they are both smoothed and striated on the upstream

side and usually craggy and plucked on the downstream side. The initial cause of steps may be one of several features: more resistant bands of rock, alternations of well- and poorly-jointed rock, preglacial rejuvenation heads, or points at which for some reason the glacier increases its erosive power, as at the junction of two glaciers or where flow is accelerated by constriction of the valley sides.

The smoothed appearance of these forms on the upstream side is due to the ice grinding its moraine over the rock, while the plucked downstream side would appear to call for a process similar to that operating on the back wall of a corrie. The problem is similar to the corrie problem in that the chief difficulty is to account for the temperature alternations needed to cause freeze-thaw on the downstream side of the step or roche moutonnée. Where the overriding glacier is thin, it is likely that it will be very crevassed where it passes over a step, so that atmospheric temperature variations and meltwater may penetrate to the face of the step, shatter the rock and so prepare it for the ice to pluck it away (Fig. 13.16). Where the ice is thick

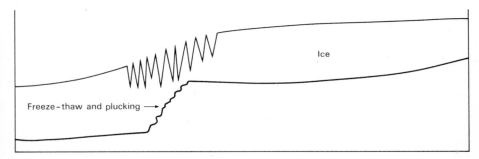

Fig. 13.16. Step beneath a thin glacier

the problem of step formation is more difficult. Just as in corries, the mechanism by which sufficiently large temperature variations may be introduced is obscure. A hypothesis involving extrusion flow has been put forward by Carol (1947) and is based on sub-glacier investigation. Measurements of the rate of movement of the ice showed that the lower layers accelerated to double the velocity of the normal flow, as they passed over a roche moutonnée (Fig. 13.17). In the lee of the rock the ice had a fluted appearance representing the moulding forced on it by projections in the rock and a subglacial cave existed. The release of pressure in this cave would slightly lower the melting point of water, so that meltwater formed in the region of high pressure on the upstream side of the rock might suddenly refreeze in the lee of the rock, thus initiating a very limited freeze-thaw process (Lewis, 1947).

PLATE 40. Roche moutonée near Loch Laggan, Inverness. The plucked side is to the left showing ice movement to be from right to left

Geomorphology

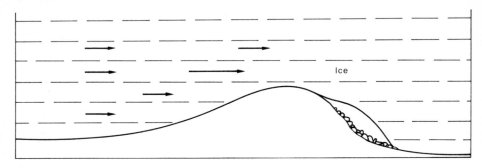

Fig. 13.17. Illustration of Carol's hypothesis of roche moutonnée formation. Arrows are proportional to velocity of ice

Although erosional forms are predominant in mountains, these regions are not devoid of depositional forms, chiefly lateral and terminal moraines. Usually lateral moraines are patchy features after glaciation, but terminal moraines (Plates 41 and 42) may be well marked and are often responsible for the impounding of lakes.

It is easy to become so obsessed with the apparent power of glacial erosion that the effect of rocks in glacial relief is forgotten. Yet the best glacial relief in Britain is found in areas of very resistant rocks. In Wales Snowdonia and the Cader Idris district have both magnificently developed (or preserved?) glacial relief on igneous rocks of Lower Palaeozoic age, while rocks of similar age but different lithologies have far less perfect forms. In the Lake District the Borrowdale Volcanic Series preserves the bulk of the best glaciated mountain scenery. In Scotland the igneous and metamorphic rocks of the Highlands have far better glacial forms than the rocks of the Southern Uplands.

There are a number of ways in which this apparent correlation of the best glacial forms with the hardest rocks may have come about.

It may be that the hardest rocks formed the highest preglacial relief. The highest preglacial relief formed the centres of local ice caps and hence glacial erosion was strongest in these areas. If this had happened the control of glacial landforms by rocks was indirect.

It could be that resistant rocks resist glacial action until a critical level is reached at which they are imprinted with glacial forms which are then preserved due to the hardness of the rock. Less resistant rocks might yield earlier to glacial denudation, but have the forms removed by later glacial action or by normal erosion in Post-glacial times.

It may simply be that the hard resistant rocks, with their usually well-developed jointing, are best adapted to produce the characteristic forms of glacial erosion.

Plate 41. Terminal moraine in Strath Oykell, Sutherland. To the right the moraine extends obliquely across the valley; to the left the smooth, cultivated floor of Strath Oykell below the moraine

388

Not only do correlations between glacial features and rocks occur on a broad general scale, but also in detail. Haynes (1969) has shown, for example, that in the Torridonian Sandstone areas of western Scotland joints dipping gently back into corries are far more likely to lead to overdeepened basins than joints dipping gently forward out of corries.

Fjords and allied features

Fjords are drowned glaciated valleys in highland regions, but their form is not due solely to the submergence of glaciated valleys. Many of them are characterised by very deep water within and by shallow water at their thresholds, a form which implies a considerable amount of glacial over-deepening. The depth of water in the closed depressions in fjords is often much greater than the known Post-glacial rise in sea level. For instance the Sognefjord in Norway reaches depths of about 1 200 m (4 000 ft), while the threshold is only of the order of 150 m (500 ft) deep, the depth even here being greater than the Post-glacial rise of the sea. In addition, an area such as Norway, which was occupied by a heavy ice cap, would rise isostatically after the ice melted, so that the Post-glacial rise of sea level in such a region would be less than on coasts not affected by glacial eustasy. In short, the Post-glacial change in sea level is a completely inadequate explanation of the major features of fjords.

Most fjords are on the west side of land masses, where high mountains lie very close to the sea, e.g. Norway, British Columbia, southern Chile, and South Island, New Zealand. The same relationships hold good on the west coast of Scotland, though it may not be wise to class the sea lochs of Scotland (Plate 43) as true fjords. As seen at the beginning of this chapter, the main accumulations of ice were probably towards the west side of land masses as they were derived from prevailing westerly winds. The glaciers from these ice caps would have had steep surface gradients to the west coasts, hence their rate of flow and their erosive ability would have been very great. Towards the sea, although it must be remembered that with lower sea levels the shore may have been considerably further out than it is at present, the rate of melting would have been rapid. Hence, we have a picture of vigorous glaciers melting away in no great distance. Where they were subject to melting, they would have moved slowly, if at all, and their erosive power near their snouts would consequently have been small, unlike conditions near their sources where erosive power was very great. Provided that the surface gradient was continuous, they would have continued to flow up a reversed slope. Therefore, it is possible to assume that they deepened pre-existing valleys greatly as they flowed quickly from the ice caps and hardly at all further out near their ends. Such is one possible

PLATE 42. Small moraine in Glen Rosa, Arran

explanation of the overdeepened forms of many fjords and their shallow thresholds (Fig. 13.18).

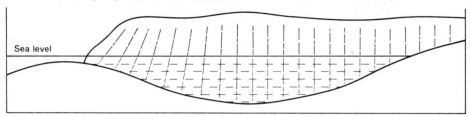

Sea level

FIG. 13.18. Sketch section of the long profile of fjord

In many areas, such as Norway, there is often a marked rectangular element in the fjord pattern and this has led to a probably erroneous attribution of these features to preglacial faulting. The true state of affairs was probably that the preglacial river valley pattern was largely determined by such things as dominant fault and joint directions, the strike of the rocks and the strike of the schistosity of metamorphic rocks. The glaciers merely scoured out pre-existing valleys and hence retained the relationship with tectonic influences.

It has been suggested (Cailleux, 1952) that one of the main aids to deep glacier erosion may be a preliminary freezing to great depths in valleys in the very cold phase before ice actually covers an area. Beneath the ice the temperature later rose and the ground shattered by deep freezing was scoured away by the glacier. Applying this idea to fjords, Cailleux suggests that the depth frozen near to the sea will be less than that inland (Fig. 13.19), so that the glacier will have less fractured material to scour

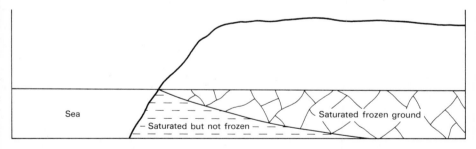

Sea

Saturated frozen ground

— Saturated but not frozen —

FIG. 13.19. Possible effect of deep freezing on fjord formation (*after Cailleux*)

away near the sea: hence, the deepest erosion would be well within the fjords.

Although deep freezing may play an important preparatory part in glacial erosion, it is unlikely that the whole of the form of fjords is due to

PLATE 43. The head of a sea loch, Loch Nevis, Inverness

such a process. The depth of some fjords seems excessively great for this explanation. In addition, fjords appear to be only a special, coastal case of what Cotton (1942) terms piedmont lakes. In South Island, New Zealand, the fjords of the west coast in the south are matched to the east by deep ribbonlike lakes. In Scotland, too, deep narrow lakes are found: Loch Ness is over 210 m (700 ft) deep, while Loch Morar on the west coast south of Mallaig is over 300 m (1 000 ft) deep and separated from the sea by land only about 15 m (50 ft) high. It seems difficult to attribute these to a seaward decrease in the depth of frozen ground, but, as they are near the centre of the ice sheet, they could be attributed to deep glacial scour caused by rapidly moving ice.

Fjords provide some of the evidence on which the idea of very selective glacial erosion may be built. It is easy to imagine that, beneath an ice sheet, all relief would be ground down. Such ideas probably originate from areas such as the Laurentian Shield, where, however, much of the grinding down may have been performed by preglacial erosion, just as it may in the Fennoscandian Shield and the Lewisian Gneiss areas of north-western Scotland. It is apparent from the relief of the Highlands of Scotland and the Scandinavian mountains, both centres of ice caps, that ice caps do not necessarily obliterate relief.

Yet in many of these areas there are curious juxtapositions of features of seemingly heavy glaciation and features which are of non-glacial origin and which can hardly have survived severe glacial erosion. The commonest observation is of the presence in the same area of true glacial U-shaped troughs and much more V-shaped valleys, which one would normally attribute to river erosion. In some areas there are signs of severe glacial erosion in valleys and a lack of it on high plateaus: an example of such a region is provided by the Cairngorms which have been discussed by Sugden (1968). The upper plateaus of these mountains preserve erosion surfaces, which are generally believed to be Tertiary in age; deeply weathered rocks which are much more likely to be preglacial than interglacial; torlike forms, which seem to have been glacially modified, though it is often reckoned that tors must be destroyed by ice sheets; sheet structures in the crystalline rocks which seem to have been formed in relation to preglacial relief and truncated by glacial features.

It is possible to argue, almost in a circle, that these features indicate that the Cairngorms have not been overrun by ice, but such an idea assumes what ice is expected to do. Other features, such as the deep troughs with signs of ice pouring into them at their heads, are hardly explicable without the presence of thick ice on the Cairngorms. Indeed, on a wider scale the intensity of glacial erosion in Scotland in general seems to imply that a thick ice sheet occupied the central highlands. In fact, Sugden advanced specific evidence in the form of the trend of the main deepened valleys, the orientation of high-level, ice-breached cols and major roches moutonnées on the plateau surface that the Cairngorms area had been overrun by ice moving from the south-west. This direction is consonant with what is inferred about the Scottish ice cap being well west of central as it was fed by prevailing winds from the Atlantic.

All this field evidence seems to support the idea of highly selective glacial erosion, an idea which involves the assumption that within an ice sheet, which may as a whole be moving slowly, there are locally streams of high velocity. High rates of ice movement demand steep surface gradients of ice sheets, conditions at present met on the edges of the Greenland and Antarctic ice sheets. Similar conditions must have occurred on the west coast of Scotland with the centre of the ice cap near the west coast and a low sea bed offshore. A like combination of circumstances would also have been present in Norway. Under these conditions we might expect generally rapid ice movement with local streams of high velocity guided by pre-existing relief. Where the valleys were aligned with the general direction of ice movement they were probably occupied by fast-moving ice streams. Where they were aligned differently, they were not so occupied and hence less modified glacially. It also seems that zones of geological weakness also tended to be scoured out by ice movement even where they were not perfectly aligned with the general direction of ice flow.

The general pattern of rock basins and sea lochs in western Scotland seems to show the two conflicting trends of valley orientation and geological weakness (Sissons, 1968). If one takes a broad view of the whole of the west coast the pattern is radial: this is consistent with the flow to be expected from an ice cap over Scotland. The pattern is most perfect in the north and the south. In the north the lochs are orientated towards the north-west in the same direction as the dykes and faults which affect the Lewisian rocks: this pattern occurs roughly north of Loch Torridon. South of Loch Linnhe the basins and lochs trend to the south-west in conformity with the Caledonian structural trend. In the central section between Loch Torridon and Loch Linnhe, ice movement would have been almost directly westwards but the structures are Caledonian. The basins and lochs are on the whole less well developed than in the northern and southern sections of the coast, while some important lochs, such as Loch Shiel, are orientated along the structural trend at an angle of about 45 degrees to the general ice movement. Similar structural control can be seen in the lochs along the Great Glen fault (Ness, Lochy and Linnhe) and in Loch Ericht farther to the east.

Most of the evidence discussed above could be most conveniently explained by selective glacial erosion, but nothing can be finally proven without much more knowledge of the behaviour of ice sheets and glaciers. At present there is some evidence for local high velocities in the steep glaciers that lead from the Greenland ice caps, through deep troughs, to the sea. Again, local high velocity ice streams are known to occur in Antarctica and have been shown to be associated with underlying troughs by seismic and gravity surveys.

Once very selective glacial erosion is initiated, presumably through being guided by a pre-existing valley, it becomes a case of positive feedback. The more the valley is deepened the steeper the gradient, the steeper the gradient the faster the flow and the greater the erosion. It has even been said that the removal of rock will bring in pressure relief effects in the valley bottom and so cause excessive jointing there and prepare the rock for glacial plucking. Many glacial troughs in Scotland are 500 m (1 650 ft) or more deep and the unloading of that thickness of rocks with specific gravity of 2·5–3·0 and its replacement by ice with a density of 0·9 must cause a considerable stress.

The cessation of the overdeepening producing rock basins, fjords or sea lochs is probably primarily due to the spreading out and retardation and melting of the ice sheet. It has sometimes been suggested that flotation of the ice in the sea may contribute to the cessation of erosion, but it can be calculated from the inferred thickness of ice in places like the Sognefjord and the relative densities of ice and water that there was just not enough depth of water available for this to be possible.

We must remind ourselves of the magnitude of the phenomenon we are trying to explain. The overdeepening of Norwegian, Chilean and Greenland fjords, i.e. the depth below the level of the submerged sill, is of the order of 1 100–1 200 m (3 500–4 000 ft). This is the height of Snowdon at least. In Antarctica overdeepening on double that scale needs to be explained. The deepening cannot have been caused by subaerial erosion. It must have been caused by stupendously great glacial erosion. It makes the 300 m (1 000 ft) of overdeepening of Loch Morar seem trivial.

Glacial breaching of watersheds

Where the glaciers are mainly confined to the valleys, the chances of diversions of the preglacial drainage occurring are great. Lakes may be dammed up between advancing glaciers and rock walls, between a main valley glacier and a tributary glacier, or by the advance of a glacier in a tributary valley across a valley not occupied by a glacier. The lake level will rise until the water can escape, unless the blocked stream can escape beneath the glacier in its path, as appears to happen in the upper Shayok valley, a tributary of the Indus, in the eastern Karakoram. This southward-flowing river was blocked in the early part of this century by the advance of glaciers in its western tributaries: at times lakes were ponded back and overflowed the ice dam, but at others the Shayok river continued to flow beneath the glacier blocking its path. Usually, however, the ice-dammed lake seems to have overflowed at the lowest available col, ripping out an overflow channel, which may be so deepened as to retain the drainage even when the ice has melted. Overflow channels are common in glaciated highlands, but examples of these features will be given in the next chapter on lowland glaciation. Most of these glacially dammed lakes are shortlived and many have left no imprint on the landscape apart from the over-flow channel. Occasionally, small terraces on the hillside mark the position of former water levels in the lakes, the best-known British example being the parallel roads of Glen Roy and adjacent valleys north-east of Fort William.

Glaciated highlands are full of anomalous drainage patterns not all of which can be explained by overflow. In many cases river capture in pre- or Post-glacial times does not provide an acceptable hypothesis, for enormous masses of very resistant rock have been cut through by very small streams and the capturing streams seem to have no advantage over the captured streams. Linton (1949, 1951) has revived the hypothesis of glacial diffluence to offer the most convincing explanations of some Scottish drainage diversions. Briefly, the hypothesis considers that if the flow of a glacier down valley is blocked, either by the entry of a side glacier or by a marked constriction in the valley form, the ice will tend to build up a sufficiently large surface gradient and consequently high velocity to ensure its flow. As it builds up, however, it may rise above the level of pre-existing cols in the valley sides and overflow into adjacent valleys. After glaciation, the drainage may revert to its former lines or it may take a new course through the glacially breached watershed.

Linton has applied these ideas to a very neat solution of the relations between the upper courses of the rivers Feshie and Geldie, between the Cairngorms and the Grampians (Fig. 13.20). It had been earlier suggested that the upper Feshie originally ran into the Geldie, but was diverted by ordinary capture following strong headward erosion of the lower Feshie (through the gorge section, A on Fig. 13.20) or by a complex chain of

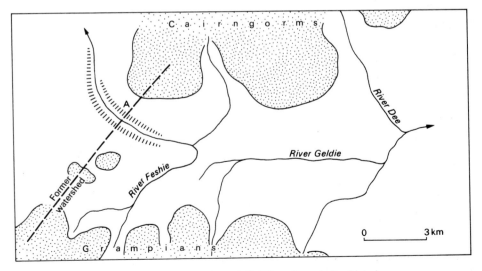

FIG. 13.20. Sketch map of upper Feshie and Geldie valleys (*after Linton*)

397

river captures and glacial diversions. The hypothesis of a glacially breached watershed is simpler and more convincing. The former continuation of the upper Feshie into the Geldie trough is apparent from remnants of an old valley floor at about 600 m (2 000 ft) extending along both sides of the Geldie trough to the head of the present Feshie, but not extending down the present Feshie beyond the gorge section (A on Fig. 13.20). During the glaciation the whole of the Geldie trough was filled with ice, which accumulated as its outlet eastwards was blocked by a powerful glacier coming down the upper Dee. The ice found a col in the former watershed at A and escaped northwards scouring out the trough now occupied by the Feshie. By projecting present slopes, Linton shows that this col was at an elevation of about 660 m (2 200 ft), i.e. 270 m (900 ft) above the height of the present Feshie valley floor at this point. The enormous deepening involved may not, of course, all have occurred during one glaciation. After the glaciation the Feshie built an alluvial fan into the Geldie trough and at various times must have flowed into both the Geldie and the present Feshie, but the outlet via the Feshie offers the steeper gradient and hence the upper Feshie became fixed in its present course. Two years later Linton (1951) applied the same type of hypothesis, with equal success, to the somewhat more complicated history of development of the rivers Tarf and Tilt, some miles south of the diversion of the Geldie.

A peculiar diversion west of Fort William on the main road and railway to Mallaig has been described by Dury (1953). Here, the Callop river does not drain eastwards into the broad open valley containing Loch Eil, but turns westwards through a steep little gorge in a ridge of very resistant rocks

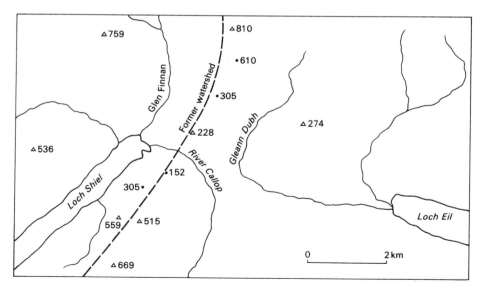

FIG. 13.21. The diversion of the Callop river, near Fort William

to enter Loch Shiel (Fig. 13.21). On the other hand, Gleann Dubh, occupying a comparable position to the Callop on the north of the Loch Eil trough, does flow eastwards. Once again, it is very difficult to account for the course of the Callop in terms of normal river capture, for there appears to be no reason why the Callop should have cut back eastwards to abstract drainage from the broad open Loch Eil trough. In Dury's reconstruction of the drainage history, he pictures a low col at 45–60 m (150–200 ft) OD in the preglacial watershed on the site of the present Callop gorge. During a retreat stage in the glaciation, a lake was dammed in the upper Loch Eil trough between the ice front to the east and the watershed to the west. This lake overflowed westwards over the col to produce the present Callop valley at about 6 m (20 ft) OD, and the gorge was later scoured by a re-advance of the ice, evidenced today by the direction of plucking and scouring on the sides of the gorge. Through this gorge, the product of glaciation, the Callop took its present course westwards. The Gleann Dubh might well have done the same, but it has become incised in the eastern side of its own fan, rather like the Feshie, impinges on a rock knob and drains eastwards to Loch Eil. If it were to recommence building up its fan it might well change its course into the Callop.

Low divides and indeterminate drainage partings are characteristic of much of the relief of the Highlands of Scotland and the examples quoted here could be multiplied again and again with more study.

To conclude, reference must be made to a most spectacular case of glacial diffluence occurring at the present time in the eastern Karakoram (de Filippi, 1932). The Rimu (Fig. 13.22), the main glacier of the eastern

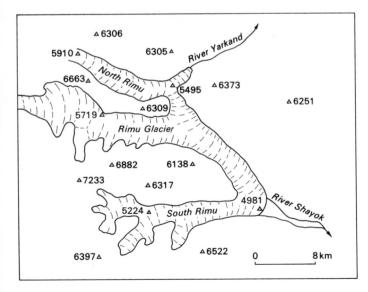

Fɪɢ. 13.22. Diffluence on the Rimu glacier (*after de Filippi*)

end of the Karakoram, consists of three branches fed by innumerable minor glaciers draining from the surrounding peaks, many of which exceed 6 000 m (20 000 ft). The northern branch, the smallest of the three, shows the remarkable example of diffluence, one lobe of the glacier going into the Yarkand valley and the other joining the main Rimu glacier lower down. Thus, this branch feeds both the Yarkand river and the Shayok river on one of the major divides in the world. The Yarkand river ultimately joins the Tarim, the combined system ending in Lop Nor east of the Taklamakan desert in one of the great interior drainage basins of central Asia. The Shayok river, on the other hand, is one of the principal tributaries of the upper Indus, which empties into the Indian Ocean. The breach where the lobe enters the Yarkand valley is about 900 m (3 000 ft) below the adjacent mountains. It would seem that conditions here are ideal for glacial diffluence: at its lower outlet the North Rimu may well have been blocked by the main Rimu glacier and overflowed a col in the watershed at the head of the Yarkand river. This valley is far less likely to have been blocked by ice, because north of the Karakoram precipitation is less and glaciers are less well developed. This is clearly seen in the superb panorama in de Filippi's book, which shows the whole of the diffluence as well as striking differences in snow conditions south and north of the North Rimu glacier. The depth of glacial scouring required is not excessive, when the intense glaciation and height of this region are considered, for it is at most little more than three times that which seems to have occurred on the Feshie. Whether the North Rimu trough, after the glaciers melt, if they ever do, would be occupied by the upper Yarkand or the upper Shayok is, of course, unknown.

14
Landforms in glaciated lowlands

In the Pleistocene the ice sheets, which covered large areas of the lowlands of northern Europe and North America, caused great modifications of the landforms there, although these effects are not as spectacular as in the highlands. The ice, which moved outwards from the mountains and other centres of dispersal, must have had a considerable velocity near the centres but there was a progressive decrease of velocity towards the margins. Thus in lowlands near the centres of dispersal eroded landforms and depositional landforms of the type caused by appreciable ice movement predominate, while near the margins the landforms are almost entirely depositional. It must be remembered, however, that during the retreat of an ice sheet depositional forms may have been left throughout the area covered by the ice, so that the simple subdivision of forms suggested above is not altogether valid. Diversions of drainage caused by the ice may also be present throughout the area formerly covered by the ice sheets.

Glacially eroded lowlands

The effect of glaciation on lowlands of ancient crystalline rocks may be the creation of an incredibly confused relief, typified by the Laurentian Shield and the Fennoscandian Shield. Both of these Pre-Cambrian shields lay near centres of ice dispersal and suffered considerable erosion. Most of the weathered rock has been stripped off by the ice and plucking and grinding have caused erosion of the solid rocks beneath. Added to this is the irregular deposition of moraine. The net result is a region of confused relief and drainage. Prior to glaciation, the drainage of such regions was probably integrated, but ice scouring and morainic deposition have resulted in the destruction of the pattern. Innumerable lakes and ponds have been formed, as a glance at a map of Finland will show. Although the number of lakes is enormous, it may have been even greater in the past, for many lakes have been filled in with various types of outwash material and later peat formation. The drainage of such areas is now effected by many small river systems terminating in lakes.

The nearest approach to this type of landscape in the British Isles is probably to be found in the Lewisian gneiss areas of north-west Scotland (Plate 44). These areas were heavily glaciated by ice moving westwards from the higher ground inland. The result of this glaciation, at its most severe, is shown in the drainage and relief pattern of west Sutherland (Fig. 14.1). The greatest elevation attained in the area mapped is only a little

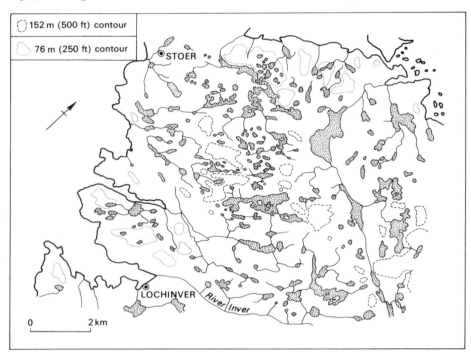

FIG. 14.1. Glaciated Lewisian gneiss country in west Sutherland. Summits shown by the 152 m (500 ft) contour inland and the 76 m (250 ft) contour near coast (*Crown Copyright reserved*)

above 240 m (800 ft) OD, and much of it is below 150 m (500 ft), but mere altitude does not give a satisfactory picture of the relief. Whether they are 100 or 250 m (300 or 800 ft) high, almost all the hills are bare, rounded, steepsided hummocks, with scanty vegetation between jointed outcrops of bare rock. Perched blocks and erratic boulders are common, many of the latter being easily detected, as they consist of red-brown Torridonian sandstone resting on predominantly grey Lewisian gneiss. Between the hills and even on flatter portions of them are numerous desolate tarns and lakes, often linked together by tortuous streams. There seems little doubt that these features are mainly due to heavy glaciation, but the effect of the rock

PLATE 44. Glaciated Lewisian gneiss from Stac Polly, Ross and Cromarty. Almost horizontal Torridonian sandstone forms mountains rising abruptly from the gneiss, Stac Polly in foreground, Suilven in middle distance and Quinag in far distance

itself must not be forgotten. The Lewisian gneiss is broken up by north-west to south-east dykes and north-east to south-west faults into a rectangular pattern, which, together with the coarse jointing of the rock, must have played a part. If the Lewisian areas are compared with adjacent areas of Torridonian sandstones at about the same elevation, the different texture of the relief becomes immediately apparent. North-west of the area shown on Fig. 14.1 the peninsula culminating in the Point of Stoer is formed entirely of Torridonian rocks and presents far smoother relief than the Lewisian areas, although it must have been equally severely glaciated. The same contrasts may be seen further south, west of Ullapool, in the neighbourhood of Gruinard Bay, where the boundaries between the Lewisian and Torridonian may be almost plotted from a boat lying offshore. These lithological controls on the effect of glaciation are probably most clearly seen at Stattic Point, between Gruinard Bay and Little Loch Broom, where the relief texture on a small inlier of Lewisian gneiss clearly distinguishes it from the smooth slopes of the surrounding Torridonian sandstones.

There is also the possibility that this Lewisian surface is a stripped surface of some sort, in which other processes effected the main denudation and glaciation merely did the stripping. Near the Torridonian outcrop it may have been exhumed from beneath those rocks and hence be an exhumed surface of Pre-Cambrian age. It is also possible that much of it might be an exhumed weathering front of Tertiary age (see Chapter 12). It was noted in discussing the selectivity of glacial erosion in Chapter 13 that apparently tropically weathered material is preserved in places in Scotland. In addition there are, intercalated in the lavas of western Scotland, which are of Tertiary age, highly aluminous soil horizons and plants which indicate at least a subtropical climate. If such weathering had acted on the Lewisian gneisses, which are fractured, jointed and split by dykes, it is highly likely that a very irregular weathering front would have been produced which has now been stripped out and slightly modified to form the apparently heavily glaciated relief in question.

Depositional landforms in lowlands

The depositional landforms found in glaciated lowlands may be subdivided in several ways. In the British Isles there is a clear division between the forms found on the Newer and Older Drifts, two useful terms designating respectively material left by the last glaciation and material deposited in earlier glaciations. In Great Britain the limit of the Newer Drift is a line extending approximately from York to Liverpool and thence south in a loop almost to the coast of South Wales (Fig. 14.2). South of that line the glacial deposits are largely featureless plateaus, typical constructional forms being present only in a very few areas such as north Norfolk, im-

Fig. 14.2. Approximate margins of Older and
Newer Drifts in Great Britain

mediately inland from Cromer and Sheringham. North of the margin of
the Newer Drift the glacial deposits still preserve in many places their
original forms. The last glaciation was not a simple advance and a retreat,
but included minor readvances in the later phase of predominant retreat:
the two most important of these are the Highland and Perth readvances.
The result is that, in general terms, the depositional forms become fresher
and fresher as the highlands are approached, especially the Lake District
and the Highlands of Scotland, because near these highlands and in valleys
within them forms associated with the last readvances are preserved. But,
although this division into Newer and Older Drifts is significant in the
relief of Great Britain, it probably reflects to a considerable extent not
merely original differences in relief forms but also differences due to subsequent erosion, either by normal subaerial or by periglacial processes.

A distinction might also be made between landforms in the zone covered
by the ice and landforms in the outwash zone beyond the ice front, but the

most useful difference seems to be that followed by Flint (1957). Two categories are recognised: forms composed of unstratified drift and those formed of stratified drift. Although the distinction is primarily a geological one based on lithology, it reflects differences in the formative processes. Forms composed of unstratified drift owe their origin almost entirely to ice action alone, while forms composed of stratified drift reflect the fact that in their construction morainic material has been resorted by water action. Most of the latter forms are found at and beyond the ice front, but not all of them, for water action may be very significant within the ice especially during the retreat and decay of an ice sheet. It would perhaps be better to class these deposits as stream deposits of a special type because their affinity to stream deposits is closer than to the unstratified drift laid down by ice sheets. They show great variability because of the highly variable discharge, which results from seasonal melting of the ice, and also probably because of the erodibility of their unconsolidated, unvegetated banks.

Forms composed of unstratified drift

The most widespread deposit of unstratified character is till, often termed boulder clay (Plate 45) in this country. Over much of East Anglia and the Midlands it consists essentially of a stony blue-grey or brown clay, which locally reaches thickness of 60 m (200 ft) or more, but is generally much thinner. It is essentially material carried by the ice sheet and deposited beneath it, by a process which is usually termed lodgment. The weight of the overlying ice ensures that the till is compacted to a considerable extent and in presentday sections it appears as a tough, roughly jointed or fissured material. Sometimes it may be overlain by a coarser, more sandy material, which represents ablation moraine formed on the surface of the ice sheet and let down on to the ground moraine as the ice sheet melted. The sandier character of this ablation moraine is due to the fact that the finer constituents have been washed out by trickles of meltwater during its formation (Flint, 1957).

The lithological character of boulder clay to a great extent reflects the rocks over which the ice has travelled. Hence, the boulder clay of the eastern parts of England owes its blue-grey or brown character largely to the broad outcrops of Lias, Oxford, Kimeridge and Gault Clays. Where the rocks traversed by the ice sheet are very different, the boulder clay may have an entirely different character. For example the boulder clay adjacent to a large granite area may be formed largely of blocks of granite in a debris of coarse sand and fine gravel.

The direction from which the ice sheet has advanced can be deduced in two different ways. The first is by tracing the erratics in the boulder clay to their sources. Many glacial erratics are hard nondescript quartzites and sandstones and it cannot be said with any certainty whence they came, but

PLATE 45. Contorted glacial till with chalk erratics, West Runton, Norfolk

there are a few particularly distinctive rocks which may be referred to their sources. In drift deposits round the Irish Sea are often found fragments of riebeckite-microgranite, which can be traced to the rock of Ailsa Craig in the Firth of Clyde. Shap granite, with its distinctive phenocrysts of pink felspar, may be recognised among the erratics in drift east of the Pennines. In East Anglia rhomb-porphyry, a syenitic dyke rock from Norway with rhomb-shaped felspar phenocrysts, is a readily recognisable erratic in the drift, while purple and grey porphyritic rocks are referable to the Devonian and Carboniferous igneous rocks of the Central Lowlands of Scotland. Many rocks, unfortunately, give such a wide choice of sources as to be of little use: thus, a fragment of garnet-mica-schist, the discovery of which in the drift of the east of England may seem exciting, offers one the choice of large parts of Scotland and Norway as its source. In addition, erratics in boulder clay may have been derived from an earlier boulder clay, so that conclusions based upon them must be guarded. Nevertheless, a study of the overall composition of the included erratics seems to be a useful means of differentiating one boulder clay from another, as has been shown by Baden-Powell (1948).

The second method of determining the direction from which the ice advanced is by measuring the orientation of the stones within the drift. It has been found that these usually have their long axes parallel to the direction of ice movement, a fact indicated by their orientation in relation to that of drumlins. Conclusions about ice movement derived from this method agree closely with those derived from a study of erratics (West and Donner, 1956).

The essence of the relief of till deposits is monotony. Where the sheets are young and sandy or gravelly they may be diversified by kettleholes, small closed depressions formed by the melting of ice blocks left in the moraine and the subsequent collapse of the latter. Where the sheets of boulder clay are old, they tend to form a plateaulike surface, which is often smoother than the relief on the solid rocks beneath. For example, between Cambridge and St Neots lies a smooth-surfaced plateau varying in elevation from about 80 m (270 ft) OD near its centre to about 55 m (180 ft) at its margins. The surface is almost entirely formed of boulder clay, which in one place in the south-east fills up a valley cut down to about 15–30 m (50–100 ft) OD so that it has no expression in the present surface relief.

The formation of till plateaus such as these presents a number of problems. They are often generally smoother than the underlying relief and it is tempting to regard them as the results of glacial deposition in the production of a smooth, streamlined form over which the ice sheet could ride easily. Yet such an explanation is difficult to square with the absence of such forms in front of major escarpments, such as those of the Chalk and Jurassic limestones, which were overridden by ice sheets. It would seem that the front of an escarpment would be the ideal place for deposition by lodgment. It could be that the smooth form is the result of normal subaerial erosion after the deposition of the till, but the degree of smoothing involved seems to be excessive for the time available. It might also be that the till was originally more mobile, because it was saturated, and subsided gently into a smooth form under its own weight. Once it had become consolidated it was difficult to erode because of its clay content.

The processes by which till is eroded from a glacier bed, carried up into a glacier, and finally deposited are being slowly clarified by painstaking observations on modern glaciers in places such as Spitzbergen (Boulton, 1970a, b). The glaciers in this area are essentially cold glaciers, ones in which the temperature is normally below the pressure melting-point so that meltwater cannot exist at depth within them; they contrast with temperate glaciers where the opposite holds good. It has been suggested that debris might be picked up from the beds of these glaciers either by being frozen into the sole of the glacier or by being thrust up from the floor in zones of compressing flow. Boulton argues that the former is the effective process in picking up debris. Where projections in the bed occur, local pressure melting releases meltwater which refreezes down-glacier from the projection. Thus debris on the bed is frozen into the glacier sole. The result is a stratification of debris layers roughly parallel to the glacier bed, the layers frozen in highest up-glacier being naturally uppermost. In suitable conditions it can be shown by erratic content, which can be related to the outcrops over which the glacier has flowed, that this has in fact happened. Where compression occurs, principally where there are obstructions in the bed, upward thrusts transfer debris from these bottom layers into the body of the glacier, but they are not thought to slice it up directly from the bed.

According to Boulton large quantities of debris carried englacially imply cold glaciers because only these freeze to their beds so providing the essentially picking-up mechanism. In temperate glaciers, in which melt-water operates at the base, most of the englacial debris must have fallen in from the valley sides because it cannot have been picked up by basal freezing. If this is correct it must mean that the Pleistocene ice sheets of Britain were cold, because they contained large quantities of debris and there was little opportunity in lowland Britain, from which most of the till was derived, for debris to be dropped in from above as the ice sheets largely obliterated the existing relief thus removing the source of the debris.

Till deposition results from both surface melt and bottom melt and in cold glaciers the former, according to Boulton, is much more effective. Basal tills are likely to be formed where the lower debris-rich layers are retarded either by a physical obstruction or through the debris content decreasing the plasticity so much that the clearer ice above flows over them. Tills deposited thus are liable to be compacted by the pressure of the ice flowing over them squeezing out their water content. Thus they are compacted before they are exposed from beneath the ice, an idea very different from that of the settlement of a mass of till under its own weight suggested above.

409

The edges of former ice sheets are often marked by terminal or end moraines, which are ridges of glacial material not usually exceeding 45–60 m (150–200 ft) in height in lowlands. In plan they often form a series of crescents, corresponding with the lobes of the ice front. They may mark not only the limits of the farthest advance of the ice but also the various positions reached during later readvances. The material in the moraines may be partly stratified and partly unstratified, for some moraines have been deposited in water, which has resulted in some stratification. Usually they are not continuous, for they have been cut through by meltwater streams and later rivers. A well-developed moraine probably indicates that the ice front occupied that position for a considerable length of time, thus allowing the accumulation of material from a considerable mass of ice. Not all former ice fronts are now marked by terminal moraines, because the moraine may have been destroyed as fast as it formed by meltwater, or the ice may have stood there for an insufficient length of time for the formation of a marked moraine, or an inconspicuous moraine may have been destroyed by later erosion. The margin of the Newer Drift is marked by an end moraine near York and by conspicuous moraines in the Neath and Tawe valleys in South Wales, but the edge of the Older Drift through Essex and north of London is not defined by a terminal moraine. In Great Britain the best examples of terminal moraines are not to be found in lowlands but in the valleys, especially in the Highlands of Scotland, where they occur at the limits of the readvances of the last ice sheet.

Push moraines are a specialised form of end moraine caused by a readvance of an ice sheet thrusting till, or some similar deposit, up into low ridges. Modern push moraines are to be found in front of present glaciers in Spitzbergen, while the Krefeld-Nijmegen moraine is a Pleistocene example from the North European Plain (Charlesworth, 1957).

Drumlins (Plate 46) are elongated, streamlined hills of drift, which have often been likened to upturned boats (Fig. 14.3). They betray the direction of ice flow in that the steeper and blunter, or stoss, end faces the direction from which the ice came. The lee end is more pointed and gentler in slope. Flint (1957) states that the ideal drumlin is 1 000–2 000 m long, 400–600 m wide and 15–30 m high (50–100 ft high). But there is considerable variation in both size and shape. The shape in particular may depart so much from the ideal form as to become little more than a rounded mound of drift. Features of drumlin type, but without the ideal form, have been called drumlinoid features.

Chorley (1959) has suggested that the degree of elongation of drumlins is really a pressure effect and hence ultimately a reflection of ice velocity. The aerofoil section of the wings of high speed aircraft is more elongate than that of slower aircraft, another manifestation of the same general

PLATE 46. Drumlins near Carnagall, Antrim. The cultivated drumlins are clearly distinguishable from the uncultivated marshy land between them

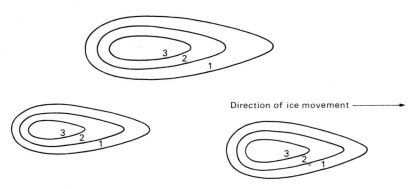

Direction of ice movement ⟶

FIG. 14.3. Idealised form of drumlins

phenomenon, while, if one wishes for a biological analogy, the eggs of small birds which lay large eggs are usually very elongate, presumably because of the pressure involved. This explanation accounts for the shape in the conditions in which streamlined forms are produced. A full understanding of the conditions in which such isolated streamlined forms are produced eludes us at present.

This is partly because there is such a range of composition in drumlins. The typical forms are composed of drift and it is these which have concerned us till now. But there are also pure rock drumlins and drumlins which consist of veneers of drift smoothing over pre-existing rock forms: many examples of both types are described by Sissons (1967) from the southern half of Scotland. Such forms are especially well developed where the strike of the rocks coincides with the direction of the ice movement, as in the areas of Carboniferous rocks in the Tweed valley around Coldstream, where the ridges have been 'drumlinised' over considerable areas. It is also suspected that some drumlins, with stratified drift in their cores, have been moulded by ice re-advances from sheets of bedded outwash or similar deposits: these are really only a special case of a rock drumlin, a soft rock drumlin.

Hence the explanation of drumlins which assumes that there was originally an irregular distribution of till in an ice sheet, that the clots of till caused local retardation of the ice sheet while the clearer areas of ice advanced more rapidly, so smoothing into streamlined shapes the clots of till, hardly seems to be of general applicability. It may be, however, that rock projections would cause the basal layers of the ice to diverge and so convert them into rock drumlins.

Whatever the explanation, the geographical distribution of drumlins, primarily in areas near highlands where it can be safely assumed that the ice sheet had relatively high velocity, indicates that speed of ice movement must be fundamental in their formation.

Usually drumlins do not occur singly but in drumlin fields, examples of which can be found in Northern Ireland around L. Neagh (Charlesworth, 1953, fig. 82). Drumlins are also common in the Eden valley between the Lake District and the northern Pennines and again south of the western half of the Southern Uplands of Scotland. The landscape of mounds and irregular drainage they present is in some ways comparable with the Lewisian gneiss country described above, but differs in being usually composed of drift and not of rock outcrops.

The feature known as crag and tail is somewhat similar to a drumlin with a rock core. Instead of an overall covering of drift, the glacial material has been deposited as a tail in the lee of the crag (Fig. 14.4). An exceptionally well developed tail stretches from Slieve Gullion, north of Dundalk, south-south-eastwards for a distance of two miles (Charlesworth, 1953).

FIG. 14.4. Crag and tail

Not all crag and tail features have the composition shown in Fig. 14.4, a fact that Sissons (1967) reminds us was observed more than a century ago by Geikie. Many of the crags and tails of the Scottish Lowlands are formed where volcanic plugs are intruded into much weaker Carboniferous sediments, for example the classic case of Castle Rock, Edinburgh. In many of these examples the tails are smooth, moulded ridges of Carboniferous sediments, with perhaps merely a patchy film of drift over them, and not the moulded mass of drift shown in Fig. 14.4. Such a composition makes them very close indeed to rock drumlins.

Forms composed of stratified drift

At the edges of an ice sheet, especially during the decline of any glaciation, there will obviously be large volumes of meltwater. These streams, many of which originate within the ice sheet, are responsible for sorting and redepositing much of the material laid down as terminal moraines and boulder clay. When it is remembered that the zone in which meltwater is dominant probably retreats back over the whole area once occupied by the ice sheet, it will become clear that considerable quantities of unstratified drift will be converted to stratified outwash (Plate 47).

Streams issuing from the ice front will be very heavily laden with debris derived from the ice, so that small decreases of gradient will provoke deposition from the streams of the coarser parts of their load. Near the ice front the deposits may be very coarse indeed, as boulders the size of packing cases are not uncommon. The material deposited will become finer

413

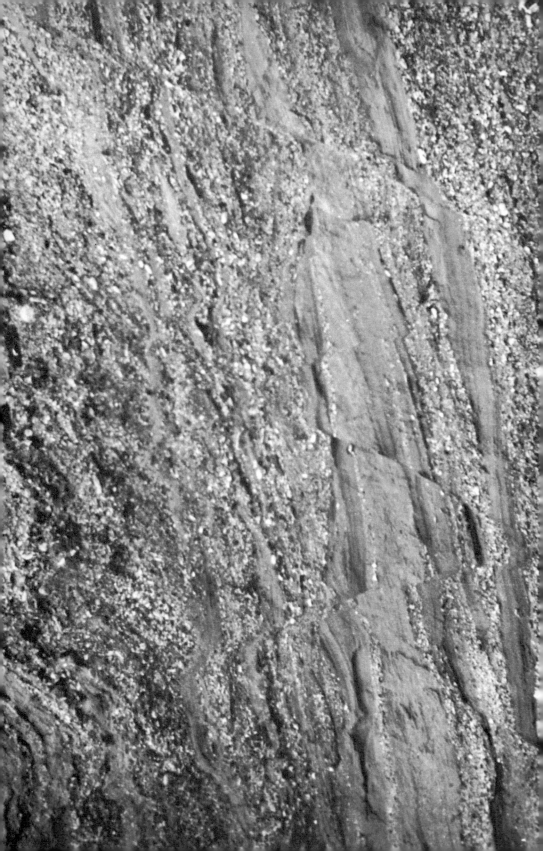

and finer away from the ice, although the finest material may well be transported to the sea unless its deposition is promoted by lakes in front of the ice. Such lakes are referred to as proglacial lakes. Most outwash material is well bedded, although current bedding is very common. It can usually be distinguished from stream gravels by a number of features. The size of many of the contained stones and boulders often indicates, especially when the gradient of the stream was obviously not very great, that a volume of water far greater than that available in interglacial and Post-glacial times was present when the gravel was deposited. Outwash often contains flattened lumps of clay, which were frozen when deposited, whereas such material would rapidly disintegrate in a normal river. Freezing may also account for the presence in outwash of Chalk districts of quantities of Chalk pebbles, which are rare in present streams there presumably because they are dissolved before they have had a chance to be moved any great distance. The surface of outwash gravels is often pitted with kettle holes, where contained blocks of ice have thawed out. Finally, many well-bedded outwash gravels are interrupted by thin beds of boulder clay, indicating that there were alternate local advances and retreats of the ice front.

The finest parts of the outwash, deposited in proglacial lakes, may form varves. These are alternations of very fine and slightly coarser deposits and represent seasonal variations in deposition. During the warmer season silt and clay will be brought into the lake, as stream erosion will be greatest then. The silt will settle immediately but the clay will remain in suspension longer, probably settling in winter when no more material is being brought into the lake. Thus, each varve, comprising the coarser and finer layer, represents one year. On this basis, a chronology for the latest phases of the Pleistocene has been established in Sweden by counting the numbers of varves in old proglacial lakes. No single lake contains a complete record in its varves, but overlapping sets of varves in different lakes can be correlated by their rhythmic pattern assuming that the climate has been constant over the area concerned. There are difficulties in interpreting varves, especially in deciding what each pair of varves means (West, 1968).

The surface form of outwash varies considerably. Obviously in the simplest example of outwash deposited by a single stream issuing from the ice front, the form will be a subdued fan. In lowland regions with practically no relief the coalescence of a series of such fans will give rise to a vast gently sloping outwash plain. Examples of such forms are rare in the British Isles, though it is probable that the Kelling and Salthouse outwash plains north of Holt in Norfolk represent somewhat smaller forms of these features (Sparks and West, 1964).

But where there is any relief outwash will tend to be confined to preexisting valleys, except where it has been deposited between lobes of ice,

PLATE 47. Contorted and faulted outwash gravel near Hitchin, Hertfordshire

which effectively masked the underlying relief. Unless the deposition of boulder clay has completely altered the relief, meltwater streams will tend to follow earlier valleys and to convert the boulder clay there into outwash deposits. Thus, in the area around Cambridge boulder clay is rare in the major valleys and outwash more common. For instance, in the valley through the Chalk south of Cambridge, which is occupied in its northern part by the Cam and in its southern part by the Stort, a tributary of the Lea, outwash deposits are common especially on the divide between the two streams, while boulder clay is rarer, and where it occurs, much thinner than on the adjacent hills.

In mountain regions, such as Scotland, outwash is very common in the valleys below the moraines. At one stage it probably formed a smooth flat floor to the valleys, but it has since been dissected by streams into flights of terraces. Outwash of this type has been termed a valley train (Plate 5). Terraces cut into valley trains of outwash are not always easily distinguished from kame terraces (see below). If, however, the bedding can be seen it should remain horizontal right to the front of the terrace, if the terrace has been cut from an outwash spread, whereas, in a kame terrace, the front of the terrace should be slumped and contorted, as it collapsed after the ice withdrew.

Outwash may be deposited from streams flowing from ice sheets in any stage, but some of the other forms of stratified drift are more likely to be associated with stagnant or dead ice sheets. It is possible for an ice front to remain stationary in two sets of conditions. The first is where the rate of advance just balances the rate of melting, so that the ice appears stationary. It is essentially active, otherwise the thawing would induce a rapid recession. The second is when the supply of ice has been cut off, so that the ice sheet is truly stagnant.

Under the latter conditions general thawing produces considerable quantities of meltwater and the formation of a bewildering mess of deposits. Marginal or lateral moraines may be formed; ablation moraine may be formed at the surface and washed down into cracks; material may be washed out of the snout of the ice sheet to form other sorts of marginal deposits; a sort of ice karst may be formed with open crevasses, solution holes and subglacial tunnels being filled with bedded, glacier stream deposits. In the final stages these various deposits will be further modified by meltwater and possibly periglacial action so that the classification of all individual forms could be very difficult, and the whole perhaps best classed as a moraine complex.

It should be noted that with moving ice, but a stationary front, holes and tunnels in the ice should be kept closed by the movement of the ice, but that with stagnant ice they will remain open. Therefore, deposits produced in holes and tunnels should, at least theoretically, be associated with stagnant ice. Deposits produced in these conditions in holes, chasms, and at the edge of the ice are usually referred to as kames of one sort or another, while deposits produced in tunnels beneath the ice are known as eskers.

Kames and kame terraces are forms composed of stratified drift developed in actual contact with the ice. Ideally, kames originate from small deltas or fans formed at the ice front. The front of these features is, therefore, a slope reflecting the angle of the foreset beds of the delta or the angle of rest of the waste fan, while the back is an irregular slope, formed by slumping when the ice withdrew and usually called the ice-contact slope. Other small ridgelike kames may have originated as the fillings of crevasses or holes in the ice sheet. The latter, which are often called perforation or moulin kames, are really only a special case of crevasse kames or vice versa. Although these forms are recognisable when their initial shape is well preserved, they may degenerate slowly, through the action of erosion, to shapeless mounds of gravel, at which stage it is very difficult to determine their origin and to distinguish them from moraine complexes.

Kame terraces originate as terraces along the sides of a tongue of ice still occupying a valley (Fig. 14.5). Heat absorbed by the rocks on the valley sides may be responsible for melting the ice touching the valley sides and a stream, depositing gravel, may occupy the trough between the ice and the valley side (Fig. 14.5A). After the ice has retreated the gravel deposited by these streams will collapse at the front, but may retain sufficient flatness adjacent to the valley sides to form a kame terrace (Fig. 14.5B). A series of kame terraces may be developed if the ice tongue in the valley wasted intermittently. A whole variety of kame features, including compound crevasse fillings, perforation kames and dissected kame terraces occurs in and near the Glaven valley, north-east of Holt in Norfolk (Sparks and West, 1964).

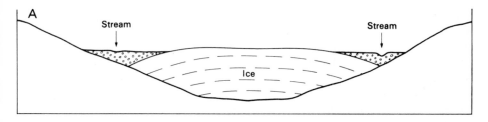

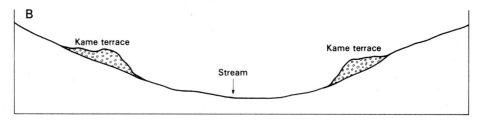

FIG. 14.5. Formation of kame terraces

417

Crevasse and perforation kames will generally tend to have accordant surface elevations if they have been formed contemporaneously within an ice sheet, because stagnant ice often has a freely communicating water table. Patterns of kames are common at the edges of the Southern Uplands of Scotland, where they are often associated with intricate patterns of melt-water channels (Sissons, 1967). Excellent examples occur in the Nith valley a few kilometres north of Dumfries.

Eskers are long gravel ridges, sometimes likened to railway embankments, which wind over the surface of glaciated areas with little regard to relief. The ridges may not be continuous, but broken by various gaps, which do not, however, destroy the essential continuity of the features. The eskers may vary in height from almost imperceptible ridges to marked forms up to 30 m (100 ft) or more in height. When they are derived from a recent glaciation the sides are steep, approaching the angle of rest of the debris composing them.

Eskers are very common in Eire where they are often so continuous as to provide the sites for routeways across the peat bogs lying between them. In Scotland they are well developed on the sides of the Moray Firth, for example around Dornoch and Golspie.

Eskers may not all have a similar origin and a number of theories have been put forward to explain them. Those which do not follow the present relief but wind uphill to cross ridges, usually at their lowest points, cannot have originated on the ice sheet as surface meltwater streams. Where eskers do not go uphill they may be explained as deposits laid down in gorges within the ice. Those formed in this way must approach crevasse kames, so that the two may not be readily distinguishable. Theoretically the kame should be short and the esker long, but there are long kames and short eskers. Again, the kame should be flat-topped, but it needs only a little subaerial degradation to round this off. It is likely that most eskers were formed by meltwater flowing in tunnels beneath the ice under hydrostatic pressure. This hypothesis explains their uphill and downhill nature and also the fact that they generally seem to use the cols over ridges, a position which would presumably be advantageous even to a subglacial stream. Had they been superimposed from meltwater streams on the surface of the ice there is no reason why they should go up and down hill and no reason why they should so often have been superimposed across cols. One other fact supports their subglacial origin: they are often associated with patterns of meltwater channels, now interpreted as subglacial (see below), in Scotland. It is still difficult to understand why these masses of coarse and readily erodible drift were not more largely washed away on the final melting of the ice sheets.

Beaded eskers, which are rarer than normal eskers, broaden out at closely spaced intervals. These may have originated in the following way. Initially there may have been an outwash cone or fan at the ice front, a feature which if left at that would now form a kame. But as the ice retreated, the cone was added to from behind so that the ice left behind a more or less continuous gravel ridge. If the recession of the ice front was punctuated by halts, those halts would be represented by local broadenings of the esker, where the stream issuing from the ice had opportunities to build up a bigger fan. In this way beaded eskers may have been formed. Some eskers are thought to have been formed by ice in proglacial lakes. The beads on these eskers represent deltas formed at the ice front during halts in its recession. It should be possible to distinguish between ideal examples of these types by a careful study of the bedding of the gravel.

Eskers, like terminal moraines, are features which extend for considerable distances, but the two may be differentiated in several ways. Terminal moraines are largely forms of unstratified or very roughly stratified drift, whereas eskers, especially if they have been formed in proglacial lakes, should be much more regularly bedded. In addition, terminal moraines lie at right angles to the direction of movement of the ice, whereas eskers are approximately parallel to it. As it is usually possible to determine the direction of ice movement, the two forms may be separated on this criterion, though certain moraines, lateral and medial moraines, may be orientated in the same direction as the eskers. The relationship between the direction of the esker train and the successive positions of the ice front, often marked by kames in this area, may be seen between Wolverhampton and Newport (Fig. 14.6) (Whitehead *et al.*, 1928).

It may be mentioned that there are deep valleys beneath the drift in places in East Anglia, which also have reversed slopes in places, and may be due similarly to streams under hydrostatic pressure. One of these valleys occurs beneath the upper reaches of the Cam and the Stort between Saffron Walden and Bishops Stortford. Excavated in chalk, it has side slopes which are of the order of 45–50 degrees in places, whereas the steepest slopes on exposed chalk relief are very little above 30 degrees. The bottoms of some of these valleys are well below sea level, the channel along the upper Stour between Clare and Long Melford in Suffolk reaching a depth of almost − 105 m (350 ft) OD. These depths, together with the steep side slopes and the locally reversed gradients, make it highly unlikely that these valleys can have been the products of normal subaerial streams. But why subglacial streams, under hydrostatic pressure, should form eskers in some places and deep valleys in others is not clear.

Glacial diversions of drainage

We have already discussed two forms of glacial diversion of drainage. In Chapter 13 the glacial breaching of watersheds was considered, while at the beginning of this chapter the disintegration of drainage which can ensue from the glacial scouring of old shield lands was illustrated from the Lewisian gneiss of north-western Scotland.

In the lowlands and the uplands the classic hypothesis of glacial diversion

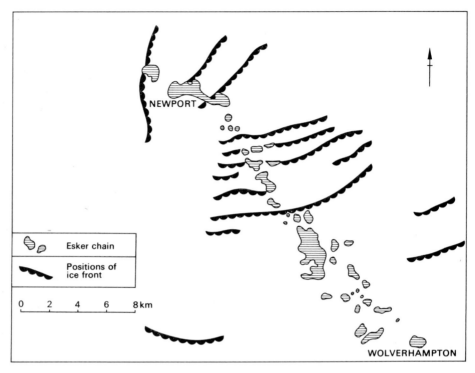

Fig. 14.6. Relation between esker train and successive positions of ice front, retreating north-westwards, near Wolverhampton (*after Whitehead et al.*)

is the formation of an ice-dammed lake and its subsequent overflow. The over-enthusiastic application of this idea has led to its being questioned in the last twenty years and replaced more and more with the idea that many meltwater channels are subglacial. Yet glacial overflow channels occur in areas not covered by the ice and these obviously cannot be subglacial. Accordingly a number of cases of the classical type are discussed first before the subglacial hypothesis is considered.

When ice blocks a valley a lake will probably be formed in that valley and rise in level until it overflows at the lowest point, which may be on the ice blocking the valley but is more usually a low point on the valley sides. The large volume of water escaping ensures that the overflow channel is vigorously deepened in a comparatively short space of time. Such overflow channels usually have steep sideslopes, forming an abrupt angle with the

PLATE 48. The glacial diversion of the Afon Soch, Caernarvonshire. The Afon Soch is thought to have flowed originally into the sea at Port Neigwl, but this outlet was blocked by ice and the impounded waters overflowed through the ridge to the left forming the gorge shown both on this plate and on Plate 49. After the ice melted the river retained its diverted course in spite of the obvious and easier exit into Porth Neigwl

Porth Neigwl

Afon Soch

Gorge

higher ground above, and flat floors (Plates 49 and 50). If they are again abandoned after the glaciation, they do not fit in with the general drainage pattern of the region, but remain as abandoned channels across watersheds and ridges. Time will soften the angles of such valleys, so that overflow channels related to glaciations earlier than the last may be subdued in form.

It has been questioned whether ice would form a solid enough dam to impound a lake, especially when it was in a stagnant condition. There seems little doubt that active ice can impound lakes, for such features have been recorded from both the Alps and the Himalayas. Sissons (1967) conceives the lakes as being often only narrow marginal features with the majority of the water stored within the voids in the decayed outer layers of ice and held there by the firm ice deep within the ice sheet (Fig. 14.7).

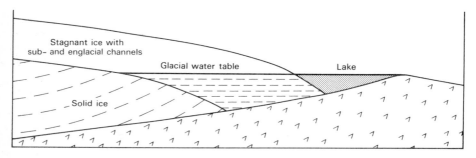

FIG. 14.7. Possible form of an ice-dammed lake (*after Sissons*)

Thus the lake is continued by the englacial water table. The effect on overflow channel formation will be the same, but the picture is probably nearer the truth than the simple one of the whole of the lake being impounded between an impermeable ice front and the containing rocks.

Ideally, such a sequence of events will leave three principal types of evidence of its existence. The first is an accumulation of lake sediments on the site of the lake. The stream entering the lake, near the edge of the ice, will probably have a considerable load of which the finer particles will settle slowly, forming laminated clays or silts. These deposits may, however, be covered by later beds, for example by boulder clay if the ice advanced after the formation of the lake, so that their existence is only known if pits or borings happen to be made in suitable areas, or the lake deposits may be removed by later erosion, either glacial or normal subaerial.

The second type of evidence to be sought is the presence of lake terraces cut into the sides of the valley occupied by the lake, but the lakes may be very temporary and the rocks very hard with the result that only a very

PLATE 49. The Afon Soch Gorge, Caernarvonshire: a glacial overflow channel (see also Plate 48)

minor terrace, if any at all, is formed. In an ideal example the terrace should end at the ice front, the position of which would be marked by a moraine. It is possible that the idea of lake terraces has been overstressed. It would take time for the smallish waves on most proglacial lakes to generate terraces, and if the lakes were temporary with steadily rising level, even though this were subject to seasonal variations, it is difficult to see why terraces should be formed, even at the highest level of the lake, because, once the overflow channel starts to form, the lake level is presumably progressively lowered. It may be that freeze-thaw effects at lake margins or ice push effects when the lakes froze in winter had something to do with the few lake terraces known. This may help to explain the fact that the clearest glacial lake terraces, the parallel roads of Glen Roy and adjacent glens north of Ben Nevis, have been formed in some of the most resistant rocks in the country.

Thirdly, the most permanent form of evidence is usually the overflow channel, which is more of a marked feature of the landscape than either of the other features.

In country of very subdued relief aggradation by streams blocked by ice may be sufficient to divert them into new courses, and lakes may be very temporary and small affairs and leave little or no evidence of their former existence.

In glaciated areas the possibility of glacial diversion offers a more likely explanation of anomalous drainage than superimposition, antecedence and capture. But the possibilities of the occurrence of these others must always be borne in mind, and it may be that the present drainage pattern is the result of the combined action of preglacial captures and glacial diversions.

The most clearly preserved records of glacial diversions of drainage in this country are in the Newer Drift areas. Systems of proglacial lakes blocked in valleys and connected by overflow channels through spurs are known down much of the eastern side of the Pennines, but the classic example of such features is to be found in the Cleveland Hills district (Kendall and Wroot, 1924).

During the last glaciation the Cleveland Hills were not glaciated, even though they rise to heights of 420 m (1 400 ft) OD. The reason probably lies in the fact that the Cleveland Hills are smooth Jurassic moorlands with no deeply marked valley heads, so that conditions were unfavourable for the accumulation of snow and the development of corrie glaciers. Instead, they were surrounded by ice from other centres which filled the North Sea and the low ground between this district and the Pennines (Fig. 14.8). Although this ice sheet stood at elevations of about 300 m (1 000 ft) OD in places, it did not override the Cleveland Hills.

PLATE 50. An abandoned glacial overflow channel on the Dee-Don watershed, Aberdeenshire

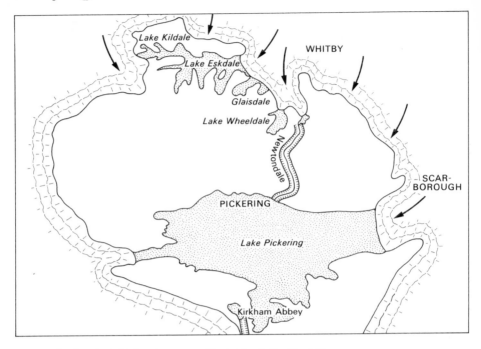

Fɪɢ. 14.8. Proglacial lakes in eastern Yorkshire (*after Kendall*)

The northern part of the hills are drained by the Esk valley running eastwards to Whitby and its numerous right bank tributaries which trench the northern face of the moors. Ice blocked the eastern end of the Esk valley while, to the west, the Leven valley (Kildale) was similarly blocked by ice in the Vale of York. In both valleys proglacial lakes developed which later united as their level rose to 225 m (750 ft) OD, considerably higher than the watershed between Kildale and Eskdale. At the same time the lake extended up all the southern tributaries of the Esk so that they were transformed into branches of the lake. This lake found the lowest point across the main Cleveland anticline at what is now the northern end of Newtondale. Over this col they overflowed southwards away from the ice. An overflow channel of this type leading directly away from the ice has been termed a direct overflow. Newtondale could be cited as the type example of an overflow channel. Its straight sideslopes form an abrupt angle with the moors above and in many places it is flat floored, though this latter feature, as in many glacial overflow channels, is to a considerable extent the result of later peat development. Its course is sinuous and might be described as meandering on a large scale.

As the floor of Newtondale was cut down, the level of Lake Eskdale fell and the spurs separating the valleys reappeared to divide the lake into a number of smaller lakes hemmed in between the ice front and the hills, this stage being shown in Fig. 14.8. With slight recessions of the ice front parts of spurs lower than the lake levels would have been exposed, thus allowing the lake waters to escape across them. At any stage a series of lakes may have been connected across the spurs, such a contemporaneous series of overflow channels being called an aligned sequence. Also as the ice receded each spur may have been trenched by an overflow channel at each stage, thus producing a parallel sequence of channels (Fig. 14.9A). Some of the overflow channels were direct cuts in the rock, while others were marginal channels between the rock on one side and the ice on the other (Fig. 14.9B). When the ice disappears such a channel is left merely as a

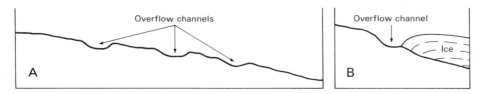

FIG. 14.9. Types of overflow channels (for explanation see text) (*after Kendall*)

shelf on the hillside. Both types are known in the Cleveland district. The later history of the Eskdale series of lakes involves periodic falls in lake levels as the ice receded and allowed overflow to take place, until finally the level of the lake at Goathland, at the northern end of Newtondale, stood at little above 150 m (500 ft) OD. At about this level the cutting of Newtondale was retarded by the presence of a bed of hard grit, and soon afterwards the ice retreated sufficiently for the Eskdale waters to escape eastwards between the ice sheet and the moors, thus causing Newtondale to be abandoned.

The evidence for the lakes in Eskdale and Kildale has been questioned by Gregory (1965), who suggested an interpretation based on stagnant ice and subglacial channels to explain much of the complex detail of lake and overflow channels implicit in the ideas of Kendall and Wroot and outlined briefly in the last paragraph. This sequence of events suggested by them for the area north of the Cleveland Hills illustrates the dangerous thinness of evidence: the position of overflow channels is used to infer the location of lakes and from the location of lakes the ice front position is inferred. We shall revert to this later. It must be stressed however that Newtondale cannot be a subglacial channel unless all the suggested ice front positions are wrong, so that the main plank of the reconstruction remains.

Newtondale discharged southwards into the Vale of Pickering (Fig. 14.8), an extensive lowland excavated in non-resistant Kimeridge Clay between the Jurassic Moors to the north and the Chalk Wolds to the south. The Vale of Pickering was blocked by North Sea ice at its eastern end and by Pennine ice at its western end. Like Eskdale, it too became a lake, in which the outflow from Newtondale built a delta at Pickering itself. The level of Lake Pickering rose until at about 75 m (250 ft) OD its waters overflowed at Kirkham Abbey and cut the sinuous gorge, by which the Derwent still drains the Vale of Pickering southwards to the Humber and not eastwards to the North Sea. The evidence for this spectacular glacial modification of drainage is still clear in the landscape in the form of overflow channels, deltas and lake sediments. But when glacial diversions of earlier age are considered the evidence becomes much less clear and at times it is only possible to infer that the drainage has been glacially modified from its anomalous pattern.

There is, however, a considerable amount of evidence for the existence of at least one proglacial lake related to the glaciation before, i.e. the Saale or Gipping glaciation. This is to be found almost in the centre of England in the upper part of the valley of the Warwickshire Avon, where an extensive lake, Lake Harrison, is thought to have been ponded up between the ice and the Jurassic escarpment (Shotton, 1953). Three lines of evidence converge in establishing the limits of this lake. In the first place, there are lake clays, the full extent of which was established after a programme of deep augering. The highest level reached by these clays is 123 m (404 ft) OD near Moreton-in-the-Marsh (Fig. 14.10), where they are thin and obviously marginal. Elsewhere the clays occur at much lower elevations and are bottom deposits of the deeper part of the lake. The highest elevation attained is closely in agreement with the heights of the gaps by which the waters of Lake Harrison could have overflowed: these are the Daventry, Fenny Compton and Dassett gaps, the first of which would have led the water into the Nene and the latter two into the Cherwell, a tributary of the Thames. Independent confirmatory evidence of the presence of the lake is provided by the presence along the Jurassic escarpment, from Moreton-in-the-Marsh to Daventry, of an erosion terrace at a little above 120 m (400 ft) OD (Dury, 1951). This feature had been interpreted as a lake terrace related to the Saale glaciation by Dury, and it is interesting to see the conclusions from the geomorphological study agreeing closely with Shotton's geological work. Lake Harrison was a proglacial lake formed during the advance of the Saale ice sheet, for it was later overridden by the ice, which moved considerably south of this area, and the lake deposits were largely hidden beneath boulder clay.

Conditions after the disappearance of the ice sheet were somewhat different from those prevailing before. Many of the present relief elements of the area were in existence before this glaciation, as can be seen from Shotton's reconstruction of the position of the ice damming the lake (Fig. 14.10). The Jurassic escarpment, Charnwood Forest, the uplands around Birmingham and Coventry, as well as Bredon Hill are clearly present, and the gaps between them blocked by ice coming from the north and northeast and along the Severn valley. The main difference lies in the trough between the Jurassic escarpment and the other uplands, which today is drained south-westwards by the Avon. Before the glaciation this broad valley sloped north-eastwards from near Bredon Hill down into the Soar

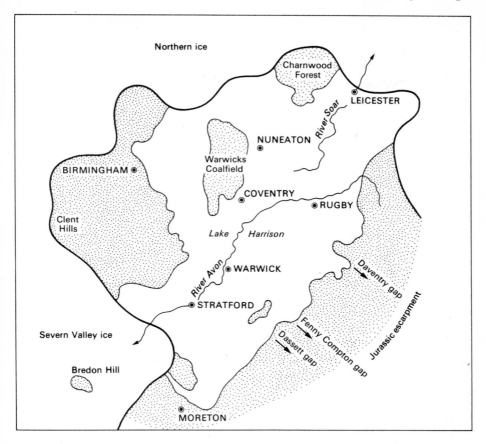

Fig. 14.10. Lake Harrison in relation to present drainage (*after Shotton*)

system at Leicester. The gradient, revealed by borehole data, is gentle but of the same order as that shown by the modern Thames and Severn between the 23 m and 7·5 m (75 and 25 ft) contours. Shotton, therefore, concludes that the main watershed of England, i.e. that between the Severn draining to the west coast and the Trent draining to the east, ran southwards through Bredon Hill, some 70 km (45 miles) south-west of its present position. Immediately after the glaciation this trough was filled with lake sediments and glacial drift to an almost level surface. The recession of the Severn ice left a face of unconsolidated glacial deposits which were quickly attacked by drainage tributary to the Severn. Further, meltwater from the north-eastern ice round Leicester would also have escaped in this direction and assisted in the establishment of the south-westerly drainage. The combined effects of these two was the formation of the Warwickshire Avon and its rapid extension north-eastwards: on this hypothesis the Avon is a comparatively recent addition to the drainage pattern, and the ice sheet was

429

responsible for a large shift of the main north-south watershed of England.

Much farther south, a complex series of marginal diversions of the Thames has been described by Wooldridge and Linton (1955) and Wooldridge and Henderson (1955). The London Basin is at the very edge of the Older Drift ice sheets, and apart from Essex and the area immediately north of London, it was not glaciated. The Thames itself, however, has not always flowed in its roughly central position through London, but followed for a long time a course along the northern edge of the basin through the area which was glaciated. Early in the Pleistocene, during a period called the Pebble Gravel phase when the land was relatively some 120 m (400 ft) higher than at present, there is geological evidence of the northerly position of the Thames. In two places north of the present Thames, between Barnet and Epping Green and again on the Laindon Hills, the Pebble Gravel contains abundant Lower Greensand chert pebbles. These must have been derived from the Hythe beds of the Wealden outcrop of the Lower Greensand, for the outcrops north of the Chilterns do not contain such material. They were brought to this area by streams flowing from the Weald across the line of the present valley of the Thames which could not, therefore, have been in existence at this stage. In fact, the Thames is believed to have flowed along what is now the Vale of St Albans and out to sea somewhere in Essex or Suffolk. Its course along the critical area of the Vale of St Albans is shown by arrow no. 1 in Fig. 14.11.

Some 18–24 m (60–80 ft) below the level of the Pebble Gravel is a terrace-like feature referred to as the Higher Gravel Train. Coming from the west this falls steadily in elevation as far east as the vicinity of Watford, but from that point its eastern counterpart, the Leavesden Gravel Train, rises eastwards. It is therefore clear that at this stage the Thames could no longer have flowed through the Vale of St Albans on its old Pebble Gravel Course, because the terrace gradients forbid this. The probable cause of its diversion was the advance across the Vale of St Albans of the ice sheet responsible for the locally derived Chiltern Drift. It is possible that the first advance across the Vale caused the Thames to loop round the ice and to form the broad Mimms depression (arrow no. 2 on Fig. 14.11). This course, however, could only have been temporary, for the gradient of the Leavesden Gravel Train falls westwards to near Watford, so that the main escape route must be sought south of this town. It is probably to be found in the two wind gaps through the Tertiary escarpment immediately south of Watford. Escaping through these gaps, the Thames is thought to have followed another broad depression, the Finchley depression, which lies parallel to and outside the Mimms depression, until it joined its old course near Ware (arrow no. 3 on Fig. 14.11). Whether these overflows were initiated by lakes ponded up by the ice is not known. By the time that the next main terrace, the Lower Gravel Train, was being formed the Thames appears to have followed substantially the same course through the Finchley depression with the exception that it took a more direct route

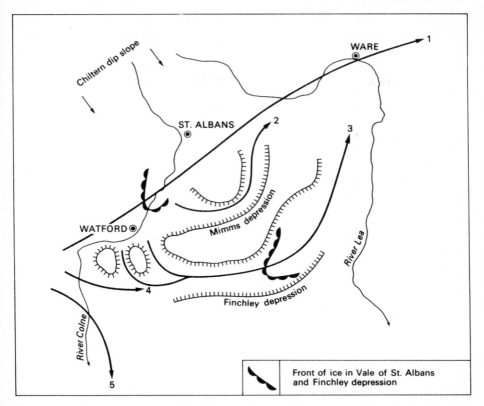

FIG. 14.11. Successive courses of the Thames between Watford and Ware (*after Wooldridge and Linton*)

into that depression, as shown by arrow no. 4 on Fig. 14.11. The effect of the advance of the Chiltern ice was, then, to cause the Thames to loop to the south, but not to take its modern course through London, for it joined its earlier course again near Ware.

At this stage the evidence for its course east of Ware must be considered. Difficulties arise because that part of Essex east of Ware is thickly covered with glacial drift, but the construction of sub-drift contours from well records has revealed the existence of a broad depression along which the Thames could have flowed. From Ware this depression extends up the lower Stort valley to Harlow, where the form of the depression can be worked out from the great number of borings put down in connection with the new town. It continues eastwards through the Rodings towards Chelmsford, and then in outline to Colchester. It seems from this that the Thames found its outlet across Essex and via one of the present east coast estuaries, though which one is uncertain.

A return must now be made to the Vale of St Albans, because the evidence of later diversion is to be found there. The mechanism of these later diversions was an advance of the Eastern Ice up both the Vale of St Albans and the Finchley depression to positions indicated on Fig. 14.11. Between the diversions caused by the earlier Chiltern Ice and those caused by the Eastern Ice there seems to have been a considerable space of time. The floor of the Vale of St Albans, along which the ice advanced from the east, falls in elevation eastwards, thus indicating that the drainage immediately before the Eastern Ice advance must have been eastwards, whereas at the time of the Leavesden Gravel Train and Chiltern Ice advance it was westwards. The eastward drainage is probably to be accounted for by a tributary of the Thames eroding headwards from Ware in the unconsolidated material filling the Vale of St Albans. Given longer time it may have succeeded in capturing the whole of the Thames into its old course, but the Eastern Ice intervened. Lobes of this ice sheet advancing along the Vale of St Albans and along the Finchley depression finally blocked completely the old eastern outlet of the Thames. Outwash from these lobes initiated the south-flowing Colne, Brent and Lea, and the Thames was forced well to the south in its basin to approximately its present course through London (arrow no. 5 on Fig. 14.11), although it has shifted uniclinally southwards in the west since then.

These diversions, like those connected with Lake Harrison, constitute a major reorientation of the drainage pattern, but glacial interference can operate on a much smaller scale, as is evidenced by certain features of the Dee valley near Llangollen (Wills, 1912). The result of glaciation here has been the plugging of two larger meanders of the earlier Dee by glacial drift and the diversion of the river across the necks of these meanders. The meander north of Berwyn (Fig. 14.12) is now short-circuited by a narrow gorge cut through hard Silurian rocks at Berwyn by the Dee at the point where the spur may well have been narrowed and lowered by the entry of the Plas-yn-Vivod stream from the south. That this stream formerly entered the Dee further to the north is shown by the fact that the tongue of drift in the Plas-yn-Vivod valley is continued north of the Berwyn gorge by an outlying small patch of drift marking the former course of the valley. A similar diversion of the Dee and the short-circuiting of a drift-filled meander occurs at Llangollen, where the modern Dee again occupies a rocky gorge.

Finally, one of those examples may be noted, which suggest some form of glacial diversion for the proof of which evidence is insufficient. It concerns the upper valleys of the rivers Cam and Stort (Fig. 14.13). The branch of the Cam in question rises some 19 km (11 miles) south of the Chalk escarpment and flows north as a stream of obsequent type before joining the main strike branch of the Cam which comes in from the west north of Whittlesford. In a number of ways it is a peculiar stream. Such a long course against the dip of the Chalk is in itself unusual, although the dip of the Chalk is

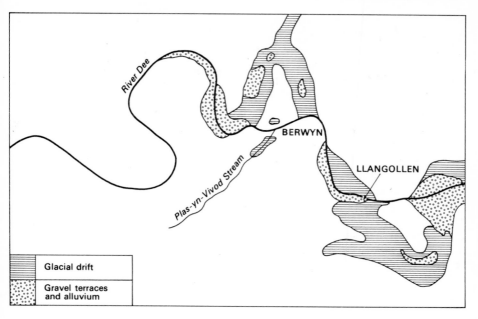

FIG. 14.12. Course of the river Dee near Llangollen (*after Wills*)

here very small. In addition, many of its tributaries flow diagonally down the dip to join it with abrupt bends. The valley itself contains none of the erosional terraces which mark the strike branch of the Cam to the north and this suggests that the valley is not as old. Finally, beneath the Cam is a deep buried channel which in general slopes southwards. This is one of those steep-sided troughs which have been described as probably being due to water beneath an ice sheet under hydrostatic pressure earlier in this chapter.

A hypothetical reconstruction of the valley's evolution may be made as follows. If the valley was originally the headstream of the Stort, as the orientation of its tributaries would seem to suggest, it may well have notched the crest of the Chalk escarpment south of Whittlesford. During the advance of the Saale or Gipping ice sheet a lake may have been ponded up between the ice and the Chalk escarpment, overflowed the crest at the notch and cut down well into the Chalk. Later, when the area was covered with ice this valley was excavated to great depths beneath the ice sheet by a subglacial stream. After the retreat of the ice the valley was choked with outwash gravels, boulder clay, and probably solifluxion fans, traces of which can still be found. These constituted an irregular surface on which the present Cam picked its way northwards.

Although these ideas would reasonably explain the present pattern, there does not seem to be sufficient evidence for conclusive demonstration. That the Cam has been reversed has been suggested several times, and it seems

433

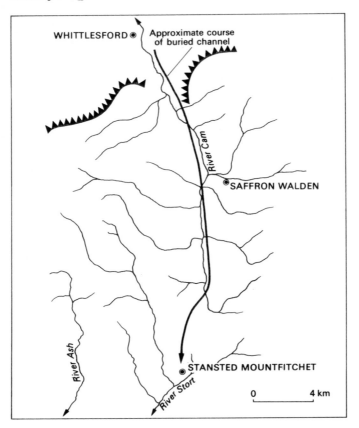

FIG. 14.13. Drainage pattern of the upper Cam and Stort valleys

that nothing more concrete than suggestion is possible. There must be many analogous cases in Older Drift areas, where glaciation seems to have affected the drainage pattern but where the evidence is merely suggestive.

Of the examples discussed above, some are inexplicable as anything other than overflow channels of glacial lakes. This seems to apply to Newtondale, which was not later covered by the ice sheet concerned. The overflows from Lake Harrison might have been subglacial features because they were later overrun, but the classic evidence for a proglacial lake is here in the form of the lake deposits, the lake terrace and the overflow channels. The Thames diversions are marginal diversions which do not involve the existence of anything more than small temporary lakes, especially as the area is lowland. The suggested Cam—Stort diversion is the weakest case. There is no evidence here other than a breach difficult to explain by normal processes. There is no evidence of a lake and really little necessity to invoke it for the same subglacial stream which later carved the deep buried channel of the Cam could well have ripped the breach in the Chalk escarpment.

The real trouble starts when the minor overflow channels of such highland regions as the northern Pennines and the Southern Uplands and Highlands of Scotland are considered. These channels are legion and if every one is used to infer the position of a proglacial lake there will be little dry land left. Further, and most important, the form of the channel is not always consistent with the overflow hypothesis.

This was clearly stated by Peel (1949) in a detailed study of two Northumbrian spillways. In these the long profile was humped, whereas in any classical lake overflow channel the slope must obviously be unidirectional. Although one can always explain things away by special assumptions, it was more honest to suggest that such channels cannot be overflow channels and to see what might arise from such an assumption.

The idea of subglacial and englacial channels had been developed by geomorphologists abroad and it has been applied steadily to many areas of Britain (Embleton and King, 1968; Sissons, 1967). In addition to humped channels, some of the channels are merely subglacial chutes gashed down the direction of steepest slope. Further, there is often an intimate association between eskers, which are generally agreed to be subglacial features, and meltwater channels. Finally, there was the scepticism mentioned earlier on about the feasibility of stagnating ice forming a dam sufficiently waterproof to hold up a lake.

The question of englacial channels was mentioned in the last paragraph. There is a certain advantage in assuming their existence, because subglacial channels might not have been able to exist due to the pressure of the overlying ice or because of its movement. Englacial channels at lesser depths could have remained open and been let down, as the ice thinned, as up and down channels across ridges. With the conversion of englacial to subglacial drainage as the ice thinned there were obviously conditions for the formation of very complex subglacial drainage patterns.

The idea of subglacial channels does not replace the idea of glacial overflow channels leading from proglacial lakes. It supplements it and saves us the necessity of inferring an embarrassingly large number of ice-dammed lakes, which will in any case not explain in every example the observable phenomena.

15
Periglacial landforms

The word periglacial has been applied both to the areas adjacent to ice sheets and to a particular form of climate involving virtually tundra vegetation and permanently frozen subsoil. It is usually used in the latter sense today and also to denote a particular set of geological and geomorphological processes that go with those conditions. This is a better meaning, because proximity to glaciers and ice sheets does not involve identity of climate. It is quite possible for vigorous confluent glaciers from high mountains to reach down into the coniferous forest zone of the surrounding areas, where the climate is certainly not periglacial. On the other hand there are large parts of Siberia, which are far from glaciers and ice sheets, because they are rather arid lowlands, but which are truly periglacial in climate.

The chief significant difference between the tundra climate and our present climate is the length of the period during which the ground is frozen. Although the climate varies with the distance from the ice front, it may be said that for approximately the winter half of the year temperatures rarely if ever exceed 0°C (32°F), while in summer the temperatures are sufficiently high to ensure the melting of approximately the top metre or so of soil. Below that level the subsoil remains permanently frozen and is often referred to as the permafrost layer. This layer, being frozen is impermeable, so that the summer thawing of the uppermost layers results in a saturated, mobile stratum at the surface.

The vegetation of the tundra region is restricted, partly as a result of the intense winter cold, partly because of insufficient summer heat and partly through the waterlogged condition of much of the soil in the summer. Thus, there is little vegetation to bind together the inherently mobile surface layer in summer.

These generalities having been expressed, one must hasten to add that a considerable variety of periglacial climate exists at the present time and probably an even greater variety existed in the past. There are the usual differences between maritime and continental climates, with associated differing amounts of snowfall and ground surface temperatures. There are differences between areas according to whether winds are mainly on or off

436

the ice caps. When one considers the Pleistocene, which was responsible for most of the periglacial features in Britain, there are more striking differences. Although the climate, vegetation and geomorphological processes seem to have been similar to their present periglacial equivalents in general terms, there must have been important differences due to latitudinal position. In Britain in the Pleistocene the length of days must have been virtually the same as now. There cannot have been the tremendous variation in day length characteristic of the Arctic. The days were shorter in Britain in summer, but the altitude of the sun was much higher, so that there may have been more effective thawing of the superficial layers and hence possibly more powerful solifluxion (see below). Although this is only conjecture, we must be prepared for differences in processes and hence for the inability to match all British Pleistocene periglacial features with forms at present occurring in the Polar tundras.

Periglacial processes and forms were clearly recognised and described in simple terms by nineteenth-century geologists, for example Reid's attribution in the 1880s of the dry valleys of the South Downs to periglacial torrents of enormous scouring and transporting power. But the main development of periglacial geomorphology has been since 1945, by the Americans and Canadians in studies of Alaska and the far north of Canada, by the French in studies of relict Pleistocene forms in their own country, and above all by the Poles. A multitude of forms has been described and separately named, although many may ultimately be shown to be elements in series of continuously variable forms. At the same time a plethora of terms has been instituted, some of them derived from Greek and Latin roots, some of them borrowings from the unfamiliar languages of the peoples who inhabit the tundras of North America and Asia.

Broadly speaking it can be said that periglacial activity leads to three main types of features:

1. The formation of new deposits which may or may not produce much relief effect.
2. The modification of the structures of existing unconsolidated deposits, largely by the formation of ground ice. Such features are largely geological, though their understanding is vital in interpreting the successions of climates under which present relief has evolved.
3. The modification of landforms by large-scale mass movements.

Periglacial deposits

The first class of deposits are those formed mainly by the fragmenting effects of freeze–thaw. The process has already been discussed in Chapter 13 as one of the main aids in the glaciation of highland regions. It is responsible for the formation of screes, provided that the lithology of the rock is also conducive to scree formation, that is, it must allow access to water along

437

discrete surfaces, usually joint planes. There is probably a joint frequency range especially favourable to scree formation. Where the joints are very widely spaced there may be little intermediate material between enormous boulders and grain-sized debris, so that good screes do not form. Such appears to be the case in the eastern, coarse part of the Arran granite. Screes are common in areas of resistant rocks in the highlands of Britain, for example North Wales, the Lake District and the Pennines. Many of them are obviously stabilised: some are virtually hidden beneath mats of turf; others are partly overgrown; others may appear unvegetated but the development of lichens on the upper surfaces of boulders betrays stability, at least for lengthy periods. A few may still be active especially where disturbed by various animals, including man, for example on the north side of Cader Idris where the Foxes Path follows a scree which is loose in places.

Freeze–thaw is most active as a process where there is the maximum number of oscillations above and below the freezing point. This does not occur in the present polar tundras where for a long period in winter the temperature probably remains below the freezing point. Freeze–thaw cycles have a much greater frequency in the local periglacial areas of the mid-latitude high mountain ranges, and the same was probably true of Britain in the Pleistocene. Accordingly scree formation in Britain in the Pleistocene was probably more comparable with its present formation in high mountains than in the Polar regions.

Various other forms of boulder accumulation occur in periglacial conditions. They are collectively known as blockfields. They may have been deposited by rock glaciers, which seem to be pure rock forms at the surface, but which have interstitial ice at depth thus allowing them to move by slow creep. Boulders may also be transported in high density mudflows and left as blockfields when the fines are later washed out of the mudflows by water action. The boulders may then start to creep downhill.

Features known as stone streams, self-evidently linear arrangements of rocks, were described from the Falkland Islands. Some of them are not confined to valleys, but seem to be related to certain outcrops because of their uniform lithology. They may represent the final degradation of those outcrops by freeze–thaw.

The second main type of periglacial deposits are solifluxion deposits. In periglacial regions a permanently frozen subsurface layer (permafrost) may form to very great depths: 600 m (2 000 ft) has been reported from northern Siberia. In summer the surface layers are thawed and constitute a temporary, saturated, mobile sheet overlying an impermeable frozen layer. Permafrost, however, is not essential for solifluxion. If deep freezing takes place in winter, there must be a stage in spring and early summer when a surface thawed layer is present over a frozen sub-surface layer, and in these conditions solifluxion may take place.

The mobility of the thawed layer (it is sometimes called the mollisol in periglacial jargon to distinguish it from the pergelisol, or permafrost, below) is due to the fact that its water content reduces its internal friction and hence its shear strength. No adequate vegetation cover binds it because the plants are mainly shallow-rooted. Therefore, it can begin to move on slopes as gentle as 2 degrees. From geological evidence in the form of various deposits classed as solifluxion it would seem necessary to assume that it can continue to flow for distances of several kilometres on slopes as gentle as 0·5 degrees. Such evidence is not absolutely conclusive, because there is always the chance of mistaking the identity of a deposit.

The word solifluxion, literally soil flowage, has been used to cover a variety of phenomena from processes as slow as soil creep to almost liquid flows of mud. In the present context it is being used to cover mostly the flows of thoroughly saturated material, but the exact spectrum included in the term must obviously be wide as becomes apparent when one starts to imagine the interdigitation of processes which may take place during summer thaws. At the dry end of the solifluxion range movement may be slow and viscous, so that there will be little opportunity for bedding to develop: the material may be distinguished from till only with difficulty. With greater fluidity rough bedding, sometimes known as tumultuous bedding, may occur. At the height of the thaw there may be enough melt-water to re-sort material and to produce intercalations of bedded material difficult to distinguish from normal or glacial river deposits. Thus the pattern of deposits formed in periglacial conditions may show rapid vertical and horizontal changes. They will also show marked areal changes, as their lithology depends very strongly on their parent material.

Solifluxion gravels of various sorts have long been recognised and mapped in Great Britain, where they are known by a variety of names, e.g. head, coombe rock, taele gravel (Plate 51), warp and trail. The word sludge commends itself as a highly appropriate word, the very sound of which suggests the process envisaged, but it is probably best to refer to the whole series as solifluxion gravels. Their nature varies widely with the character of the parent material, but they are all essentially unsorted and either unbedded or very poorly bedded. Many of the dirty and rubbly gravels, containing stones in a matrix of silt or clay, which are unwelcome to the gravel industry, fall into the category of solifluxion gravels.

Many solifluxion gravels do not form surface features, but extend as featureless spreads at the foot of higher ground. For example, south of the South Downs a vast spread of coombe rock and brickearth, the latter probably largely a windborne sediment, covers the coastal plain of West Sussex, while occasional patches of angular solifluxion gravel occur on the Weald north of the South Downs escarpment. At the foot of the Chalk escarpment in Cambridgeshire are discontinuous spreads of Taele gravel, which have essentially the same origin.

Occasionally solifluxion gravels do form ill-defined surface features, especially when the flow is in the form of a waste fan, the material of which

439

has not spread out over lower ground. Certain very subdued spurs in the upper Cam valley near Little Chesterford, Essex, have been cut through by gravel pits and railway cuttings and seem to be wholly composed of gravel. Such features may be more common than realised, for they can only be shown to be waste fans if large and deep workings, of the order of 6–10 m (20–30 ft), are opened in them. Faint convexities may also be observed at the mouths of dry valleys, for instance in the South Downs, and such forms have been interpreted as low-angle fans (White, 1924).

In the present tundra region and in comparable areas high on mountains in more temperate latitudes, small-scale terraces may be formed on slopes of 10–30 degrees by solifluxion, especially when the surface is well covered with herbaceous vegetation. These minute terraces, with heights and widths of the order of a metre (3 ft), seem to be due to solifluxion acting slowly. They may be distinguished from the familiar minute terraces caused by animal tracks on steep slopes by a number of features: they are always parallel and horizontal, unlike animal tracks which may go obliquely upslope and merge with each other; they are on the whole on a smaller scale than animal tracks; and their distribution is related to height and climate, whereas animal tracks are independent of these factors.

Thirdly, wind action may be an important agent of deposition in periglacial regions. The conditions are ideal; great areas of bare unvegetated waste left by retreating ice sheets and further areas bared by the devegetating activities of processes such as solifluxion. The period of incomplete vegetation cover may have been longer in the more continental climates of parts of the interior of Europe than it was in Britain.

Two grades of material are movable by wind action: sand is the coarser and moves mainly by saltation (see Chapter 11), while the finer silt and clay particles are carried mainly in suspension. Thus, the areas of wind-blown sands should lie between the glacial deposits and the areas of wind-blown silts. In Europe this is generally true. The sands are known as cover-sands and occur widely in the Netherlands to a depth of a few metres (West, 1968). Their aeolian origin is suggested by mechanical analyses of grain size and the frosted appearance of many grains, a characteristic of wind-blown sands. Traces of dune-bedding occur here and there, while further east actual sand dunes are found in Poland.

In Britain much of the sand of the Breckland on the margin between East Anglia and the Fens must have been affected by wind action under periglacial conditions. Sand blowing is also thought to have been responsible for the facetting and polishing of many of the flints found in these deposits. Areas of degraded, vegetated small dunes occur around Lakenheath aerodrome and wind blowing still affected unvegetated sand there until recent years. It is quite obvious at the present time that it needs little

PLATE 51. Taele gravel, near Thriplow, Cambridgeshire: unbedded, calcareous solifluxion gravel, deeply weathered at surface

to disturb ecological equilibrium and to start sand blowing again. The areas marginal to the Breckland are mostly arable areas, growing poor land crops such as rye, and wind action on them can readily cover the local roads to depths of several centimetres with sheets of sand, while in exceptionally severe cases windbown sand can reduce the visibility to that of a moderately dense London fog.

More important than cover sands is the deposit known as loess. This is primarily a silt deposit, characterised by a very sharp mode in its grain size. It may be formed anywhere where dust-laden winds have their suspended material precipitated either by a drop in wind velocity or by an increase in precipitation (see Chapter 11). It is, therefore, not necessarily periglacial. For example, the loess of northern China was formed by northwesterly winds from the winter Siberian anticyclone blowing over unconsolidated deposits in the deserts of Mongolia.

Periglacial loess in Europe was presumably formed by anticyclonic winds winnowing fine material from the sheets of deposits left by the Scandinavian ice sheets. It extends as a belt of diminishing importance from the Ukraine to southern Britain, thus reflecting presumably an association with continental climatic conditions. It expands in lowlands such as upper Silesia, the Leipzig area, and where the Rhine valley opens out north of the Rhine block around Bonn and Cologne. It narrows in between, where it forms a fringe between the Hercynian blocks and the north European lowland. In northern France it is known as limon: in Sussex and the Thames valley it is called brickearth, although this purely utilitarian term may include any brick-making raw material as well as loess.

Although mainly aeolian, much loess must have been deposited in a zone of areas of dry ground separated by marshy slacks. Its snail faunas are famous and clearly point to such a combination. Some of it may even have been reworked by local downhill processes of mass movement. Locally water action may have occurred during deposition, because pebbles are known to exist in it in places in southern Britain.

Yet, essentially, loess is unbedded and subject to a rough vertical fissility, which enables it to stand in cliffs. It is plastered over various parts of the landscape ranging from river terraces to the Chalk hills of northern France. In western Europe it probably largely conforms to a pre-existing relief pattern, but farther east, where it is thicker, it may blanket the relief.

Fourthly and finally, the deposits of periglacial rivers must be considered. These are very difficult to distinguish from outwash. Outwash is formed by the melting of ice: periglacial deposits from the melting of snow. Streams from both carried large loads over relatively gentle gradients because of their high peak discharges. Even so the gradients may have been steeper than the gradients of interglacial or Post-glacial streams, so that periglacial streams graded to low glacial sea levels may have profiles which cross those of interglacial streams (see Chapter 9).

Periglacial structures

Two trends dominate the formation of periglacial structures in existing soil and waste mantles. They are, firstly, the tendency for ice masses to be segregated and, secondly, the tendency for patterns, usually polygons and stripes, to be formed on a variety of scales. Ice wedges provide the natural link between the two classes, for they are ice segregations which often form polygonal patterns.

Apart from the ice which causes freeze–thaw shattering and is a form of segregation, needle ice or pipkrakes may develop immediately below the surface of the soil or weathered rock. The crystals grow at right angles to the face and prise off weathered material as well as assisting in the general downhill creep of material (see Chapter 4).

On a larger scale is the general tendency for ice segregation to occur in the soil or waste mantle. Segregation depends on two main factors, the rate of freezing and the lithology of the material, the latter having a very important effect on water content.

If freezing is very rapid the water in the soil has not time to migrate and to build up discrete masses of ice and is frozen in its normal disseminated state within the mantle. Under these conditions ice segregations may only build up due to the thawing of meltwater in contraction cracks (see below).

If freezing is slower ice segregation can occur, but it is much better developed in some types of sediments than in others. In clays it seems that the pore spaces are so small that movement is too restricted to allow active segregation. In sands the spaces are too large and the columns of water under tension tend to break. In silts the pore spaces seem to be of the optimum size and segregations are best developed in this material. There is also a tendency for ice segregations to be formed under large stones, because the thermal conductivity of rock is greater than that of sediments with large pore volumes.

Thermokarst

Thermokarst or cryokarst (frost karst) is a term used in somewhat different ways by different authors. The landscape resulting from ground ice development is one of depressions of irregular size and shape, which have some degree of similarity to karst depressions. It is alleged to be caused by the irregular melting of ground ice layers. Probably it requires the development of ground ice layers and sheets without the regular geometric forms which are responsible for pingos and allied features (see below). If irregular ground ice development causes the surface to be heaved up irregularly, it seems that during summer thawing solifluxion will probably transfer material from the raised sections above major ice segregations to the intervening hollows. As a result, when the ground ice itself thaws, a series of irregular depressions will form on the sites of the former ice masses. The

443

exact form of the depressions will depend on a number of factors: the original distribution of the ice, the extent to which subsurface movement of water in thaws causes washing out, and whether the thermokarst hollows become filled with water. Water-filled hollows tend to perpetuate themselves and to become regular in shape because the banks are thawed at water level. In addition, where the water freezes in winter ice pressure and thrusting may also tend to smooth the banks. Above a critical size the water surface may be large enough for the development of wind-generated waves, which can further enlarge and regularise the hollows by mechanical action. At this stage they become transformed into thaw lakes, which cover extensive areas in Arctic North America (Hopkins, 1949). Fossil Pleistocene thermokarst has been reported from New Jersey and the Paris Basin, and it is possible that some of the landscape of gentle depressions on the Chalk between the Fens and the Breckland in East Anglia is of this type. The Breckland meres, lakes many of which have a diameter of a hundred metres or more and which occur mainly to the north-east of Thetford, might conceivably be examples of thaw lakes.

Ice mounds

In many places ground ice forms much more symmetric lenses than are required for thermokarst development. The general development of an ice mound feature is illustrated in Fig. 15.1. The ground above the ice lens is forced up, solifluxion shifts material off it, so that when the ice melts one is left with a hole often containing a pond and surrounded by a rampart, which may be breached if the pond is drained and which has been formed from the waste sludged off the ice mound. There are many variations on

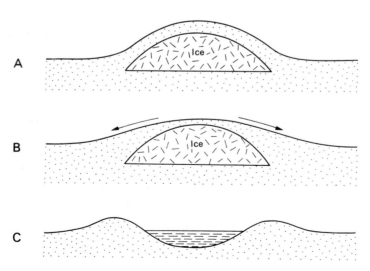

Fig. 15.1. Ideal development of an ice mound

this type of feature—the literature has been summarised by Maarleveld (1965)—and they range in size up to about 500 m (1 600 ft) in diameter and 50 m (160 ft) in height. If one makes allowance for the fact that they are unlikely to be perfectly regular cones, this means maximum sideslopes of 25 degrees or more, making the larger of them truly impressive features.

In practice there is a wide range of features of this type from small mounds a few metres large, which may be produced in a single winter to the large ones which are semipermanent features of the landscape and may take hundreds of years to form. A variety of names have been used for them. The larger ones are usually called pingos, an Eskimo word, in Arctic North America and Greenland. The Russians use the term, bulgunnjakh, while hydrolaccolith has also been used from the analogy with a lens-shaped igneous intrusion. Small forms usually arising in peat bogs have been called pals (plural = palsas). There is undoubtedly a considerable range of forms and transitions from these lenses to the more general type of ground ice distribution which produces thermokarst on thawing.

The larger pingos are found in northern Siberia, east Greenland, the Mackenzie delta of northern Canada and in Alaska. Two main hypotheses of origin have been advanced: by Mackay for the Mackenzie delta forms and by Müller for the Greenland forms.

The Mackenzie delta type is associated with the sites of former lakes: similar features in northern Siberia are reported to occur in the same type of situation. It is believed (Mackay, 1963) that the lake will prevent permafrost from forming below it for some while (Fig. 15.2A), but that ultimately filling and shallowing of the lake will allow permafrost to develop on its floor. At this stage the permafrost at the sides of the lake starts to extend inwards and exerts strong lateral pressure on the mass of unfrozen, saturated material beneath the lake (Fig. 15.2B). The pressure set up causes bulging at the weakest point, i.e. beneath the old lake where the permafrost is thin. The result is that the ice lens formed from the water in the unfrozen pocket is emplaced there (Fig. 15.2C). This type of pingo has been referred to as a closed system pingo, because it is formed from an enclosed, unfrozen pocket trapped by advancing permafrost.

The East Greenland type of pingo (Müller, 1959) is virtually an artesian feature. Pressure of water either within or beneath the permafrost may build up so much in an artesian situation that the permafrost may be buckled up and the uprising water frozen as a hydrolaccolith. Occasional eruptions of water, mud and gas have been reported from such features. Pingos are finally destroyed when the enclosed ice lens is exposed to thawing and are often replaced by a form of crater lake. Pingos of the East Greenland type are sometimes called open system pingos.

Fossil pingos or pingo-casts should consist of a circular rampart, probably breached in one place, enclosing often a marsh or pond. Such forms have been reported from a variety of localities in north-western Europe, from central Wales and from the margins of East Anglia and the Fens (many of

445

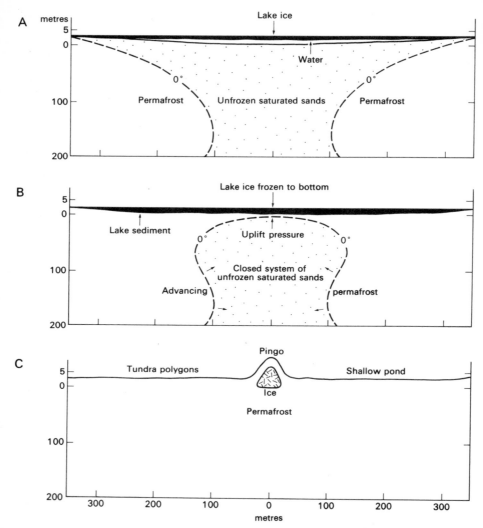

FIG. 15.2. Development of a pingo (*after Mackay*)

the quoted examples are summarised in Sparks, Williams and Bell, 1972). However, there are important differences in size and frequency between present pingos and their alleged fossil counterparts. At the present time, Mackenzie delta pingos have a maximum density of about 60 per 100 sq km (160 per 100 sq miles), while Alaskan forms reach a density of only 6 per 100 sq km approximately (16 per 100 sq miles). The smaller fossil European forms have densities up to 30–50 per sq km (80–130 per sq mile) i.e. fifty times the density of the present Arctic forms. Their size ranges up to about 120 m (400 ft) in diameter, which is very much less than the maximum

446

size of modern pingos. Forms similar to the British forms have been described from waterlogged, fluvioglacial material in the far north of Norway by Svensson (1969), who regards them as intermediate in type between the pingo and the pals. Svensson suggests that the enclosing rampart is partly due to sludging of material and partly due to the pressure exerted by seasonal ice developed in the pond in the decay stages of the ice mound feature, a process suggested above as a possible contributor to the development of thermokarst.

Forms of this type are very well developed at Walton Common, some 12 km (8 miles) east of Kings Lynn, and a map of part of this area is shown in Fig. 15.3. In this locality it seems that a weak artesian state may have led

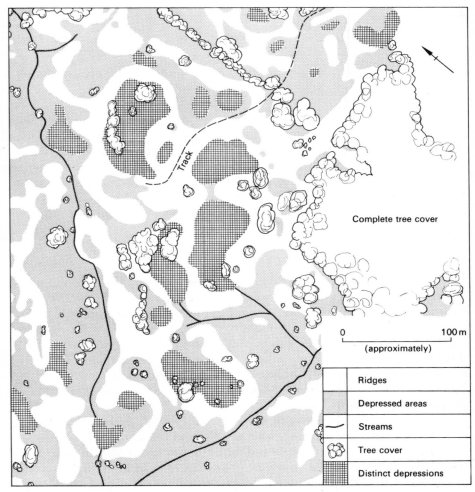

Complete tree cover

0   100 m

(approximately)

	Ridges
	Depressed areas
	Streams
	Tree cover
	Distinct depressions

FIG. 15.3. Pleistocene ice mound features on Walton Common, Norfolk (*after Sparks, Williams and Bell*)

447

to the location of the features where springs and rises still occur at the foot of the Chalk between that rock and the impermeable, interglacial, Nar Valley Clay to the west (Fig. 15.4). This does not mean that the features must be considered to be East Greenland type pingos: they are more likely to have been formed like naleds (see below) but below the surface because of the weak artesian state.

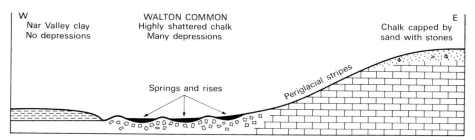

FIG. 15.4. Geological location of Walton Common ice mounds (*after Sparks, Williams and Bell*)

If the artesian condition is strong enough water may be forced through to the surface and form surface ice masses of varying size, known as naleds to the Russians and Aufeis in German. Such features may occur where weakness occurs in permafrost or where water continues to flow under surface ice and then breaks through. Cases are known where heated houses have so weakened permafrost as to localise breaches of it, with the result that the houses have quickly become filled with solid ice much to the discomfort of the inhabitants.

Ice wedges

Ice wedges form a natural link between ice segregation features and polygonal pattern features. When unconsolidated deposits freeze they first expand as ice is formed and then contract as the temperature falls to the region of -20 to $-40°C$ (-4--$-40°F$). When they contract they tend to split into a polygonal pattern, very much like a drying playa lake floor or a cooling sheet of basalt. The pattern of polygonal cracks is usually fairly coarse, the diameter of the polygons ranging up to about 50 m (150 ft or more). In the cracks so formed, and reopened year after year, wedges of ice slowly accumulate.

The normal development of an ice wedge involves the formation of contraction cracks when the seasonally thawed surface layer is frozen in winter (Fig. 15.5A). In summer and autumn meltwater fills the crack and freezes to ice in the permafrost zone below the thawed layer, in which the ice wedge does not develop (Fig. 15.5B). Repeated winter freezing opens up the same cracks and allows a thick ice wedge to develop in time (Fig. 15.5C). The pressure exerted by the seasonal expansion of the material

448

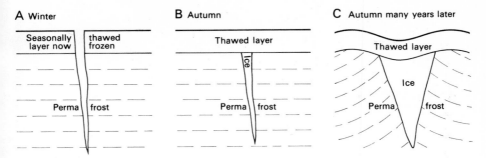

FIG. 15.5. Formation of ice wedges (for explanation see text)

between the ice wedges results in that material being upturned at the edges of the wedges, which are themselves marked by surface gutters.

Fossil ice wedges, or ice wedge pseudomorphs, are often found in British periglacial deposits. As the permafrost thaws and the ice disappears, irregular slumping leads to the development of the sort of structures shown in Fig. 15.6 and Plate 52. The very irregular form shown in Fig. 15.6B is

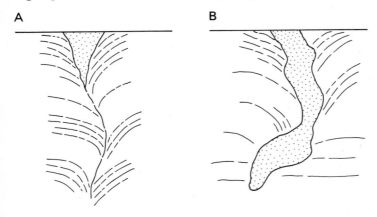

FIG. 15.6. Ice wedge pseudomorphs

possibly due to water action modifying the infill of the wedge as it develops. Ice wedge pseudomorphs are very useful features in reconstructing palaeo-geography in former periglacial areas, especially when several generations of ice wedges in gravels illustrate the alternation of periglacial climates and more temperate intervals. They can be recognised in sections in gravel pits, from which they were often reported as channels or replacements of tree roots in earlier literature, and also as polygonal patterns on aerial photographs of those deposits.

It is possible for features similar to ice wedge pseudomorphs to develop without the formation of ice wedges themselves. A similar pattern of narrow contraction cracks may form and be filled with ice and hoar frost in winter.

449

During the summer thaw silt and sand may be washed into the thawed crack. In the next winter the crack might well be reopened on the same line as it is now a potential line of weakness. A great number of repetitions of this cycle could lead to the formation of wedgeshaped intrusions of material difficult to tell from those replacing former ice wedges. On the whole this seems a less likely process though such cracks have been observed in parts of Siberia and North America. If the cracks extend far into the permafrost it is difficult to see how they can be seasonally thawed and avoid becoming true ice wedges. Further, it has been observed that such cracks only seem to occur in areas which are both vegetation-free and snow-free in winter (Péwé, Church and Andresen, 1969).

Polygonal patterns

The ice wedges discussed above form polygons on a large scale, but polygons are also formed on intermediate and small scales. A descriptive classification of these forms of patterned ground was instituted by Washburn and is now generally used. The following classes may be distinguished:

(*a*) Circles are isolated features. A typical example is a circle of coarser stones surrounding a centre of fine materials.
(*b*) Nets are really concentrations of circles, which have not reached the stage of interfering with each other. Some concentrations of ice mound features might be described as nets.
(*c*) Polygons occur when the whole area is covered with contiguous features, the interference producing the polygonal pattern. It is interesting that the formation of sheets of polygons must indicate the contemporaneity of the pattern, whereas ice mound features seem to overlap and eliminate each other, thus showing that their formation is not strictly synchronous. Polygonal ice mounds do not seem to develop.
(*d*) Steps may occur on gentle slopes where polygons become elongated into small terracelike features. They do not always occur and in some places polygons change directly into stripes. In the sorted type of step a bank of coarser stones forms the front of the terrace.
(*e*) Stripes are elongations of polygons and steps on slightly steeper slopes.

Any of these classes of features may be divided into sorted and unsorted types. The sorting is defined in terms of calibre of material. In sorted features, such as stone circles and stripes, the coarser material is in the margins and the finer in the centres. In non-sorted features there may be lithological differences but not grain size differences. Typical unsorted polygons are ice wedge polygons. The bedding of the material occupying the sites of former wedges may be different from the material between the wedges, but there are not significant grain size differences.

PLATE 52. Ice wedge in glacial gravel, Furze Hill, Cambridgeshire

The most common polygons and stripes are the small-scale sorted type. These polygons, which usually have diameters of less than one metre (3 ft) and consist of stones at the margins and finer material in the centres, occur in most tundra regions and also high on mountains where similar climates prevail. They are also found fossil in areas which suffered similar conditions in the Pleistocene. According to Cailleux and Taylor (1954) the polygons are regular on slopes up to 2 degrees, but become elongated on slopes of 3 to 7 degrees and change into stone stripes on slopes of 8 to 26 degrees (Fig. 15.7). Stone stripes are alternations of strips of stones and strips of finer

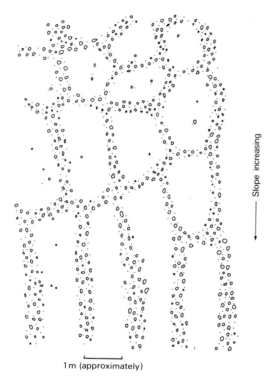

Slope increasing

1 m (approximately)

FIG. 15.7. Relation between slope and stone polygons and stripes

material corresponding with the edges and interiors of polygons. All these features are usually held to have a common origin, the differences being due merely to the effects of slope. But there is not final agreement as to the method of formation.

It must be emphasised that small-scale sorted polygons do not require permafrost conditions, but merely severe winter freezing. They form, and re-form after they have been disturbed, in a matter of a few years in the southern French Alps, the Pyrénées, and the higher parts of the Lake District and Scotland.

One hypothesis of their origin invokes convection currents arising in the following manner. During the summer thaw the water in the topmost layers of the soil reaches a temperature of 4°C (39°F) before that at the contact of thawed and frozen layers below. As water attains its greatest density at 4°C, the surface water sinks and the lighter colder water below rises to replace it. Thus, convection currents are set up. Experiments made on pasty material have demonstrated that such currents do occur, and that a polygonal pattern develops with rising currents in the centres of the polygons and descending currents at their margins. It seems doubtful, however, whether such currents would be sufficient to move boulders weighing in extreme cases over 50 kg (1 cwt) such as have been observed in some Greenland polygons (Cailleux and Taylor, 1954). An extension of this idea invokes differences in density within the soil caused by drying. The surface layers would dry first in summer and hence become denser and tend to sink, while the lower layers, full of meltwater caused by the summer thaw, would be lighter and tend to rise. Again, a polygonal pattern has been demonstrated by experiment to follow this type of density variation.

Although the convection hypothesis outlined above is not now generally believed, it was an attempt to provide a basic explanation of the pattern, whereas many hypotheses only really explain the continuation and perfection of a pattern once it has been formed. Just as polygonal patterns of igneous rock joints can be derived from random patterns of centres of cooling, it may be that random distributions of patches of coarse and fine material might lead to polygons, but the distributions are obviously not random in many cases. Perhaps the downward penetration of freezing in autumn and the differing susceptibility of different beds to freezing allows, for example, slow-freezing beds of waterlogged silt to be injected upwards, rather in the manner suggested for involutions (see below). Thus, a more random distribution of coarser and finer material might be created from originally stratified materials.

Once a more or less random distribution has been achieved the coarser patches will freeze faster than the finer patches, because their thermal capacity is smaller than that of the fine patches which have a high water content. Thus pressure will be exerted from the coarser patches towards the finer patches, doming them up. This doming would be accentuated when the finer patches, with their higher water content, froze and expanded. Any stones which then came to the surface of the finer patches would migrate downhill to the coarser margins so leading to the accentuation and perpetuation of the pattern.

The upward movement of stones through freezing and thawing surface layers is quite characteristic. As freezing works downwards the stones are firmly frozen into their correct relative position in the soil when their upper parts are frozen in. When thawing occurs, also operating from the surface, they are held by the bottom in an elevated position in the frozen soil while the thawed levels collapse around them. Repetition of such annual cycles result in the stones rising to the surface.

It must be emphasised that the origin of small sorted polygons is still obscure and it may well be that various other processes help to maintain the sorted state. Among them may be the tendency for any meltwater to follow the lower stony margins of polygons and wash the finer material away.

Good examples of unsorted polygons and stripes are provided by the Breckland type (Williams, 1964). They are intermediate in size between stone and ice wedge polygons (Fig. 15.8). The average distance between

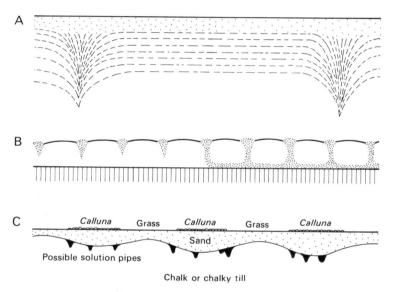

FIG. 15.8. Types of polygons. A. Ice wedge polygons. B. Stone polygons, 'floating' to the left and 'rooted' to right. C. Breckland polygons

polygon centres is 10·5 m (35 ft) and between the axes of stripes, which replace polygons on slopes exceeding 1–2 degrees, is 7·5 m (25 ft). It seems unlikely that the stripes are polygons elongated by solifluxion caused by gravitational effects on slopes, for they are immediately replaced where the slope flattens by polygons. Had solifluxion been operative, there would presumably have been a zone of flattening slope on which polygons slowly replaced stripes.

These features are clearly revealed on aerial photographs and are readily observable at ground level, for example between the disused Thetford–Bury railway and the Elveden–Bury road. On Thetford Heath and other places they show up as vegetation patterns with heather and other acid plants occupying the margins, where the depth of sandy material is greater, and more calcicolous plants the centres of polygons and stripes, where the Chalk or chalky till is much nearer the surface (Fig. 15.8c). In some the sandy material is confined to the margins; in others it is merely much thicker at the margins. These polygons are bounded by shallow troughs and they are essentially non-sorted being formed primarily from an uneven distribution of sandy material.

Their origin is not understood. They have been compared to tussock-birch-heath polygons at present forming in Alaska (Watt, Perrin and West, 1966). They are more or less confined to the Chalk and chalky tills in East Anglia (with an outlier in Thanet). These are among the most frost-susceptible materials in Britain, but, as polygons are not found on the more westerly Chalk outcrops, it seems that a continental situation and presumably climate in the Pleistocene must also have been very important in their formation. It seems that the East Anglian chalky areas, to judge from the prevalence of patterned ground and ice mound features, suffered the most intense periglacial modification in lowland Britain during the last (Weichsel) glaciation. There are, of course, Chalk areas on the continent which should have had the right sort of climate in the Pleistocene for similar features to be formed, but nothing seems to have been reported from them yet.

Involutions

Many superficial materials which have been subjected to periods of periglacial conditions have involuted or festooned structures (Fig. 15.9). These may be of various types. Fig. 15.9A shows the pattern often observed when an underlying bed of gravel or weathered rock is involved. The vertical attitude of the stones is probably due to the freezing and thawing which aids their migration to the surface. But involutions can also affect beds of silt or clay lying beneath gravels and cause them to be injected

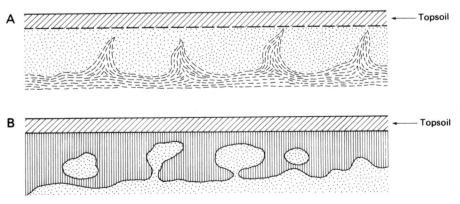

FIG. 15.9. Periglacial involutions (for explanation see text)

towards the surface (Fig. 15.9B). In plan many such structures are circular or polygonal and the processes leading to involutions may be almost identical with those leading to polygons. The basic causes of involutions, as of many other periglacial features, remain obscure, but two hypotheses are generally advanced.

The first is that the mobile layer is trapped as winter freezing starts from the surface, either between the surface freezing layer and the permafrost or between the surface freezing layer and an impermeable substratum. One can imagine that in these conditions pressure and movement would result in a generally chaotic structure being imparted to the mobile layer, although it is not easy to understand why the pattern produced should have elements of regularity.

The second hypothesis is that irregular frost-susceptibility of different materials leads to contortion. The tendency for ice segregations in silts, which would expand rapidly on freezing, would lead to injection of these into surrounding material. Again one would expect an irregular pattern and the problem is very similar to that of accounting for the regularity of small-scale stone polygons.

Both hypotheses regard involutions as consequences of the freezing of the active layer, so that the depth to which involutions go is useful in reconstructing the depth of annual thawing in areas of former periglacial climates.

Landform modification in periglacial climates

Most of the periglacial features discussed above are structural features of superficial deposits, very useful in interpreting past climatic history, but not prominent landforms. It is probably better to speak of periglacial modification of landforms rather than complete landscapes of specifically periglacial forms. Peltier (1950) outlined a periglacial cycle of erosion, which largely involved frost shattering and solifluxion, but it is very doubtful whether there was enough time for such a cycle ever to occur in the Pleistocene, however violent the processes of erosion may have been. The Pleistocene was a period of a couple of million years of fluctuating climate, only a quarter of which at an absolute maximum is likely to have been periglacial. Yet the imprint of periglacial denudation may be strong on the British landscape, because the last period of strong erosive activity may well have been the post-Weichselian periglacial phase.

Many of the processes of mass movements on slopes described in Chapter 4 were probably accelerated during Pleistocene periglacial conditions. The rotational slips affecting the Lower Greensand escarpment of the Weald are thought to have been much more common and on a larger scale than they are now. Similarly, the cambering mentioned in the same chapter and the breaking up of the feather edge of the camber by joints parallel to the contours, known as gulls, was probably accentuated in periglacial conditions.

In Chapter 7 the probable effects of periglacial conditions on the origin of dry valleys in Chalk areas was discussed, in particular the strong evidence for considering that much of the formation of the small valleys near Brook in Kent was periglacial. Other small valleys of the same type occur on the north side of the Cromer moraine in Norfolk, on Royston golf course on the Cambridgeshire–Hertfordshire border and in various other places. They are usually called periglacial valleys and were probably formed by a combination of snow melt runoff and spring activity. Possibly initiated by surface runoff over a frozen subsurface, though not necessarily permafrost, layer, complete thawing would later have saturated the rocks and spring flow would have been localised in the valleys and so enlarged them. The enlarged valleys more effectively channelled later melt runoff and so a typical positive feedback situation probably arose.

Waste mantles formed by earlier more tropical weathering may well have been removed by solifluxion in periglacial conditions, so exposing tors and other features of the weathering front as bare rock outcrops.

Various terracelike forms have been attributed to periglacial conditions. They include the stonebanked steps, intermediate between polygons and stripes, discussed above. Comparable in size are turf-banked terraces and lobes with fronts 2 m (6 ft) or so high and treads up to 5 or 6 m (15 or 20 ft) wide, though much larger features of the same type have been reported from outside Britain. These are essentially solifluxion features in which the turf banks serve to restrain the movement rather like a cooling skin of lava can restrain the flow of more mobile material within. Examples are known from the highlands of both Wales and Scotland. In favourable cases an overridden humus layer can be traced beneath such features and, if this can be dated, estimates of the rate of downhill movement of these terraces can be made. Overall rates of about 20 cm (8 in) per century have been obtained from Norwegian examples, though these average rates must include lengthy periods of stability. Jahn (1961) reports annual movements of solifluxion lobes of 10 cm (4 in) per year in Alaska and Spitzbergen, while extreme rates of 50 cm (20 in) per year have been observed.

On a larger scale are altiplanation terraces. These are irregular features found in highland regions, ranging up to 1 000 m (1 000 yd) or so in length, several hundred metres in width and 5 or 10 m (15 or 30 ft) high at the frontal edge. They cannot be explained by rejuvenation and are thought to be nivation features. It seems unlikely that they are due to the levelling of waste by periglacial action, as was once suggested, for many of them are cut into solid rock. They are much more likely to be due to nivation acting round snowbanks settling in existing hollows and on steps. Solifluxion may help to remove the debris formed by nivation. It is curious that nivation around snowbanks has been suggested as an explanation of both altiplanation terraces and of the early stages of corrie formation. Alti-

457

planation terraces are said to exist in the highest areas of south-west England.

Periglacial processes may also be responsible for some asymmetric valleys, which have been discussed in Chapter 5.

When one looks at the widespread distribution of polygonal and striped patterns and of presumed ice mound features in East Anglia, it seems that there has been little serious attack on the landscape since they were formed. It then begins to seem that many of our slopes may be in equilibrium with past, periglacial, solifluxion conditions and not with present processes, especially in areas where slopes are very gentle. Such a conclusion can only be reasonably firm where periglacial patterns are extensively preserved. But, theoretically, it is likely to apply widely, because stability on slopes affected by solifluxion is likely to be achieved at much lower angles than stability in relation to the now prevalent soil creep. Just as our highland landscape may be a relict glacial feature, so may much of our lowland landscape be a relict periglacial feature.

Williams (1968) tried to estimate the possible amounts of periglacial erosion in various parts of southern Britain. This is a difficult task, because ideally one must be able to measure the volume of material moved and also know the area from which it had been derived. A suitable situation is where a raised beach is backed by a degraded cliff, so that the material shed from the cliff accumulates over the beach. In other areas the sources of streams of boulders may be fairly closely definable. The results of these calculations give stripping varying from about 3 m (10 ft) thickness in parts of the Dartmoor granite to about 9–10 m (30–35 ft) on parts of the Chalk of central southern England. The difference may reflect the greater susceptibility of Chalk to this form of denudation and/or the more severe periglacial activity in southern than in south-west England. It is probable that both contributed, because any greater resistance of rock in the south-west should have been offset by the steeper slopes there encouraging denudation through solifluxion.

There seems to have been increasing mobility of material eastwards in southern Britain. In Dartmoor material moved about 1 km (1 000 yd) on slopes ranging from 6 to 12 degrees; in the Chalk of the Marlborough area 4 km (4 300 yd) over slopes of 1·5 to 3 degrees; in the Hythe Beds of west Kent over 6 km (7 000 yd) on a slope of 1 degree; in the Chilterns nearly 9 km (9 700 yd) over a slope of 1 degree; and in the Cambridge area over 10 km (11 300 yd) over a slope as low as 0·5 degrees.

16

Erosion surfaces and their interpretation

Strictly speaking, any surface which is not an original structural or constructional surface is an erosion surface. The term would thus include such forms as hill and mountain slopes and sea cliffs. But in practice it is applied essentially to surfaces of faint relief, the end products of complete or incomplete cycles of erosion.

Various processes already discussed in previous chapters are theoretically capable of forming such surfaces. Ordinary subaerial erosion, as envisaged by Davis, marine planation, pediplanation and possibly others lead ultimately to erosion surfaces, which should vary in form with the process concerned and in completeness with the length of time for which the process has been operating. Because the earth's crust has been unstable in the recent geological past, the remnants of successive attempts to wear the land's surface down may be preserved in the present landscape. They will be most readily recognisable where warping of the crust has not occurred or has been very limited in extent. Where great dislocations have taken place, the recognition and correlation of such surfaces will be practically impossible.

The fact that the remains of a series of surfaces can often be observed in a present landscape offers a way of interpreting the landscape differently from that suggested by Davis in terms of youth, maturity and old age. Instead, the landscape is described in terms of the various cycles or stages, through which it has passed. Such interpretations dominate present attempts to analyse the erosional history of the landscape. By studying the height and origin of the erosion surfaces, it is possible to describe approximately what base levels and what processes of erosion have been in operation, while the study of gentle warping of these surfaces sometimes adds to knowledge of recent tectonic movements, a fact clearly demonstrated by the study of Post-glacial shorelines discussed in Chapter 13.

Yet, in practice, problems arise, because of the great difficulty of distinguishing between erosion surfaces of various types, the poor state of preservation of many of them, and the possibility of other explanations.

459

Characteristics of various types of erosion surface

Peneplains

The peneplain, the end product of Davis's cycle of subaerial erosion, has already been briefly considered in Chapter 2. It is essentially a surface formed by the intersection of the minimum slopes needed to ensure that waste is evacuated from the surface of the land. The main controls will be the gradients of the major rivers, which cannot be less than those needed to ensure that the rivers will transport the load supplied to them. The minor streams will be of steeper gradient, because they will be less efficient transporting agents than the main streams, even though the calibre of the debris supplied to them may be the same as to the main rivers. Finally the slopes of the land will be somewhat steeper still and designed to ensure the transport of fine waste to the sluggish drainage system. Thus a peneplain is a lowland of faint relief, the divides being broad, gentle swells separating wide river valleys. It is not an easily defined form: it cannot be said that when slopes reach a certain gentleness, a peneplain comes into being. It marks the late stages of a cycle of subaerial erosion, but is not the final form, as it is always subject to very slow and very slight further degradation. It is neither a perfect plain, nor do the summits of its broad swells reach a common elevation, for the rivers that have been the essential controls in its development must preserve some gradient, so that the general elevation of the peneplain rises inland. Theoretically, in the case of an extensive peneplain, there is no reason why it should not rise to an elevation of several hundreds or even a thousand or more metres well into the interior. In addition, the state of peneplanation will be first achieved near the coast, where the drainage pattern first reaches a level below which it can only cut with the greatest difficulty. The peneplain will extend progressively inland and merge gradually with areas of very late maturity, without there being any clear dividing line between them.

If peneplains were ever observed in the degree of perfection described above, there should be little difficulty in recognising them, but good examples of peneplains are extremely rare and some would say that they do not exist. Davis himself reckoned that the time required for their formation would be that of a major geological period, that is something of the order of the length of the whole of the Tertiary or the Cretaceous periods. It is not likely that any part of the earth's crust has been sufficiently stable for a sufficient length of time to allow for the formation of a widespread peneplain, but there is the possibility of finding examples of what may be termed partial peneplains.

These may be recognised theoretically by a number of criteria some of which are of more value than others. The form, a system of gentle slopes, is rarely of any use, for surfaces which are thought to be peneplains are rarely found at the level of their formation. Instead, they have been uplifted and dissected and their remains are found only on ridge tops and summits, where it is very difficult, if not impossible, to decide whether the gentle slopes there are part of the peneplain or have been formed since.

The second criterion by which they may be recognised is by a consideration of the deposits on them. These should consist of a thick mantle of deeply weathered waste, but it is possible that similar thick waste mantles would also be found on some tropical forms, such as pediplains (see below), although it might be possible to distinguish between the often lateritic waste of a pediplain and that of a peneplain formed in a cool temperate humid climate. Unfortunately, owing to subsequent uplift and dissection it is only infrequently that erosion surfaces betray their origin by their waste mantles. Too often, they have merely a thin skin of amorphous waste, which is extremely difficult to interpret. A further difficulty may arise from our imperfect knowledge of the rate of weathering, for it is not known how long would be required for an uplifted and dissected plain of marine denudation to acquire a mantle of subaerial waste.

Finally, it is characteristic of a peneplain that its drainage should be very well adapted to the structure of the underlying rocks. The consequent streams of the early stages of the cycle of erosion are slowly replaced by the subsequent drainage characteristic of the later stages, so that by the time the stage of peneplanation is reached, the whole or the vast majority of the drainage should be adjusted to the rock structures. It is true that in old age the streams may become slightly out of adjustment due to lateral shifting on the thick waste mantle, but this will cause anomalies only in detail and not in the general plan. Such an adjustment to structure is not typical of an uplifted plain of marine denudation, although, again, only estimates of the rate at which streams will adapt themselves to structures on an uplifted surface of this type are available. On the other hand, an uplifted and deformed peneplain might have its drainage partly superimposed in peculiar directions from the waste mantle on to the underlying rocks, and its recognition could be difficult. On the whole, though, an uplifted and dissected surface with a drainage pattern adapted to the detail of the structure is more likely to be a peneplain than any other form of erosion surface, except perhaps a panplain.

Although the accordance of summit and ridge elevations over large areas, coupled with a well-adapted drainage pattern, could well be interpreted as a peneplain, the concept is less easily applied to the staircase pattern of erosion surfaces so common in many parts of the world. These surfaces, which are commonly no more than a few kilometres wide, are separated from each other by areas of more abrupt slopes (Fig. 16.1A). It is difficult to imagine this state of affairs being brought about by a series of incomplete peneplains. As suggested above, a peneplain gradually becomes less and less perfect towards the interior until, unless peneplanation is complete, it is no longer recognisable as a peneplain, so that the junction between two peneplains would be extremely ill-defined and probably not

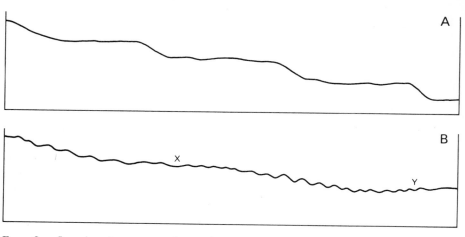

FIG. 16.1. Junctions between erosion surfaces: A. Staircase of erosion terraces; B. Junction of two incomplete peneplains, X and Y

recognisable after dissection (Fig. 16.1B). Fig. 16.1 lacks the third element of the actual picture, for in practice the breaks of slope in a staircase of erosion surfaces often run fairly straight across country. The junctions between two peneplains should not have this form. Just as peneplanation is first achieved near the sea, so inland it is extended headwards along rivers with the result that the lower of two peneplains should finger into the higher one along its valleys. The junction between the two would, therefore, be extremely obscure. Further, it is very doubtful whether any form of peneplanation would be complete in the narrow strips of a few kilometres width which form the erosion surfaces of many coastal areas.

The general accordance of summit levels, often interpreted as the remnants of a peneplain or other form of erosion surface, may possibly be explained in another way by recourse to the Gipfelflur hypothesis, which has been discussed by Baulig (1952). This hypothesis was largely derived from studies in the Alps, where there is an approximate equality in summit levels, so that a surface tangent to the summits represents roughly the general form of the mountain range. The interpretation of this as a summit peneplain obviously involves an enormous amount of uplift, and as other and even higher mountain ranges probably show comparable features, there would seem to be almost no limit to the height to which a peneplain might be uplifted, and warped, and still remain recognisable.

The Gipfelflur hypothesis would treat these accordant summits and crests as the intersections of slopes belonging to adjacent valleys. If the valleys are equally spaced, cut down to approximately the same depth, and flanked by slopes of the same angle, the accordant summit levels might be interpreted as a feature without significance (Fig. 16.2). Although these are somewhat specialised conditions, it might be possible for them to be

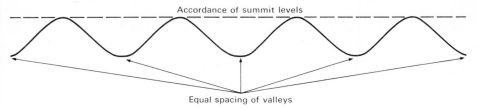

FIG. 16.2. The Gipfelflur hypothesis of accordant summits

realised in practice. The equal spacing of river valleys might be achieved in areas of homogeneous structure, especially in areas of massive crystalline rocks or in areas of highly folded, thinly bedded sediments. In these the spacing of the rivers would not be strongly subject to structural control and would possibly reflect the rainfall. With large amounts of precipitation the master streams would probably soon dominate the area and would probably be approximately equally spaced. Their equal size could ensure that they cut down at approximately the same rate, while, if they were all graded, they might be at approximately the same height above base level in any given part of the mountain region. The homogeneity of the structure would tend to ensure that valley slopes were at roughly the same angle provided that they were all being actively eroded at their bases by the rivers.

Another possible subsidiary cause was added by Richter to the original idea which was put forward by A. Penck. He observed that below the snow-line the slopes were to a considerable extent protected by vegetation, while above it they were protected by snow, but at the snowline freeze-thaw action was most violent. Hence there would be a tendency for mountain massifs to be planed off at this level. However, as Baulig points out, the snowline varied by more than 900 m (3 000 ft) in the Pleistocene, so that this idea cannot readily be applied to the Gipfelflur.

The very number of the special assumptions which have to be made renders the hypothesis less acceptable, but there are, in addition, certain points of direct evidence against the idea. A number of workers in the Alps have shown that Alpine relief probably consists of glacially modified relief of preglacial origin. The relief is essentially polycyclic and the slopes between valley bottoms and mountain tops are not uniform, but broken by terracelike features, which are sometimes sufficiently well preserved for the series of stages in which the valleys have been cut to be recognised. Thus, the crests do not represent the intersections of active valley slopes, but are the intersections of slopes now fossilised and divorced from the effects of present river erosion. In fact the Alpine crests seem to represent in detail the remnants of a number of local and sometimes imperfect planations, and not one broad surface still in process of being lowered.

Nevertheless, the idea remains for consideration, even though it may not seem to apply to the Alps. With more irregular spacing of rivers and

463

less constant angles of slope, the equality of summit heights would become less and less, but might still be sufficiently close to tempt one to postulate a dissected peneplain. There is no definition of what constitutes accordant summit levels and the term is interpreted subjectively by each investigator. Sometimes, the summits attributed to a surface, at 600 m (2 000 ft) for example, may vary by 100 m (330 ft) or more. It is then extremely difficult to decide whether they do represent an erosion surface or whether they might be due to the lowering of crests by the intersections of the slopes of more or less equally spaced valleys.

A variation of this idea has been revived by Hack and Goodlett (1960) in a discussion of the relief of a part of the central Appalachians. Many workers have explained the relief of this region as one of a series of partial planations, but it is possible to explain it in a different way. The ridge crests are approximately parallel to the valley bottoms and separated by equilibrium profiles which vary from rock to rock. Therefore, if the spacing of the streams is fairly regular and the typical valley sideslope angles vary with lithology, the result will be an apparent accordance of summit levels on each rock outcrop, thus simulating a staircase of erosion surfaces. The correspondence with lithology and structure should be so obvious that the plausibility of such a hypothesis in an area of alternating lithology ought to be capable of being tested merely by inspection.

Panplains

Certain difficulties in connection with the idea of peneplanation led Crickmay (1933) to postulate an alternative mechanism of normal subaerial erosion, namely lateral planation by rivers ultimately leading to the coalescence of floodplains into a feature which he termed a panplain. This idea, he considered, would more nearly reconcile observed facts with suggested hypotheses. For example, the fairly common occurrence of wide flat valley floors separated by steeper slopes from flat interfluves is not easy to explain by normal peneplanation. The gently undulating country, thought to be characteristic of a peneplain, occurs usually where a lithological control might be suspected, for example in broad clay lowlands. Again, many uplifted features, interpreted as peneplains, appear to be too flat to have been formed in that way, while the concept of a staircase of peneplains is not acceptable for reasons given above. Finally, the residuals on peneplains, including the type example of Mount Monadnock, seem to be too steep-sided to be accounted for by peneplanation.

Accordingly, Crickmay suggested his hypothesis of lateral river planation, which in essence is very like the lateral planation hypothesis of pediment formation (see Chapter 11). He observed that many floodplains are erosion features, thinly covered with alluvium, and that the extension and coalescence of these would lead to a flatter form with steep residuals more in accord with many observable features. There seems to be little doubt that river planation is an important agency of erosion, as is evidenced by the formation of erosional river terraces, but whether these can extend indefinitely to form a panplain seems more doubtful. Obviously, the meander belt of a river cannot exceed a certain width without cutoffs straightening it out again. The greatest width would be attained where the meanders are incised, but on the subdued relief of a broad floodplain cutoffs should readily occur, so that the meander belt would be considerably narrower. To effect the formation of a panplain, then, it would be necessary for the whole of the meander belt to migrate across country. Where it impinged on the steeper slopes at the edge of the floodplains, a very great amount of material would descend from the slopes into a river with a very low gradient and hence small transporting power. This would seem to be likely to decrease the erosive power of the river to an enormous extent. There remains, too, the theoretical difficulty, considered in Chapter 5, of attributing great powers of lateral planation to a river which has reached almost to the end of its powers of downcutting. Nevertheless, the idea remains as an alternative to peneplanation.

A panplain should be recognisable by its form, which would be very different from a peneplain, although possibly capable of confusion, if indifferently preserved, with a plain of marine denudation. Deposits, if present, should consist largely of stratified river alluvium and not of a mantle of weathered waste, as on a peneplain, nor of marine deposits, as on a plain of marine denudation.

Plains of marine denudation

Many geomorphologists have held the view that the sea, if left to act over a sufficient length of time, would carve out an erosion surface of considerable width. It has been seen in Chapter 8 that some would dispute this view on the grounds that, as the erosion platform increases in width, the power of the waves is expended on that platform and not on the cliffs behind it. Nevertheless, the idea of marine planation, which was very popular among English workers in the nineteenth century, is still frequently used by English geomorphologists, although it is usually applied to marine terraces of limited width rather than to erosional surfaces covering wide areas.

Davis was less favourably disposed to the idea of widespread marine planation and suggested that many might be marine-trimmed peneplains, formed during a period of slow submergence. To test such a hypothesis one would need to consider the nature of any sediments preserved beneath the general cover of marine deposits. If some of these sediments were obviously the products of subaerial weathering, then it would be highly likely that the feature was a marine-trimmed peneplain. On the other hand, a plain of marine denudation very slightly uplifted and then dissected by subaerial erosion would be difficult to distinguish from a peneplain, so that the argument cuts both ways. In addition, Baulig has pointed out that if the feature were really a plain of marine denudation it should have as the basal member of its covering rocks a conglomerate representing the material eroded from the cliffs by the sea during its task of planation. Some plains seem to be buried beneath marine deposits indicating quiet water conditions, as though the sea merely flooded over and buried a plain of another origin.

However, whether a surface is a true marine erosion feature or a marine-trimmed peneplain is of principal interest to those assessing the relative erosive power of rivers and of the sea. The geomorphological effect of both is almost identical. The marine-trimmed peneplain will be flatter than the peneplain and not readily distinguishable from a true marine surface, while drainage will be superimposed from the marine sediments overlying it, just as with a true marine plain.

Compared with a peneplain, the marine surface will be much flatter, though not without appreciable gradient. The gradient, which will be steeper in the inshore portion, will vary with the nature of the rocks and the exposure, and may well be as great as that of rivers producing a panplain, thus making it difficult to distinguish between the two, though the landward margin of a plain of marine denudation, representing the old cliffs, should be considerably straighter than that of any subaerial feature. When coastal staircases of terraces are being considered the hypothesis of marine planation obviously offers the most likely explanation, as an intermittently falling sea would naturally produce features of this sort.

Differentiations on form alone are often difficult because of the amount of dissection which has taken place. Unfortunately deposits are often, even usually, lacking from erosion surfaces attributed to the sea. Where they are present, the extreme rounding of the pebbles or marine fossils may serve to betray their origin.

There is, however, one criterion which is of much greater value than any of the above: it is the relationship of the drainage to the structure. On a marine erosion surface with deposits the drainage will be dependent on the slope of those deposits and will then be superimposed on to the rocks below. On a marine erosion surface without deposits it is likely that the new streams will either flow down the original beach gradient or in the direction in which the surface is tilted. Both lead to a superimposed pattern of drainage, which is usually considered to persist for a considerable length of time, at least till the original plain of marine denudation has been reduced to nothing more than accordant ridges and summits. It is possible that superimposition might occur on a subaerial surface if considerable aggradation had occurred, followed by a slight tilt, but the number of such cases is probably very small. Of course, the superimposed drainage, derived from a marine surface, will slowly become adapted to structure, so that some measure of adaptation is not against the interpretation of the surface as marine. The rate at which such adaptation occurs would probably depend on the vary-

ing resistance of the rocks, being quickest where there are large lithological variations (see Chapter 6).

Pediplains

Essentially pediplains (Chapter 11) are areas of gentle gradients dominated by steepsided residual hills. Whether they are due to lateral planation or to scarp recession, they cannot differ greatly in form from panplains.
They may even resemble marine plains, but the gradient of the latter
should be away from the former coastline, while the gradients of pediplains
will be away from the residuals or the places where those residuals existed,
and will therefore vary widely over the surface. In areas where semi-arid
conditions prevail and are known to have prevailed for a long time in the
past, widespread erosion surfaces are usually interpreted as pediplains. In
humid temperate regions one needs geological evidence in the form of such
beds as playa deposits or desert sandstones before attributing any surface
to pediplanation. Such surfaces are usually exhumed surfaces (see below)
for desert conditions do not seem to have existed in England since the
Triassic period, although surfaces as late as early Tertiary have been
attributed to desert erosion in France. Some of these are discussed later in
this chapter. In the same period in Britain tropical or semitropical weathering seems to have prevailed.

Glacial and periglacial erosion surfaces

Both glacial erosion (Chapter 13) and periglacial erosion (Chapter 15) are
powerful forces and various attempts have been made to develop cycles of
erosion under both glacial and periglacial conditions. Although it is possible to deduce theoretical cycles, it is doubtful whether there is ever
sufficient time for them to attain a sufficient stage of completion for a widespread erosion surface to be developed. The Pleistocene period probably
occupied about two million years but by no means the whole of that was a
period of cold conditions. If we assume that half was occupied by interglacial
conditions, we are left with roughly half a million years each for purely
glacial and for periglacial conditions in areas marginal to the ice sheets
such as the British Isles. However powerful the agents of erosion, this time
is so short compared with estimates for peneplanation or panplanation of
anything from 15 to 200 million years that it is unlikely that glacial or periglacial processes can lead to wide surfaces of low relief. In addition, the
sequence of glacial to interglacial climates ranged over each area in turn,
so that it might be very difficult to attribute any erosion surfaces to one
particular process.

One way in which planation might be effected is by the recession of
cirques into a mountain (Fig. 13.10). The formation of arêtes and pyramidal peaks has been attributed to this process in Chapter 13 and it is possible
that further development headwards would eliminate these features and

reduce a mountain to an irregular surface at approximately constant altitude with jagged, irregular ridges separating the greatly enlarged cirques. If, with increasing glaciation, the whole mountain block were to be occupied by an ice cap, these jagged ridges might be smoothed off (Hobbs, 1922). Such a degree of cirque erosion would only occur where the Pleistocene had been mainly a period of glaciation, i.e. nearer to the Pole than to the margins of an ice sheet, for only there would the snow-line and cirque level have remained sufficiently constant over a long enough period of time.

Occasionally, glacially scoured lowlands of the Laurentian shield type have been attributed to glacial erosion, but there is considerable evidence, summarised in Baulig (1952), that glaciation only modified in detail a pre-existing erosion surface due to other causes. In places preglacial clays and sands have not been destroyed by the ice sheet, while great thicknesses of rock weathered in the preglacial or interglacial periods have not been entirely scoured away by the ice. These facts are not consistent with a hypothesis of intensive glacial erosion.

A periglacial cycle has been outlined by Peltier (1950) and involves frost-shattering and solifluxion as its principal processes. A dissected upland would be attacked by frost-shattering, both on its flattish upper surfaces and on the valley slopes. Frost-shattered rock fragments would tend to remain undisturbed on the upper surfaces, but on the slopes there would arise a frost-shattered cliff at the top, with a scree slope below it. With time all rock soaked with water would be shattered and graded slopes formed. The material shattered by frost would be transported down these by solifluxion processes and meltwater action and accumulate in the valleys, unless the latter were periodically scoured out by streams. Peltier envisages these slopes slowly flattening until finally a surface of faint relief with slopes of less than 5 degrees is formed solely through periglacial processes.

Although such processes do operate in periglacial conditions and are extremely powerful agents of erosion, it seems unlikely, at least in the British Isles, that they have done more than modify an existing relief.

Exhumed surfaces

Most of the erosion surfaces discussed so far have never been buried beneath any great thickness of sediments, although they may have been thinly covered with weathered material or shallow marine deposits. There are, however, erosion surfaces buried beneath hundreds and in places thousands of metres of solid rock, some of them of very great age. While still buried they are simply unconformities in the geological sense. All unconformities, however, do not fall within the definition of erosion surfaces used at the beginning of this chapter, for some of them are by no means plane but consist of highly irregular surfaces buried beneath later sediments. As long as such an unconformity is not greatly deformed there is obviously the possibility of its being stripped by erosion and appearing in the landscape as an exhumed surface. In addition, in areas of horizontal or nearly horizontal sediments there is the possibility of softer material being stripped off a hard bed and that bed appearing as a marked terracelike feature.

Hard beds forming ledges or plateaus are often referred to as structural surfaces, but they are not truly structural as they do not represent the original surface of a series of beds. They have been exposed to view by erosion and hence are, in a sense, erosion surfaces. It is possible that downward erosion would be arrested by a hard bed and the surface of that bed stripped by some form of lateral erosion. In this case the ledge or plateau formed on the hard bed should coincide exactly with the surface of that bed. The impression of exact coincidence is usually lost when the feature is studied in more detail. The terrace is often seen to be slightly transgressive across various horizons of the hard bed, a fact which can only be established by a careful study of the stratigraphy. In examples of this sort it is much more likely that the hard bed happened to coincide more or less with the level at which an erosion surface was being formed. Where that erosion surface was formed on softer beds it was more readily destroyed and its long preservation on the hard bed heightens its apparent structural origin. Good examples of apparent structural surfaces, which are seen to be erosional when studied in detail, have been described from the Ingleborough district of the Pennines (Sweeting, 1950). Here the Carboniferous Limestone consists of massive limestone below and the alternating series of limestones, shales and sandstones known as the Yoredales above. A pronounced erosion surface at about 390 m (1 300 ft) OD locally coincides markedly with the massive limestone, but in other places it cuts across rocks below the Carboniferous Limestone and across the Yoredales above, while, where the detail of the Carboniferous Limestone bedding can be seen, it can be shown that the surface cuts across the bedding.

The exhumation of a gentle, undeformed unconformity is probably effected in a similar way to the stripping of a structural surface. The chances of an unconformity being revealed are greater when there is a series of non-resistant beds overlying very resistant beds, for example unconsolidated Tertiary sediments overlying granites or metamorphic rocks. Where the rocks on either side of the unconformity are approximately equal in resistance to erosion there is little chance of it being exhumed.

The exhumation will rarely if ever be perfect for the drainage pattern cutting through, for example, weak sediments on to an underlying crystalline complex will probably still go on cutting down. The rate of downcutting may, however, be slowed down and allow more lateral planation to shift away the remains of the weak sediments from the interfluves. In this case the exhumed surface will be represented on interfluves and summits and not necessarily be greatly different from any other uplifted and dissected surface. Its recognition will depend primarily on the presence of outliers and on the fact that the general form of the surface will prolong the plane of unconformity (Fig. 16.3).

In other areas, and especially near the coast, it may happen that an unconformity is approximately exhumed by the development of a series of erosion terraces (Fig. 16.4). If the beds above the unconformity are weak, the sea or river might quickly erode them and be considerably retarded when it meets the more resistant rocks below. Thus a series of terraces may almost completely destroy the weak beds and barely nick the more resistant

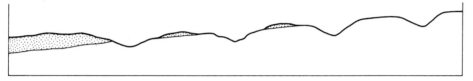

FIG. 16.3. Section of an exhumed erosion surface

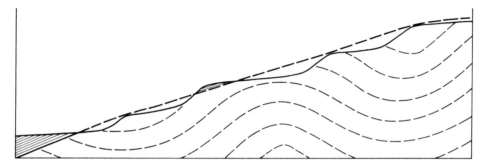

FIG. 16.4. Approximate exhumation of an unconformity by a series of erosion terraces

beds below. Whether one regards this as a series of erosion terraces or an exhumed surface depends largely on which aspect of the landscape one is most interested in. Approximate exhumations of sub-Eocene surfaces are to be found in some of the Chalk areas of southern England and northern France, where Eocene outliers are locally common, thus indicating the possibility of exhumed surfaces, while the abundance of Pliocene and Pleistocene terraces offers at the same time a mechanism to explain the stripping and an alternative view of the landscape.

This analysis of the various possible forms of erosion surface is by no means complete, but the question has been discussed critically by Baulig (1952) to whom reference should be made for further detail. It is time to turn to a discussion of actual landscapes to see the way in which they have been interpreted in terms of erosion surfaces. For this purpose, three contrasting areas will be taken: the Central Plateau of France, where ancient exhumed surfaces are of great significance; south-eastern England, where a whole succession of partial planations has occurred; and parts of Africa, which have been interpreted in terms of a series of cycles of pediplanation.

The Central Plateau of France

The Central Plateau of France is complicated in its geological detail, but the essentials of its form are not difficult to grasp. It is one of the largest of Europe's Hercynian blocks, the original folds of which were almost planed down and covered by a transgressing series of Jurassic deposits. At the time of the Alpine folding the Central Plateau was uplifted and fractured. The uplift was greatest on the south and east sides facing the Alps and the great escarpments of the Cevennes (Fig. 16.5) contrast strongly with the north-

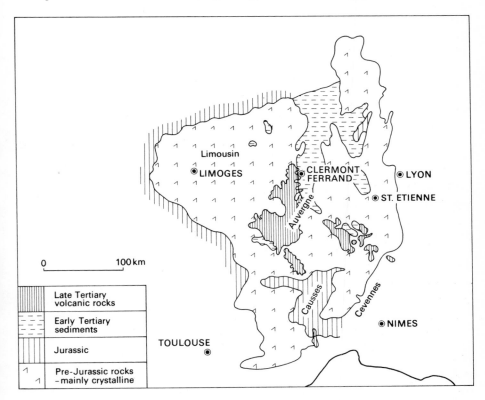

FIG. 16.5. Generalised geology of the Central Plateau of France

western part of the plateau, the Limousin, where the old rocks of the plateau disappear beneath the surrounding Jurassic sediments. At the same time fracturing took place within the plateau, the most notable results being the series of downfaulted troughs preserving Tertiary sediments in Auvergne. These Tertiary filled troughs are known as limagnes, the greatest of them being that followed by the river Allier to the east of Clermont Ferrand. After the fracturing considerable volcanic activity broke out in the later part of the Tertiary period and continued into the Quaternary. This was responsible for the great series of lava flows and volcanic plugs of the central parts of the plateau.

Within the plateau four major types of rocks occur. Firstly, there are the great series of igneous and metamorphic rocks, which pre-date the Hercynian folding and occupy the largest area (Fig. 16.5). Small areas of Carboniferous, Permian and Triassic rocks are locally preserved through downfolding and downfaulting, but these are not separated on Fig. 16.5. Secondly, there are the Jurassic rocks, mostly limestones or dolomites, which cover the plateau in the south-west to form the karst region, known as the Causses, and outcrop almost continuously on the western and northern borders of the plateau. Thirdly, there are Eocene and Oligocene sediments, lacking in resistance to erosion and preserved in downfaulted troughs mainly in Auvergne. Finally, the late Tertiary and Quaternary volcanics pile up to great thicknesses on the crystalline base and form the highest ground in the plateau.

Each major region of the plateau may be considered as being formed of one or more of these elements. Thus, Limousin is essentially an area of plateau developed on crystalline rocks, while the Auvergne owes its distinction to the juxtaposition of extensive volcanic areas and Tertiary filled basins. The Causses are the Jurassic limestone regions within the plateau while the eastern zone between the Loire and the Rhône consists of a series of crystalline blocks with a north-east to south-west trend often separated by Permian or Carboniferous basins.

But, although the lithological groups of the plateau thus give it regional differentiation, the surface is often a monotonous plateau, obviously due to erosion. In a very detailed analysis of the area Baulig (1928) has carefully distinguished the various erosion surfaces prominent in the present configuration of the landscape. These are of two main types, exhumed surfaces of great extent and considerable age, and more limited Pliocene and Pleistocene terraces.

The manner in which the Jurassic rocks of the northern margins lap on to the plateau clearly suggests the possibility of finding exhumed sub-Jurassic surfaces on the Central Plateau. But this surface is not to be found in Limousin, for the plateau surface there cuts across both the crystalline rocks and the Jurassic sediments and must be younger than both: it is in fact an early Tertiary surface, which will be discussed below. Fragments of the sub-Jurassic surface, which is known as the post-Hercynian surface, do occur on the northern borders of the plateau adjacent to the Jurassic rocks, and are also found in some of the less disturbed parts of the interior. The origin of the surface is betrayed largely by the deposits resting on it.

As Baulig observes, the post-Hercynian surface in the Central Plateau, as in many other of the Hercynian regions of Europe, is remarkable for its uniformity, which is too great for it to be attributed to a normal peneplain. This conclusion is reinforced by the nature of the deposits on it. The surface is fossilised by a transgressing series of deposits, ranging from Trias to Lias in age, of uniform lithology. This is not a surface of marine abrasion, for there is no basal marine conglomerate denoting long continued and powerful wave action and, even where the basal deposits are marine, they are often limestones indicating deposition in quiet water. The basal beds seem to consist of a slightly reworked layer of decomposed rock varying with the nature of the underlying strata, which are themselves often rotted and stained. This fossil waste mantle is succeeded by a series of red sandy clays, lenticular limestones, mottled clays and gypseous marls, which are overlain in turn by bedded sandstones and conglomerates. The lithology clearly suggests semi-arid tropical weathering and erosion, with some variation in the amount of precipitation. At times the climate must have been sufficiently humid to cause intense rotting of the underlying rock, while at others the presence of lenticular limestones and gypseous beds indicates semi-arid conditions with playa lakes. In fact, this is an exhumed erosion surface, which bears a remarkable resemblance to a pediplain. It was buried beneath the deposits of the transgressing Jurassic sea and parts of it, exhumed at a much later date, are still visible in the landscape.

The formation of this surface was brought to an end by uplift at the end of the Jurassic period. A new cycle of erosion was then initiated and ran on till the whole Central Plateau was uplifted and fractured during the Alpine folding in the middle of the Tertiary period. Although this cycle produced a surface less perfect than that resulting from the post-Hercynian cycle, it was nevertheless another semi-arid cycle. This is indicated by the xerophytic nature of the fossil plants and by the lithology of the deposits. Many of these deposits were formed by weathering: they include laterites, arkoses and, on the Mesozoic limestones, clays formed by decalcification. These may be replaced both laterally and vertically by gypseous and calcareous deposits of lake or lagoon origin. The whole complex is known as the sidérolithique and evidently denotes a long period of tropical weathering, during which this second, or Eogene, erosion surface was formed.

The erosion surface was of course deformed by the mid-Tertiary uplift and fracturing and the role it plays in the present landscape varies with the degree of disturbance of the area concerned. In Limousin, which was the least disturbed area of the whole of the Central Plateau, it covers vast areas and still bears patches of unstratified, residual material. These deposits can be traced laterally on to the Mesozoic rocks to the west, where they are residual below and bedded above, and these in turn can be shown to pass into marine deposits of Eocene and Oligocene age in the Aquitaine Basin. Thus the deposits betray not only the origin of the surface but also its age.

The Auvergne, on the other hand, was greatly dislocated by the Alpine movements and the Eogene surface does not play nearly so important a part in the landscape. In the basins downsinking was taking place at the same time as sedimentation, so that the Eogene surface is to be found at the bottom of the limagnes beneath a considerable thickness of fresh or brackish water deposits. Even there the underlying crystalline rocks are rotted and laterites locally present. In the volcanic regions the Eogene surface lies beneath the volcanic rocks, which were erupted on to an already dislocated Eogene surface, and buried both the surface and the deposits on it. In the block-faulted crystalline area between the Loire and the Allier, the Eogene surface is in places preserved on the tilted surface of the blocks, e.g. Forez and Livradois.

Whereas the post-Hercynian and Eogene surfaces cover considerable areas of plateau, the much later Pliocene and Pleistocene surfaces are found within the valleys and especially on the Mediterranean border of the Central Plateau. These, unlike the earlier exhumed surfaces, are essentially horizontal for no appreciable earth movements have occurred since their formation. In the area between the Cevennes and the Mediterranean, and especially well preserved on the limestones there, is a series of terraces at elevations of 380, 280, 250, 180 and 150 metres (1 250, 930, 830, 600 and 500 ft) above sea level. Of these the 380 m, 280 m and 180 m are the most important. They remain at constant elevations round the Gulf of Lions and even extend about 100 km (60 miles) up the Rhône valley. Although it would be tempting to regard them as marine terraces, Baulig states that the evidence points to a subaerial origin. The surfaces of the terraces are not flat enough for marine terraces and seem to possess a pattern of shallow valleys organised for the drainage of the area. Lower terraces penetrate along valleys into country mainly occupied by higher terraces, while the terraces are perfectly well developed in areas sheltered on the seaward side by barriers of higher land. If they had been marine, Baulig argues that they should be less well developed there because marine erosion would have been reduced. The argument is not completely convincing, for it has been seen in Chapter 8 how well raised beaches are developed in extremely sheltered areas of the west of Scotland. However, there are no traces of cliffing and no marine deposits on the terraces, which bear only a thin residual layer of terra rossa. Thus, they would seem to be of subaerial origin, but developed in close proximity to the sea, an assumption necessary to explain their uniform heights. They are correlated with terraces at similar elevations near Algiers, where the 280 and 180 m terraces cut across inclined Lower Pliocene strata and are thus probably Upper Pliocene in age.

Thus the Central Plateau is seen to betray the presence of two major cycles of erosion and a number of later partial cycles. In detail more than two exhumed surfaces may be recognised, for very locally a sub-Permian surface occurs on the margins of Permian basins. The interpretation of the general history of the development of the landscape is typical of work of this sort. It is not, of course, a complete interpretation of landscape, but defines the major stages in its evolution. Superimposed on it are all the diverse effects of lithology on relief and the intervention of processes such as glacial and periglacial action, which exerted a profound local modification of the landforms but did not result in the formation of any erosion surfaces. These effects are usually much more apparent in the landscape than the erosion surfaces, the recognition of which demands the interpretation of fragmentary data in the field and its reconstruction on a map.

Southeastern England

The area occupied by the Hampshire Basin, the London Basin and the Weald has probably been more closely studied than any other in Britain, at least as far as erosion surfaces are concerned. In some ways it is well suited to these investigations, for it contains large areas of Chalk on which erosion surfaces are well preserved owing to the permeability of the rock. Further, the area south of a line east–west along the foot of the North Downs escarpment is corrugated with minor east–west folds of Alpine age, so that the relations between structure and drainage are readily observed. On the other hand, the eastern part of the region falls within the area affected by the downsinking of the North Sea basin, so that towards the east coast, especially in the north of the area, erosion surfaces are warped and in places no longer recognisable.

The fact that summit levels of south-eastern England are approximately accordant has long been recognised, but the exact nature of the surface thus represented has been disputed. Earlier English opinion tended to adopt the idea of a plain of marine denudation, but in 1895 Davis argued that it was an uplifted peneplain. His argument was based upon the degree of adaptation of the drainage to the structure, which he considered to be so complete as to require for its formation a complete cycle of erosion, which ended in the development of the surface now represented by the accordant summits, together with part of a second cycle, which has dissected that surface to its present state.

But, in fact, the drainage is not nearly as perfectly adapted to the structure as Davis maintained. Locally, it corresponds closely, but in other areas such as the Hampshire Basin and the South Downs it is at right angles to the folds and obviously superimposed. Such considerations led Bury (1910) to reject the idea of a peneplain and to revive the idea that the upland surface was of marine origin. This surface, at least in the Weald, was thought to have been warped slightly about both north–south and east–west axes, i.e. the Weald was very slightly domed up again. From it the present drainage was superimposed and those streams at present adapted to the structure have achieved this adaptation in the present cycle of erosion through the rapidity with which subsequent streams develop in rocks with such contrasting lithologies as those of the Weald.

It appears that both these ideas contain an element of truth, for in the later hypothesis of Wooldridge and Linton (1955) the upland surface is regarded as a peneplain, the edges of which in the Hampshire and London

Basins and along the South Downs have been trimmed up into a plain of marine denudation.

The surviving portions of the peneplain are small, for it has only been preserved on the more resistant rocks of the region, principally the Chalk. At the present time the largest area of peneplain is to be found in southern Wiltshire on the Great Ridge south of the Wylye valley. There is no doubt that it is an erosion surface for it truncates the folds of southern England, which are of Alpine, mid-Tertiary age, and a number of lines of evidence suggests that it is subaerial in origin. It possesses low residual eminences and the direction of the gradient, where it can be observed, seems to be towards the present drainage lines. Further, it is suggested that the true residual Clay-with-flints is best developed on the surface. These features are all consistent with the interpretation of the feature as a subaerial peneplain, but the strongest evidence is based on the relations of drainage to structure. Those areas not trimmed up by a later phase of marine planation generally show a longitudinal drainage pattern well adjusted to the structure (Fig. 16.6), a feature typical of a prolonged period of subaerial denudation. This is perhaps best seen in parts of the Weald, where the tributaries of the upper Medway and the streams draining eastwards towards Dungeness are approximately parallel to the folds. The same is true of the west Sussex Rother, the principal tributary of the river Arun. It must not be thought that within this area remnants of the peneplain may be observed on the ridge tops in all areas, for it is probably true that none are high enough except the Lower Greensand (Hythe Beds escarpment) and the higher parts of the Chalk in the northern and western Weald.

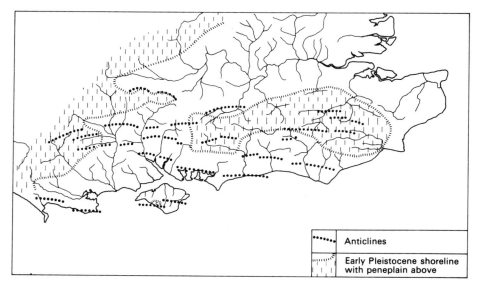

••••••	Anticlines
⁙⁙⁙⁙	Early Pleistocene shoreline with peneplain above

Fɪɢ. 16.6. Fold trends and drainage in south-east England (*after Wooldridge and Linton*)

Elsewhere all that remains of the peneplain is the drainage pattern inherited from it.

The area outside the peneplain has a drainage pattern less closely adapted to the structure, as can be seen by the way in which the rivers of the Hampshire Basin and the South Downs transect the folds (Fig. 16.6), but there is other evidence for regarding this area as a marine erosion surface. For this it is necessary to turn to the London Basin, which is not a good area for studying the relations between drainage and structure, as it is outside the area of east–west Alpine folding. At three places fossiliferous marine deposits are known: piped into the Chalk at Lenham in Kent; resting on the Chalk dipslope at Netley Heath in Surrey, and in relict blocks at Rothamsted in Hertfordshire. Elsewhere unfossiliferous material of the same type, and distinguishable from Eocene outliers by its heavy mineral assemblage, occurs at a number of localities at similar elevations. Often these deposits occur on a terrace cut into the Chalk dipslopes of the Chilterns and North Downs. Even where the deposits are not found, the terrace still occurs, in places cut into the dipslopes and in others bevelling the crest of the North Downs (Fig. 16.7). Thus in the ideal example there is an area of peneplain on the crest of the Downs (Fig. 16.7A) separated by a degraded cliff from the marine terrace lower down the dipslope. This arrangement is probably best seen between Reigate and Sutton in Surrey (Fig. 9.16).

The Lenham deposits appear to be the natural continuation and to contain the same type of fauna as the Diestian beds of Belgium, while the Netley Heath deposits are equivalent to the Red Crag of the East Anglian succession. This indicates that the sea responsible for their deposition was a transgression, for the Red Crag is younger than the Diestian. Both these beds used to be classed as Lower Pliocene in age, but, due to revisions of the horizon at which the Pliocene–Pleistocene boundary is drawn, they now represent the base of the Pleistocene. The peneplain above must be older, but it cannot be older than the folding of the area, since it is not deformed.

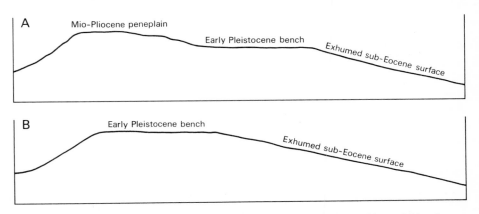

Fig. 16.7. Forms of Chalk dipslope on the North Downs (*after Wooldridge and Linton*)

This leaves the Miocene and Pliocene periods for its formation: hence, it is termed the Mio-Pliocene peneplain.

Although early Pleistocene deposits are absent from the Hampshire Basin and virtually so from the South Downs, remnants of the terrace on which they are thought to have been deposited can still be found, especially in the Hampshire Basin. In the South Downs the terrace has been reduced to a number of accordant summit levels. But the presence of marine planation may be inferred from the superimposed drainage pattern, which has been discussed in Chapter 7 and illustrated in Fig. 6.27. Thus the picture of a peneplain with well-adjusted drainage and its marine-trimmed edge with superimposed drainage is complete throughout south-eastern England.

A third surface of a very different type occurs on the dipslope of the Chalk in places below the level of the early Pleistocene bench. It is the stripped sub-Eocene surface of the Chalk (Fig. 16.7A and B). It can also be seen on the northern side of Fig. 9.16 below the early Pleistocene bench. A study of the 1-inch Geological sheet, No. 271, of Dartford is also to be recommended: west of the river Cray a solid mass of Thanet Sands, Blackheath Beds and London clay obscures the Chalk, but east of that stream outliers of Thanet Sands and Blackheath Beds indicate that the intervening areas of Chalk must approximate to the sub-Eocene surface.

The early Pleistocene bench at approximately 195–165 m (650–550 ft) OD represents one of the longest and most clearly datable stages in the evolution of south-eastern England. Below it a whole series of terraces, both simple and composite, continue down to present sea level. Some of these later stages in the evolution of the relief of the area have already been considered in the discussion of the glacial diversions of the lower Thames (Chapter 14). Unfortunately, these are river terraces in the broad sense and they are terraces of a river flowing towards a downwarped area. Furthermore, although they are well developed in the middle part of the course of the Thames, approximately west of the latitude of London, they are not well developed to the east, where the sea must have been, so that it is difficult to be sure of the sea levels to which they were graded.

On the other hand, the south coastal areas, which would seem to be the obvious place to look for lower and later terraces, are also not entirely satisfactory. In the South Downs (Sparks, 1949) there appears to be a series of marine terraces at 145, 131, 116, 100 (composite), 70 and 56 m (475, 430, 380, 330 (composite), 230 and 180 ft) OD. But the whole area is very dissected and the terraces are no more than flattenings on ridges, some of them almost destroyed and very indistinct. In the Hampshire Basin (Everard, 1954) some of the same stages may be recognised, though they are not all marine, while in east Kent (Coleman, 1952) the same series, except the upper two, appear to be present and to have been downwarped 6–10 m (20–30 ft). It is interesting that the upper terraces of east Kent have been downwarped most, a feature entirely in agreement with the progressive sinking of the North Sea basin.

It seems that after the two early major episodes of the Mio-Pliocene peneplain and the early Pleistocene marine transgression, south-eastern England was affected by an intermittently falling base level. Near the sea, this resulted in the cutting of a series of marine terraces, as in the South Downs and parts of the Hampshire Basin, while elsewhere, for example in the Thames and east Kent, a corresponding series of river terraces is found. None of these attained any great width. Our knowledge of them is somewhat fragmentary for their correlation poses difficult problems. They are so closely spaced that correlations from one area to another almost require tracing the terraces continuously on the ground. This is impossible in some areas, for example in the interior of the Weald. Although the terraces may be traced on the South Downs and although terraces have been observed on the high ground formed by the Hastings Beds in the centre of the Weald, it is very difficult to correlate the two areas, for the Weald Clay between has been excavated to such a depth that few if any terraces remain.

It can be seen, then, that the denudation of the Weald has been effected by a number of partial cycles of erosion, most of which have done no more than form terraces at most a few kilometres wide. In detail, the history of the evolution of relief is much more complex and more problematical than depicted here, for it ends with the oscillating base levels of Pleistocene times and the features formed in relation to those base levels. Even in the earlier history there are problems, but they do not seriously affect the general conclusion that the evolution of relief of south-eastern England is best treated in terms of a terrace succession.

Southern Africa

When one turns from south-eastern England to southern Africa, the scale of things changes stupendously. Instead of a small part of a small country one is dealing with a third of a vast continent, while the multitude of small Pliocene and Pleistocene erosion surfaces is replaced by a few vast erosion surfaces formed during the whole period from the Jurassic to the present day. Some of these surfaces are composite, while in places they are warped considerably. Thus, though this is another problem of cyclic geomorphology, it must be approached with an attitude of mind somewhat different from that adopted when dealing with small parts of Europe which have been the subject of intensive research for long periods.

The vast plateaus of southern Africa have received considerable attention and syntheses covering the whole area have been put forward by King (1950, 1951, 1955 and 1967). Although exhumed surfaces, not necessarily flat, have been discovered, these are subordinate to vast pediplains, formed under arid or semi-arid climates. King specifically rejects peneplanation as a mechanism, for the African surfaces are separated by clearly defined escarpments, which could not have survived the flattening and degrading processes associated with peneplanation. It is very difficult if not impossible to visualise peneplanation producing a vast staircase of erosion surfaces: this is the same problem discussed earlier in this chapter in connection with the much smaller scale staircases of erosion surfaces in England.

The age of the pediplains is not always easy to assess for a number of reasons. Datable sediments lying on them are distinctly rare and, King argues, these may not give the exact date of the main phase of pediplanation, but only the date of a slight reworking of the surface. Pediplanation involves scarp retreat and pediment formation and progresses inward towards interfluvial areas from incised rivers. Thus, a given block of upland may be planed off at one time by a pediment extending back from one river and at a later date, and possibly at a slightly lower elevation, from another pediment extending back from another river. Conditions of this sort are illustrated in Fig. 16.8, where the pediment from river B is extending at the

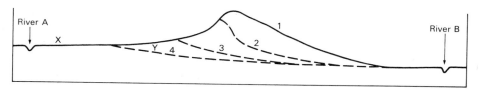

Fig. 16.8. Extension of a pediment across an earlier pediment

expense of that developed from river A. It is obvious that on surface 4 the sediments at Y may be considerably younger than those at X. Pediments, too, may be expected to suffer reworking as the rivers which control their level are slowly incised and as the material on them becomes progressively finer in the late stages of the cycle. Thus, the deposits of a pediment, the main cutting of which may have occurred in the Mesozoic, may be quite recent.

The only areas of southern Africa with a fairly full record of Mesozoic and Tertiary sedimentation are the coastal fringes, and King uses breaks in sedimentation there as indications of the period when the various cycles of pedimentation began and ended. Sedimentation began in the early Jurassic, was broken about the time of the Middle Chalk of England and again broken in the Miocene.

Very local traces of exhumed surfaces may be found in southern Africa where overlying sediments are being stripped from unconformities. Some of these are extremely old, being Pre-Cambrian in age, but the most important seems to be the pre-Karroo surface. The Karroo beds range in age from the Carboniferous to the Jurassic, so that the surface buried beneath them is not all of one age. In the south it is in places an ancient glaciated surface, formed by the great Permo-Carboniferous glaciation of the southern hemisphere. The sub-Karroo surface, which is locally deformed, has a considerable relief and such features as fossil valleys and solution holes have been found. But, in spite of its interest, it occupies but a very small part of the total landscape, compared with the later cyclic surfaces.

The oldest and highest surface is known as the Gondwana surface. It is thought to have been formed while all the southern land masses, Africa, the plateau region of South America, India and Australia, all formed part of a single continent, Gondwanaland, which was later disrupted by continental drift. Its age is considered to be Jurassic and early Cretaceous and it was the most perfect planation in southern Africa. It corresponds very closely in age with the post-Hercynian surface of the Central Plateau of France.

When the old continent of Gondwanaland broke up the African plateau, with a general elevation of approximately 600 m (2 000 ft), was downwarped at its edges, where marine sediments were deposited. A new cycle, the post-Gondwana cycle, started to erode inwards from the coast and from the rejuvenated rivers, but towards the end of the Cretaceous further warping occurred. This is attributed by King to isostatic uplift of the continental margin as material was eroded from it. It resulted in the Gondwana and post-Gondwana surfaces being tilted back towards the interior of the continent and the initiation of another cycle of erosion.

The next cycle, the African cycle, ran on through the Tertiary to the late Oligocene or early Miocene periods, when further uplift of the margins of the continent occurred. At the present time this surface occupies the greatest area in southern Africa, although it is not such a perfect pediplain as the Gondwana surface (cf. the Eogene surface of the Central Plateau of France, which resembles it closely in age and in being less perfect than the preceding surface).

The succeeding Late Tertiary cycles have merely trimmed the edges of southern Africa except in northern Natal and Mozambique, where they cover considerable areas. They are of Pliocene and Pleistocene age, and nowhere produce such perfect plains as the earlier cycles. Added to them are the late Pleistocene changes, which have produced minor coastal features, which are insignificant by comparison with the others.

Once again, the landscape is seen as the product of a series of cycles of varying degrees of completeness, cycles of pediplanation in this example. The reconstruction of the history of south African relief is bold and it must be added that it involves hypotheses that are very disputable. The subject of Continental Drift has only to be mentioned to arouse violent disagreement among geologists, geophysicists, zoologists and others interested in various aspects of the history of the earth. Similarly, the idea of isostatic upwarping of continental margins as material is removed from them by erosion does not command universal agreement. Some of the processes of pediplanation are little understood, but the same applies to most other processes thought to be capable of forming erosion surfaces.

The examples of the analysis of the history of evolution of relief described above have all used deposits, lying on the erosion surfaces or correlated with

481

the erosion surfaces, in the interpretation of those surfaces. There are examples from other areas where there is little evidence beyond the staircase of erosion surfaces. In such cases one can do little more than quote a catalogue of elevations at or near which base level is suspected to have been. When the erosion surfaces are fragmentary in the extreme there is room for scepticism, specially in examples where they are alleged to go up to the summits of the mountains, though there is a sufficient number of clear examples known to make it foolhardy to dismiss such hypotheses out of hand. It is mentally more satisfying to the investigator to have more complex arguments to toy with, and this has led people, if they remain interested in the history of the evolution of relief, to attempt to integrate this type of geomorphological study with Pleistocene studies, where specialised techniques abound. Precision is gained and truth is more likely to be revealed when converging lines of evidence lead to a conclusion, but studies on this basis usually demand collaboration if the full benefits of expertise are to be reaped. Even then one must not expect too much, for it is only too easy to assume that other techniques are infallible, an attitude more common to the outsider than to the practitioner. As an illustration of the methods and their uses let us consider the area around Cambridge.

The Cambridge area

The valley of the river Cam, which is merely a tributary of the Great Ouse, has always had a lot of attention paid to its Pleistocene deposits, because successive professors of geology at Cambridge were interested in the deposits which were extensively exposed in the nineteenth century. The relief, which is subdued if not monotonous, has never received as much attention. It consists essentially of two blocks of subdued uplands, separated by a broad strike valley occupied by a puny stream (Fig. 16.9).

It lies too near the downwarped area of the southern North Sea for it to be too readily assumed that one will find the terrace pattern of other parts of southern England developed at the same elevations here. The Chalk is nowhere high enough for the preservation of Wooldridge and Linton's early Pleistocene terrace at its original 195 m (650 ft) elevation, for example: one does not know whether it has been eroded away or warped down.

Yet on the geomorphological side there are traces of erosion terraces at 100, 70, 56 and 40 m (330, 230, 180 and 130 ft) (Fig. 16.10), and, as far as one can tell, they remain horizontal through the area. There are no deposits on them, at least no deposits likely to be synchronous with their formation. So, one is left merely with what can be gleaned from their form and distribution to explain their origin. Their horizontality suggests a marine origin, and this idea receives some support from other features. The 100 m terrace is located in front of the Chalk escarpment in the west near Ashwell and on the Chalk escarpment near the Saffron Walden Cam.

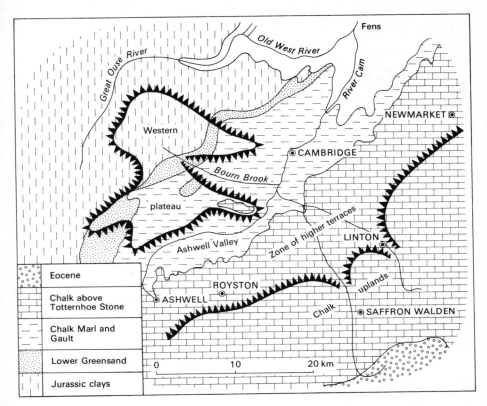

Fig. 16.9. Relief regions and solid geology of the Cam valley (*after Sparks and West*)

East of that, the general surface elevation is of the order of 115 m (380 ft) on boulder clay, but beneath the boulder clay a plot of the elevation of the Chalk in wells and outcrops on to a generalised dip section reveals a dissected landscape with a remarkable summit uniformity of 100 m (Fig. 16.11). Because it is all on one rock, this might be dismissed in terms of the Hack and Goodlett explanation mentioned earlier in the chapter, but in fact the back of the 100 m terrace must run diagonal to the strike of the Chalk and it seems very unlikely that subaerial processes would produce an escarpment oblique to the strike, whereas marine processes might.

Again, the 40 m terrace also seems to be horizontal. This can be traced intermittently round the Fens at constant elevation. If the line along which it meets higher ground is regarded as an old degraded shoreline, it is interesting that two of the rivers flowing into the Fens, the Great Ouse and the Nene, have abrupt bends on that line, at Huntingdon and Wansford respectively. They cross the shoreline at a very oblique angle, but are then aligned at right-angles to the shore as though they were flowing down the direction of maximum gradient of an emerged sea floor.

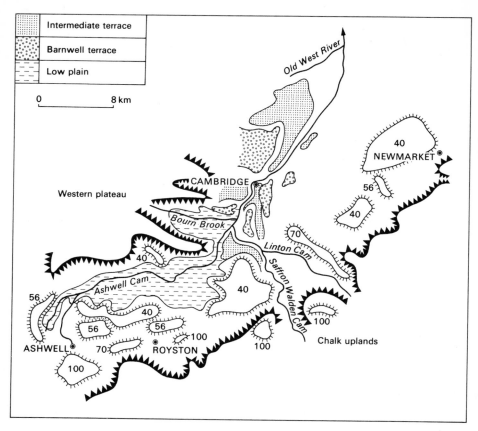

FIG. 16.10. Terraces of the Cam valley (*after Sparks and West*)

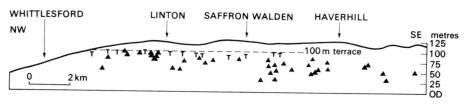

FIG. 16.11. Concealed 100 m terrace beneath the boulder clay in the Cambridge region. Heights of Chalk in wells (▲) and outcrops (**T**) (*after Sparks and West*)

Beyond this one can only argue by analogy. Lower-level interglacial and Post-glacial marine invasions of the Fens are evidenced by fossiliferous marine deposits, so that earlier, higher-level invasions cannot be excluded as improbable.

Lower terraces in the Cam valley are gravel features around Cambridge, while an eroded valley floor in the upper part of the Ashwell Cam seems to be their equivalent (Fig. 16.10). These features are probably later than the boulder clay which forms the Western Plateau and caps the Chalk Uplands.

The relation of that boulder clay to the general geomorphological evolution of the Cam valley is important, because it can provide a firm marker in the erosion history of the region. In this simple account we shall have to ignore the detailed, local question whether there are in fact two boulder clays of different ages. It might be argued, because the Cam valley is cut into a plateau capped with boulder clay, that the valley must be later than the boulder clay. Very little, if any, indisputable boulder clay is found in the valley of the Ashwell Cam. Yet instinct cries out against this deduction: the Cam valley is 10 km (6 miles) or so wide and, if one assumes it to be post-boulder clay in age, a quite fantastic amount of erosion is called for since the Gipping (last but one) glaciation.

One needs to look at the indirect geological relations between the boulder clay and the valley deposits (Fig. 16.12). Around Barrington pure

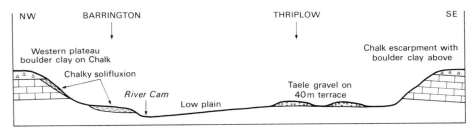

FIG. 16.12. Geological relations between boulder clay and Cam valley

chalky solifluxion occurs in little filled gullies in the hillside and spreads out over the terrace on which Barrington is sited. As the chalky solifluxion does not appear to contain boulder clay material, it has usually been alleged to predate the boulder clay: in passing, it should be stressed that such negative evidence is never wholly satisfactory. As the solifluxion conforms to the present general slope of the valley side, that slope must predate the solifluxion and hence the boulder clay. This argument could be turned by assuming some rather curious physical distributions of material; it has been stated here dogmatically only for the sake of brevity.

On the other side of the valley near Thriplow the taele gravel, a solifluxion gravel shown by its erratic content to be derived from the boulder clay, again seems to conform closely to the present regional slope. In age it must just postdate the boulder clay, so that again it seems highly likely that the relief was there before the boulder clay, unless we assume a catastrophically rapid valley formation between the boulder clay and the solifluxion.

Further, in the Western Plateau and the Chalk Uplands borings show that the base of the boulder clay goes down to the general level of the floor of the Cam valley. In the former at least the boulder clay is stuffed into valleys tributary to the Cam, which must therefore have been in existence.

485

So, the major elements of the relief, on the basis of geological arguments, are shown to predate the main glaciation of the area, which stone orientation shows to be Gipping on the Chalk Uplands, although it may be one glaciation earlier on the Western Plateau.

As the gravel terraces of the Cam valley contain Gipping till erratics they cannot be earlier than last (Ipswichian) interglacial in age, but more about their age and conditions of deposition can be learned from their detailed study. There are certainly three terraces (four according to the Geological Survey).

3. The Barnwell Station beds: these hardly form a true terrace as they occur beneath the Post-glacial floodplain mainly.
2. The Intermediate terrace: elevation about 9–10 m (30–35 ft) in Cambridge.
1. The Barnwell terrace: elevation about 15–17 m (50–60 ft) in Cambridge.

Where the Barnwell terrace was most closely investigated on the road to Histon north of Cambridge, it consists of about 7–8 m (25–28 ft) of silt and gravel beds, which represent the latter half of the last interglacial period. This is demonstrated by pollen analysis, which shows hornbeam forest at the base, succeeded upwards by pine and birch woodlands with much open country, the succession ending in frost-heaved gravels. Local environmental conditions, as revealed by a study of fruits, seeds and non-marine Mollusca, can be regarded as those of a river meandering over its own aggradation plain. The continued aggradation is somewhat surprising, as we know that the highest base level of the interglacial occurred in the hornbeam forest zone, yet here aggradation went on after this when base level must have been falling. It seems that the fall in sea level merely extended rivers over the gentle North Sea floor without providing a rejuvenation: any tendency to cut down may have been offset by the greater supply of debris which the streams had to remove from the less forested landscape.

Finally, for some reason unknown, rejuvenation occurred and the Intermediate terrace was formed. It consists of about 5 m (16 ft) of aggradation and was closely investigated in west Cambridge. The general look of the flora and fauna is glacial. In this and similar age deposits around the edge of the Fens bison, reindeer and mammoth are the main vertebrate fossils, while the upper parts of these deposits, which are almost certainly to be correlated with the Intermediate terrace, show fine periglacial ice wedge and ice lens features. The deposits occur, where the rivers enter the Fens, at almost uniform heights of about 4 m (13 ft), as though a large shallow lake was blocked by ice at the time. The lake, if it existed, would presumably have been slightly earlier than the ground ice features which cut its deposits, though several generations of ice wedge features suggest a complex glacial history.

Finally another wave of rejuvenation occurred, possibly when the mouth of the Wash was freed of its ice sheet, and the Cam cut down to deposit the Arctic Barnwell Station beds in what is now a buried channel beneath the Post-glacial floodplain caused by aggradation consequent on a rise of sea level.

This summary account is designed to illustrate the synthesis of different types of evidence. The full details of the reconstructions and the degree of confidence in them can be found in the references quoted by Sparks and West (1965). Difficulties have been glossed over and certain of the conclusions presented here are bound to be modified by later work.

These areas illustrate the way in which the development of landscape tends to be treated by geomorphologists today. It is no longer described in terms of youth, maturity and old age, probably because nearly all landscapes seem to be polycyclic. The stages recognised are the end products of various cycles and partial cycles of erosion of various types and the alternating climates of the Ice Age. In these interpretations many assumptions are made about processes involved, for geomorphologists interested in this type of work infer processes from the presumed results of those processes, many of which are imperfectly understood. The stages recognised by the study of erosion surfaces form the skeleton of the landscape, but it must be remembered that that skeleton may be almost completely obscured by lithological effects and by dissection related to modern base levels.

References[1]

Chapter 2   The Davisian geographical cycle

CHORLEY, R. J., DUNN, A. J. and BECKINSALE, R. P. (1964) *The History of the Study of Landforms, I: Geomorphology before Davis*. Methuen. This work is an invaluable summary of the development of ideas, including those of Davis's immediate predecessors on which much of his own generalisation was based. An analysis of Davis and his work is promised for a later volume.

DAVIS, W. M. (1909) *Geographical Essays*. Boston, Ginn (reprinted 1954, Dover).

DAVIS, W. M. (1912) *Die erklärende Beschreibung der Landformen*. Leipzig, Teubner.
These are the only collected accounts of Davis's thought, the second being much more comprehensive.

GILBERT, G. K. (1880) *Report on the Geology of the Henry Mountains*, 2nd edn, United States, Department of the Interior, Washington.

GLOCK, W. S. (1932) 'Available relief as a factor in the profile of a land form', *J. Geol.* **40**, 74–83.

Chapter 3   Weathering

BARTON, D. C. (1916) 'Notes on the disintegration of granite in Egypt', *J. Geol.* **24**, 382–93.

BIROT, P. (1960) *Le cycle d'érosion sous les différents climats*, Universidade do Brasil, Faculdade nacional de Filosofia, Centro de Pesquisas de Geografia do Brasil.

BLACKWELDER, E. (1925) 'Exfoliation as a phase of rock weathering', *J. Geol.* **33**, 793–806.

BLACKWELDER, E. (1933) 'The insolation hypothesis of rock weathering', *Am. J. Sci.* **26**, 97–113.

CAILLEUX, A. (1962) 'Études de géologie au détroit de McMurdo', *Comité national français pour les recherches antarctiques, Publication* no. 1.

[1] Abbreviations used are listed on p. 508.

DURRANCE, E. M. (1970) Letter. *Geol. Mag.* **106**, 496–7.

EVANS, I. S. (1970) 'Salt crystallisation and rock weathering: a review', *Revue Géomorph. dyn,* **19**, 153–177.

GOLDICH, S. S. (1938) 'A study in rock weathering', *J. Geol.* **46**, 17–58.

GOUDIE, A., COOKE, R. U. and EVANS, I. S. (1970) 'Experimental investigation of rock weathering by salts', *Area (Inst. Br. Geogr.),* **2**, 42–48.

GRAWE, O. R. (1936) 'Ice as an agent of rock weathering: a discussion', *J. Geol.* **44**, 173–82.

GRIGGS, D. T. (1936) 'The factor of fatigue in rock exfoliation', *J. Geol.* **44**, 781–96.

KELLER, W. D. (1957) *The Principles of Chemical Weathering,* rev. edn, Columbia, Missouri, Lucas. This is a comprehensive and useful summary rather like the survey by P. Reiche.

LINTON, D. L. (1955) 'The problem of tors', *Geogrl J.* **121**, 470–87.

MERRILL, G. P. (1921) *Rocks, Rock-weathering and Soils,* New York.

OLLIER, C. D. (1969) *Weathering,* Oliver & Boyd. This book is indispensable for anyone wishing to go further into the subject of weathering, while its comprehensive list of references leads directly into the original literature.

RASTALL, R. H. (1927) *Physico-chemical Geology,* London.

READ, H. H. (1936) *Rutley's Elements of Mineralogy,* 23rd edn, London.

REICHE, P. (1950) 'A survey of weathering processes and products', *University of New Mexico Publications in Geology,* no. 3. A comprehensive and useful summary.

SMITH, L. L. (1941) 'Weather pits in granite of the southern Piedmont', *J. Geomorph.* **4**, 117–27.

SPARKS, B. W. (1971) *Rocks and Relief,* Longman. A more detailed discussion of some of the subjects dealt with in this chapter is contained in this book.

TAMM, O. (1924) 'Experimental studies on chemical processes in the formation of glacial clay', *Sveriges Geologiska Undersökning,* series C, no. 333.

TYRRELL, G. W. (1926) *The Principles of Petrology,* London.

WILLIAMS, J. E. (1949) 'Chemical weathering at low temperatures', *Geogrl Rev.* **39**, 129–35.

WILLIAMS, R. B. G. (1969) 'Periglacial climate and its relation to landforms', unpublished Ph.D. thesis, University of Cambridge.

Chapter 4   The development of slopes

BAKKER, J. P. and LE HEUX, J. W. N. (1946) 'Projective-geometric treatment of O. Lehmann's theory of the transformation of steep mountain slopes', *Proc. K. ned. Akad. Wet.,* ser. B, **49**, 533–47.

BAKKER, J. P. and LE HEUX, J. W. N. (1947) 'Theory on central rectilinear recession of slopes', *ibid.,* **50**, 959–66 and 1154–62.

BAKKER, J. P. and LE HEUX, J. W. N. (1950) 'Theory on central rectilinear recession of slopes', *ibid.,* **53**, 1073–84 and 1364–74.

BAKKER, J. P. and LE HEUX, J. W. N. (1952) 'A remarkable new geomorphological law', *ibid.*, **55**, 399–410 and 554–71.

BAULIG, H. (1940) 'Le profil d'équilibre des versants', *Annls Géogr.* **49**, 81–97.

BAULIG, H. (1950) *Essais de géomorphologie*, Paris, Les Belles Lettres. This book is invaluable as it contains most of Baulig's best essays on processes and geomorphological methodology reprinted with notes and commentaries on other research published after the dates of the original essays.

BIROT, P. (1949) *Essais sur quelques problèmes de morphologie générale*, Lisbon, Centro de Estudos Geográficos.

BLACKWELDER, E. (1942) 'The process of mountain sculpture by rolling debris', *J. Geomorph.* **5**, 325–8.

BRUNSDEN, D. (Ed.) (1971) 'Slopes: form and process', *Inst. Br. Geogr.*, Special publication No. 3.

CARSON, M. A. (1969a) 'Models of hillslope development under mass failure', *Geographical Analysis*, **1**, 76–100.

CARSON, M. A. (1969b) 'Soil moisture', Ch. 4 (II) in R. J. Chorley, ed., *Water, Earth and Man*, Methuen.

CARSON, M. A. and PETLEY, D. J. (1970) 'The existence of threshold hillslopes in the denudation of the landscape', *Trans. Inst. Br. Geogr.* **49**, 71–96.

CARTER, C. S. and CHORLEY, R. J. (1961) 'Early slope development in an expanding stream system', *Geol. Mag.* **98**, 117–30.

COMMON, R. (1954) 'A report on the Lochaber, Appin and Benderloch floods', *Scott. geogr. Mag.* **70**, 6–20.

COMMON, R. (1966) 'Slope failure and morphogenetic regions', in G. H. Dury, ed., *Essays in Geomorphology*, Heinemann.

COTTON, C. A. (1952) 'The erosional grading of convex and concave slopes', *Geogrl J.* **118**, 197–204.

DIJK, W. VAN and LE HEUX, J. W. N. (1952) 'Theory of parallel rectilinear slope-recession', *Proc. K. ned. Akad. Wet.*, ser. B, **55**, 115–29.

ENGELN, O. D. VON et al. (1940) 'Symposium on Walther Penck's contribution to geomorphology', *Ann. Ass. Am. Geogr.* **30**, 219–80.

FENNEMAN, N. F. (1908) 'Some features of erosion by unconcentrated wash', *J. Geol.* **16**, 746–54.

FISHER, O. (1866) 'On the disintegration of a chalk cliff', *Geol. Mag.* **3**, 354–6.

GIFFORD, J. (1953) 'Landslides on Exmoor caused by the storm of 15th August, 1952', *Geography* **38**, 9–17.

GILBERT, G. K. (1909) 'The convexity of hilltops', *J. Geol.* **17**, 344–51.

GOSSLING, F. (1935) 'The structure of Bower Hill, Nutfield (Surrey)', *Proc. Geol. Ass.* **46**, 360–90.

GOSSLING, F. and BULL, A. J. (1948) 'The structure of Tilburstow Hill, Surrey', *ibid.*, **59**, 131–40.

GROVE, A. T. (1953) 'Account of a mudflow on Bredon Hill, Worcestershire, April 1951', *Proc. Geol. Ass.* **64**, 10–13.

HOLLINGWORTH, S. E., TAYLOR, J. H. and KELLAWAY, G. A. (1944) 'Large-scale superficial structures in the Northampton ironstone field', *Q. Jl geol. Soc.* **100**, 1–44.

HORTON, R. E. (1945) 'Erosional development of streams and their drainage basins: hydrophysical approach to quantitative morphology', *Bull. geol. Soc. Am.* **56** (1), 275–370.

KENNEDY, B. A. (1969) 'Studies of Erosional Valley-side Asymmetry', unpublished Ph.D. thesis, University of Cambridge.

KIDSON, C. (1953) 'The Exmoor storm and the Lynmouth floods', *Geography* **38**, 1–9.

KIRKBY, M. J. (1967) 'Measurement and theory of soil creep', *J. Geol.* **75**, 359–78.

KIRKBY, M. J. (1969) 'Erosion by water on hillslopes', Ch. 5 (II) in R. J. Chorley, ed., *Water, Earth and Man*, Methuen.

LAWSON, A. C. (1915) 'The epigene profiles of the desert', *University of California Dept. of Geology Publication No. 9*, 23–48.

LAWSON, A. C. (1932) 'Rain-wash erosion in humid regions', *Bull. geol. Soc. Am.* **43**, 703–24.

LEHMANN, O. (1933) Morphologische Theorie der Verwitterung von Steinschlagwänden. *Vierteljahrsschrift der Naturforschende Gesellschaft in Zürich*, **87**, 83–126.

LEOPOLD, L. B., WOLMAN, M. G. and MILLER, J. P. (1964) *Fluvial Processes in Geomorphology*, San Francisco, Freeman.

LINTON, D. L. (1930) 'Notes on the development of the western part of the Wey drainage system', *Proc. Geol. Ass.* **41**, 160–74.

MELTON, M. A. (1957) 'An analysis of the relations among elements of climate, surface properties, and geomorphology', *Technical Report No. 11, United States Office of Naval Research*.

MELTON, M. A. (1960) 'Intravalley variation in slope angles related to microclimate and erosional environment', *Bull. geol. Soc. Am.* **71**, 133–44.

MELTON, M. A. (1965) 'Debris-covered hillslopes of the southern Arizona desert—consideration of their stability and sediment contribution', *J. Geol.* **73**, 715–29.

PENCK, W. (1924) *Morphological Analysis of Landforms*, English translation by H. Czech and K. C. Boswell, London, Macmillan, 1953.

SAVIGEAR, R. A. G. (1952) 'Some observations on slope development in South Wales', *Trans. Inst. Br. Geogr.*, no. 18, 31–52.

SCHEIDEGGER, A. E. (1961) *Theoretical Geomorphology*, Berlin, Springer.

SCHUMM, S. A. (1956a) 'Evolution of drainage systems and slopes in badlands at Perth Amboy, New Jersey', *Bull. geol. Soc. Am.* **67**, 596–646.

SCHUMM, S. A. (1956b) 'The role of creep and rainwash in the retreat of badland slopes', *Am. J. Sci.* **254**, 693–706.

SCHUMM, S. A. (1964) 'Seasonal variations of erosion rates and processes on hillslopes in western Colorado', *Z. Geomorph.*, Suppl. 5, 215–38.

SCHUMM, S. A. (1966) 'The development and evolution of hillslopes', *J. geol. Educ.* **14**, 98–104.

SCHUMM, S. A. and CHORLEY, R. J. (1966) 'Talus weathering and scarp recession in the Colorado Plateaus', *Z. Geomorph.*, n.s. **10**, 11–36.

SHARPE, C. F. S. (1938) *Landslides and Related Phenomena*, Columbia University Press.

SKEMPTON, A. W. (1964) 'Long-term stability of clay slopes', *Géotechnique*, **14**, 77–101.

SOUCHEZ, R. (1966) 'Slow mass-movement and slope evolution in coherent and homogeneous rocks', *Bull. Soc. belge Géol.* **74**, 189–213.

STRAHLER, A. N. (1950) 'Equilibrium theory of erosional slopes approached by frequency distribution analysis', *Am. J. Sci.* **248**, 673–96 and 800–14.

STRAHLER, A. N. (1956) 'Quantitative slope analysis', *Bull. geol. Soc. Am.* **67**, 571–96.

WOOD, A. (1942) 'The development of hillside slopes', *Proc. Geol. Ass.* **53**, 128–40.

WOOLDRIDGE, S. W. (1950) 'Some features in the structure and scenery of the country around Fernhurst, Sussex', *Proc. Geol. Ass.* **61**, 165–90.

WURM, A. (1935) 'Morphologische Analyse und Experiment Hangentwicklung, Einebnung, Piedmonttreppen', *Z. Geomorph.* **9**, 57–87.

YOUNG, A. (1960) 'Soil movements by denudational process on slopes', *Nature*, **188**, 120–2.

Chapter 5   The nature of a river valley

The following two references are of the utmost importance:

LEOPOLD, L. B., WOLMAN, M. G. and MILLER, J. P. (1964) *Fluvial Processes in Geomorphology*, San Francisco, Freeman.

MORISAWA, M. (1968) *Streams: their dynamics and morphology*, McGraw-Hill.

The following should also be consulted for more details:

BATES, R. E. (1939) 'Geomorphic history of the Kickapoo region', *Bull. geol. Soc. Am.* **50**, 819–80.

BAULIG, H. (1948) 'Le problème des méandres', *Bull. Soc. belge Etudes géogr.* **17**, 103–43.

BAULIG, H. (1950) *Essais de géomorphologie*, Paris, Les Belles Lettres.

BIROT, P. (1949) *Essais sur quelques problèmes de morphologie générale*, Lisbon, Centro de Estudos Geográficos.

BIROT, P. (1952) 'Sur le mechanisme des transports solides dans les cours d'eau', *Revue Géomorph. dyn.* **3**, 105–41.

BLACHE, J. (1939, 1940) 'Le problème des méandres encaissés et les rivières lorraines', *J. Geomorph.* **2**, 201–12, and **3**, 311–31.

COLEMAN, A. M. (1952) 'Some aspects of the development of the lower Stour, Kent', *Proc. Geol. Ass.* **63**, 63–86.

DAVIS, W. M. (1909) *Geographical Essays*, Boston, Ginn (reprinted 1954, Dover).

DURY, G. H. (1965) 'Theoretical implications of underfit streams', *Prof. Pap. U.S. geol. Surv.* no. 452-C.

DURY, G. H. (1966) 'The concept of grade', in G. H. Dury, ed., *Essays in Geomorphology*, Heinemann.

DURY, G. H. (1969a) 'Hydraulic geometry', Ch. 7 (II) in R. J. Chorley, ed., *Water, Earth and Man*, Methuen.

DURY, G. H. (1969b) 'Relation of morphometry to runoff frequency', Ch. 9 (II), in *ibid.*

FRIEDKIN, J. F. (1945) *A Laboratory Study of the Meandering of alluvial rivers*, U.S. Waterways Experiment Station, Vicksburg, Miss.

GILBERT, G. K. (1880) *Report on the Geology of the Henry Mountains*, 2nd edn, United States, Department of the Interior, Washington.

GILBERT, G. K. (1914) 'The transportation of debris by running water', *Prof. Pap. U.S. geol. Surv.* no. 86.

HACK, J. T. (1965) 'Postglacial drainage evolution and stream geometry in the Ontonagon area, Michigan', *Prof. Pap. U.S. geol. Surv.* no. 504-B.

HAPP, S. C., RITTENHOUSE, G. and DOBSON, G. C. (1940) *Some Principles of Accelerated Stream and Valley Sedimentation*, U.S. Department of Agriculture, Washington.

HJULSTRÖM, F. (1935) 'Studies in the morphological activity of rivers', *Bull. Geol. Inst., Upsala*, 25.

HORTON, R. E. (1945) 'Erosional development of streams and their drainage basins: hydrophysical approach to quantitative morphology', *Bull. geol. Soc. Am.* **56** (1), 275–370.

KENNEDY, B. A. (1969) 'Studies of Erosional Valley-side Asymmetry', unpublished Ph.D. thesis, University of Cambridge.

KESSELI, J. E. (1941) 'The concept of the graded river', *J. Geol.* **49**, 561–88.

KIDSON, C. (1953) 'The Exmoor storm and the Lynmouth floods', *Geography*, **38**, 1–9.

KIRKBY, M. J. (1963) 'A Study of the Rates of Erosion and Mass Movements on Slopes, with special reference to Galloway', unpublished Ph.D. thesis, University of Cambridge.

KIRKBY, M. J. (1969) 'Infiltration, throughflow, and overland flow', Ch. 5 (I) in R. J. Chorley, ed., *Water, Earth and Man*, Methuen.

LEIGHLY, J. (1934) 'Turbulence and the transportation of rock debris by streams', *Geogr. Rev.* **24**, 453–64.

LEWIS, W. V. (1936) 'Nivation, river grading and shoreline development in south-east Iceland', *Geogrl J.* **88**, 431–47.

LEWIS, W. V. (1944) 'Stream trough experiments and terrace formation', *Geol. Mag.* **81**, 241–53.

LEWIS, W. V. (1945) 'Nick points and the curve of water erosion', *ibid.*, **82**, 256–66.

MACKIN, J. H. (1948) 'Concept of the graded river', *Bull. geol. Soc. Am.* **59**, 463–512.

MATTHES, G. H. (1941) 'Basic aspects of stream-meanders', *Trans. Am. geophys. Union*, **22**, 632–6.

MELTON, M. A. (1960) 'Intravalley variation in slope angles related to microclimate and erosional environment', *Bull. geol. Soc. Am.* **71**, 133–44.

RASTALL, R. H. (1944) 'Rainfall, rivers and erosion', *Geol. Mag.* **81**, 39–44.

SCHEIDEGGER, A. E. (1961) *Theoretical Geomorphology*, Berlin, Springer.

SCHUMM, S. A. (1956) 'The role of creep and rainwash in the retreat of badland slopes', *Am. J. Sci.* **254**, 693–706.

TAILLEFER, E. (1950) 'Le versant atlantique des Pyrénées et son avant-pays', *Revue Géomorph. dyn.* **1**, 101–22.

Chapter 6   The development of drainage systems

ARKELL, W. J. (1956) 'Geological results of the cloudburst in the Weymouth district, 18th July, 1955', *Proc. Dorset nat. Hist. archaeol. Soc.* **77**, 90–5.

BAILEY, E. B. (1939) 'Tectonics and erosion', *J. Geomorph.* **2**, 116–19.

BAULIG, H. (1938) 'Questions de terminologie', *J. Geomorph.* **1**, 224–9. This paper is reprinted with comments in *Essais de Géomorphologie*, 1950.

BROWN, E. H. (1960) *The Relief and Drainage of Wales*, University of Wales Press.

DAVIS, W. M. (1909) *Geographical Essays*, Boston, Ginn (reprinted 1954, Dover).

ENGELN, O. D. VON (1942) *Geomorphology*, New York, Macmillan, Ch. xvi, summarises the Appalachian drainage and gives references.

GREEN, J. F. N. (1949) 'The history of the river Dart, Devon', *Proc. Geol. Ass.* **60**, 105–24.

HORTON, R. E. (1945) 'Erosional development of streams and their drainage basins', *Bull. geol. Soc. Am.* **56** (1), 275–370.

HOWARD, A. D. (1938) 'A case of autopiracy', *J. Geomorph.* **1**, 341–2.

JOHNSON, D. W. (1932) 'Streams and their significance', *J. Geol.* **40**, 481–97.

JONES, R. O. (1939) 'The evolution of the Neath–Tawe drainage system, South Wales', *Proc. Geol. Ass.* **50**, 530–66.

JONES, O. T. (1952) 'The drainage systems of Wales and adjacent regions', *Q. Jl geol. Soc.* **107**, 201–25. Many references to other papers.

LEOPOLD, L. B., WOLMAN, M. G. and MILLER, J. P. (1964) *Fluvial Processes in Geomorphology*, San Francisco, Freeman.

MARR, J. E. (1906) 'The influence of the geological structure of English Lakeland upon its physical features', *Q. Jl geol. Soc.* **62**, lxvi–cxxviii.

MORISAWA, M. E. (1968) *Streams: their dynamics and morphology*, McGraw-Hill.

SHREVE, R. L. (1966) 'Statistical law of stream numbers', *J. Geol.* **74**, 17–37.

SHREVE, R. L. (1967) 'Infinite topologically random channel networks', *ibid.* **75**, 178–86.

SPARKS, B. W. (1949) 'The denudation chronology of the dip-slope of the South Downs', *Proc. Geol. Ass.* **60**, 165–215.

SPARKS, B. W. (1951) 'Two drainage diversions in Dorset', *Grography*, **36**, 186–92.

SPARKS, B. W. (1953) 'Erosion surfaces around Dieppe', *Proc. Geol. Ass.* **64**, 105–17.

SPARKS, B. W. (1953) 'Stages in the physical evolution of the Weymouth lowland', *Trans. Inst. Br. Geogr.* **18**, 17–29.

STRAHLER, A. N. (1968) 'Quantitative geomorphology', in R. W. Fairbridge, ed., *Encyclopedia of Geomorphology*, New York, Reinhold, pp. 898–912.

WAGER, L. R. (1937) 'The Arun river drainage pattern and the rise of the Himalaya', *Geogr. J.* **89**, 239–49.

WOOLDRIDGE, S. W. and LINTON, D. L. (1955) *Structure, Surface and Drainage in South-east England*, 2nd edn, London, Philip.

WOOLDRIDGE, S. W. and MORGAN, R. S. (1937) *The Physical Basis of Geography*, Longmans.

ZERNITZ, E. R. (1932) 'Drainage patterns and their significance', *J. Geol.* **40**, 498–521.

Chapter 7   The effects of rocks on relief

AUB, C. F. (1964) 'Karst problems in Jamaica, with particular reference to the cockpit problem', *20th International Geographical Congress, Karst Symposium* (duplicated).

BÖGLI, A. (1960) 'Kalklösung und Karrenbildung', *Z. Geomorph.* Suppl. **2**, 4–21.

BÖGLI, A. (1964) 'Mischungskorrosion—Ein Betrag zum Verkarstungsproblem', *Erdkunde*, **18**, 83–92.

BULL, A. J. (1936) 'Studies in the geomorphology of the South Downs', *Proc. Geol. Ass.* **47**, 99–129.

BULL, A. J. (1940) 'Cold conditions and landforms in the South Downs', *ibid.* **51**, 63–71.

CHANDLER, R. H. (1909) 'On some dry chalk valley features', *Geol. Mag.* **46**, 538–9.

CORBEL, J. (1959) 'Erosion en terrain calcaire', *Annls Géogr.* **68**, 97–120.

COTTON, C. A. (1944) *Volcanoes as Landscape Forms*, Christchurch, New Zealand, Whitcombe and Tombs.

FAGG, C. C. (1923) 'The recession of the Chalk escarpment', *Trans. Croydon nat. Hist. scient. Soc.* **9**, 93–112.

FAGG, C. C. (1939) 'Physiographical evolution in the North Croydon survey area and its effect upon vegetation', *Proc. Trans. Croydon nat. Hist. scient. Soc.* **2**, 29–60.

FAGG, C. C. (1954) 'The coombes and embayments of the Chalk escarpment', *ibid.* **12**, 117–31.

FAGG, C. C. (1958) 'Swallow holes in the Mole gap', *South-Eastern Naturalist and Antiquary*, **62**, 1–13.

FORD, D. C. and STANTON, W. I. (1968) 'The geomorphology of the south-central Mendip Hills', *Proc. Geol. Ass.* **79**, 401–28.

GARRELS, R. M. and CHRIST, C. L. (1965) *Solutions, Minerals and Equilibria*, Harper & Row.

HODGSON, J. M., CATT, J. A. and WEIR, A. H. (1967) 'The origin and development of Clay-with-flints and associated soil horizons on the South Downs', *J. Soil Sci.* **18**, 85–102.

HUNT, C. B. (1953) 'Geology and geography of the Henry Mountains region, Utah, *Prof. Pap. U.S. geol. Survey*, no. 228.

JUKES-BROWNE, A. J. (1906) 'The Clay-with-flints: its origin and distribution', *Q. Jl geol. Soc.* **62**, 132–64.

JUKES-BROWNE, A. J. and WHITE, H. J. O. (1908) 'The geology of the country around Henley-on-Thames and Wallingford', *Mem. Geol. Survey, U.K.*

KERNEY, M. P., BROWN, E. H. and CHANDLER, T. J. (1964) 'The Late-glacial and Post-glacial history of the Chalk escarpment near Brook, Kent', *Phil. Trans. R. Soc.* B, **248**, 135–204.

KIRKALDY, J. F. (1950) 'Solution of the Chalk in the Mimms valley, Herts.', *Proc. Geol. Ass.* **61**, 219–23.

MONROE, W. H. (1964) 'Lithologic control in the development of a tropical karst topography', *20th International Geographical Congress, Karst Symposium* (duplicated).

NORTH, F. J. (1930) *Limestones*, London, Murby.

NORTH, F. J. (1949) *The River Scenery at the Head of the Vale of Neath*, 3rd edn, National Museum of Wales, Cardiff.

PITTY, A. F. (1968) 'The scale and significance of solutional loss from the limestone tract of the southern Pennines', *Proc. Geol. Ass.* **79**, 153–78.

PRESTWICH, J. (1855) 'On the origin of the sand- and gravel-pipes in the Chalk of the London Tertiary district', *Q. Jl geol. Soc.* **11**, 64–84.

PRINGLE, J. (1948) 'British Regional Geology: the South of Scotland', 2nd edn, *Mem. Geol. Survey, U.K.*

REID, C. (1887) 'On the origin of dry chalk valleys and of coombe rock', *Q. Jl geol. Soc.* **43**, 364–73.

REID, C. (1899) 'The geology of the country around Dorchester', *Mem. Geol. Survey, U.K.*

SAUNDERS, E. M. (1921) 'The cycle of erosion in a Karst region (after Cvijić)', *Geogrl Rev.* **11**, 593–604.

SMALL, R. J. (1961) 'The morphology of Chalk escarpments: a critical discussion', *Trans. Inst. Br. Geogr.* **29**, 71–90.

SMALL, R. J. (1964) 'The escarpment dry valleys of the Wiltshire Chalk', *ibid.* **34**, 33–52.

SMALL, R. J. and FISHER, G. C. (1970) 'The origin of the secondary escarpment of the South Downs', *ibid.* **49**, 97–108.

SMET, R. E. DE and SOUCHEZ, R. (1964) 'Evolution comparée de deux massifs dolomitiques: Catinaccio et Sella', *Revue belge Géog.* **88**, 157–86.

SPARKS, B. W. (1949) 'The denudation chronology of the dip-slope of the South Downs', *Proc. Geol. Ass.* **60**, 165–215.

SPARKS, B. W. (1957) 'The evolution of the relief of the Cam valley', *Geogrl J.* **123**, 188–207.

SPARKS, B. W. (1971) *Rocks and Relief*, Longman.

SPARKS, B. W. and LEWIS, W. V. (1957) 'Escarpment dry valleys near Pegsdon, Hertfordshire', *Proc. Geol. Ass.* **68**, 26–38.

SWEETING, M. M. (1950) 'Erosion cycles and limestone caverns in the Ingleborough district', *Geogrl J.* **115**, 63–78.

SWEETING, M. M. (1958) 'The karstlands of Jamaica', *ibid.* **124**, 184–99.

SWEETING, M. M. (1966) 'The weathering of limestones, with particular reference to the Carboniferous limestones of northern England', in G. H. Dury, ed., *Essays in Geomorphology*, Heinemann.

SWEETING, M. M. (1968) 'Some variations in the types of limestones and their relation to cave formation', *Proceedings of the 4th International Congress of Speleology in Yugoslavia*, (Ljubljana), iii, 227–32.

THOMAS, T. M. (1954) 'Swallow holes in the South Wales coalfield', *Geogrl J.* **120**, 468–75.

WELLS, A. K. and KIRKALDY, J. F. (1967) *Outline of Historical Geology*, 6th edn, Murby.

WILLIAMS, P. W. (1966) 'Limestone pavements with special reference to western Ireland', *Trans. Inst. Br. Geogr.* **40**, 155–72.

WILLIAMS, R. B. G. (1969) 'Periglacial Climate and its Relation to Landforms', unpublished Ph.D. thesis, University of Cambridge.

WOOLDRIDGE, S. W. and KIRKALDY, J. F. (1937) 'The geology of the Mimms valley', *Proc. Geol. Ass.* **48**, 307–15.

WOOLDRIDGE, S. W. and LINTON, D. L. (1955) *Structure, Surface and Drainage in South-east England*, 2nd edn, London, Philip.

Chapter 8 Coastal features

The literature on shoreline forms is not as scattered as that on some of the other geomorphological subjects and a considerable amount is available in book form.

BIRD, E. C. F. (1968) *Coasts*, Massachusetts Institute of Technology Press. This provides a wide descriptive treatment of coasts suitable for the university student at an early stage.

FAIRBRIDGE, R. W., ed. (1968) *Encyclopedia of Geomorphology*, Reinhold. Many interesting and informative articles on specific coastal subjects are contained in this work.

GUILCHER, A. (1958) *Coastal and Submarine Morphology*, Methuen. This work gives a comprehensive account of the general problems involved in coastal geomorphology. The English edition is a revision as well as a translation.

JOHNSON, D. W. (1919) *Shore Processes and Shoreline Development*, Wiley.

Johnson's account is the classic on shorelines and includes comprehensive discussions of processes and theoretical development, although some of the ideas are now rather out of date, including the cycles of shoreline development.

KING, C. A. M. (1959) *Beaches and Coasts*, Arnold.

STEERS, J. A. (1953) *The Sea Coast*, Collins, New Naturalist series. This remains one of the best general books on the coast for undergraduate reading.

STEERS, J. A. (1960) *The Coast of England and Wales in Pictures*, Cambridge University Press.

STEERS, J. A. (1964) *The Coastline of England and Wales*, 2nd edn, Cambridge University Press.
These two books now provide the standard treatment of British coasts. Further details may be obtained from the references included therein.

STEERS, J. A. (1969) *Coasts and Beaches*, Oliver & Boyd. A simple account by an authority on the subject.

ZENCOVICH, V. P. (1967) *Processes of Coastal Development*, Oliver & Boyd. This massive work is a mine of information, especially as it mainly deals with examples and authors not readily accessible to most of us in the original Russian.

References to other books and papers mentioned in the chapter are as follows:

CHALLINOR, J. (1949) 'A principle in coastal geomorphology', *Geography*, **34**, 212–15.

COTTON, C. A. (1952) 'Criteria for the classification of coasts', *Proceedings 17th International Geographical Union Conference*, Washington.

COTTON, C. A. (1954) 'Tests of a German non-cyclic theory and classification of coasts', *Geogrl J.* **120**, 353–61.

DALY, R. A. (1934) *The Changing World in the Ice Age*, Yale University Press.

DAVIS, W. M. (1928) *The Coral Reef Problem*, American Geographical Society.

GUILCHER, A. (1953) 'Essai sur la zonation et la distribution des formes littorales de dissolution de calcaire', *Annls Géogr.* **62**, 161–79.

GUILCHER, A. (1963) 'Estuaries, deltas, shelf, slope', in M. N. Hill, ed., *The Sea*, New York, Interscience, vol. iii, ch. 24.

HARDY, J. R. (1964) 'The movement of beach material and wave action near Blakeney Point, Norfolk', *Trans. Inst. Br. Geogr.* **34**, 53–70.

JOHNSON, D. W. (1925) *The New England–Acadian Shoreline*, Wiley.

JUTSON, J. T. (1939) 'Shore platforms near Sydney, New South Wales', *J. Geomorph.* **2**, 237–50.

KING, C. A. M. and WILLIAMS, W. W. (1949) 'The formation and movement of sand bars by wave action', *Geogrl J.* **113**, 70–85.

KUENEN, PH. K. (1950) *Marine Geology*, Wiley.

LAKE, P. (1949) *Physical Geography*, 2nd edn, Cambridge University Press.

LEWIS, W. V. (1931) 'The effect of wave incidence on the configuration of a shingle beach', *Geogrl J.* **78**, 129–48.

LEWIS, W. V. (1932) 'The formation of Dungeness foreland', *Geogrl J.* **80**, 309–24.

ROBINSON, A. H. W. (1953) 'The storm surge of 31st January–1st February, 1953', *Geography*, **38**, 134–41.

ROBINSON, A. H. W. (1955) 'The harbour entrances of Poole, Christchurch and Pagham', *Geogrl J.* **121**, 33–50.

RUSSELL, R. C. H. and MACMILLAN, D. H. (1952) *Waves and Tides*, Hutchinson.

RUSSELL, R. J. (1967) *River Plains and Sea Coasts*, University of California Press.

SHEPARD, F. P. (1963) *Submarine Geology*, 2nd edn, Harper & Row.

STODDART, D. R. (1969) 'Ecology and morphology of recent coral reefs', *Biological Reviews, Cambridge Philosophical Society*, **44**, 433–98.

WENTWORTH, C. K. (1938–9) 'Marine bench-forming processes', *J. Geomorph.* **1**, 6–32, and **2**, 3–25.

WILSON, G. (1952) 'The influence of rock structures on coast-line and cliff development around Tintagel, North Cornwall', *Proc. Geol. Ass.* **63**, 20–48.

Chapter 9   Movements of base level

BAULIG, H. (1935) 'The changing sea-level', *Institute of British Geographers*, Publication no. 3.

BAULIG, H. (1950) *Essais de géomorphologie*, Paris, Les Belles Lettres.

BLACHE, J. (1939–40) 'Le problème des méandres encaissés et les rivières lorraines', *J. Geomorph.* **2**, 201–12, and **3**, 311–31.

BROWN, E. H. (1952) 'The river Ystwyth, Cardiganshire: a geomorphological analysis', *Proc. Geol. Ass.* **63**, 244–69.

DARBY, H. C., ed. (1938) *The Cambridge Region*, Cambridge University Press.

EVERARD, C. E. (1954) 'The Solent river', *Trans. Inst. Br. Geogr.* **20**, 41–58.

FAEGRI, K. and IVERSEN, J. (1964) *Text-book of Modern Pollen Analysis*, 2nd edn, Oxford, Blackwell.

GEORGE, T. N. (1955) 'British Tertiary landscape evolution', *Science Progress*, **43**, 291–307.

GODWIN, H. (1938) 'The origin of roddons', *Geogrl J.* **91**, 241–50.

GODWIN, H. (1941) 'Pollen-analysis and Quaternary geology', *Proc. Geol. Ass.* **52**, 328–61.

GODWIN, H. (1956) *History of the British Flora*, Cambridge University Press.

HANSON-LOWE, J. (1935) 'The clinographic curve', *Geol. Mag.* **72**, 180–4.

HOLLINGWORTH, S. E. (1938) 'The recognition and correlation of high level erosion surfaces in Britain', *Q. Jl geol. Soc.* **94**, 55–74.

JONES, O. T. (1924) 'The upper Towy drainage system', *Q. Jl geol. Soc.* **80**, 568–609.

LEWIS, W. V. (1944) 'Stream trough experiments and terrace formation', *Geol. Mag.* **81**, 241–53.

MILLER, A. A. (1935) 'The entrenched meanders of the Herefordshire Wye', *Geogrl J*. **85**, 160–78.

MILLER, A. A. (1939) 'Attainable standards of accuracy in the determination of preglacial sea levels by physiographic methods', *J. Geomorph*. **2**, 95–115.

PEEL, R. F. (1949) 'Geomorphological fieldwork with the aid of Ordnance Survey maps', *Geogrl J*. **114**, 71–5.

SPARKS, B. W. (1953) 'Effects of weather on the determination of heights by aneroid barometer in Great Britain', *Geogrl J*. **119**, 73–80.

SPARKS, B. W. and WEST, R. G. (1963) 'The interglacial deposits at Stutton, Suffolk', *Proc. Geol. Ass*. **74**, 419–32.

SPARKS, B. W. and WEST, R. G. (1970) 'Late Pleistocene deposits at Wretton, Norfolk. I, Ipswichian interglacial deposits', *Phil. Trans. R. Soc*., B, **258**, 1–30.

STAMP, L. D. (1927) 'The Thames drainage system and the age of the Strait of Dover', *Geogrl J*. **70**, 386–90.

WEST, R. G. (1968) *Pleistocene Geology and Biology*, Longmans.

WEST, R. G. and SPARKS, B. W. (1960) 'Coastal interglacial deposits of the English Channel', *Phil. Trans. R. Soc*., B, **243**, 95–133.

WOOLDRIDGE, S. W. and LINTON, D. L. (1955) *Structure, Surface and Drainage in South-east England*, 2nd edn, London, Philip.

Chapter 10   The importance of changes of climate

BAULIG, H. (1928) *Le plateau central de la France*, Paris, Colin.

BÜDEL, J. (1963) 'Klima-genetische Geomorphologie', *Geographische Rundschau*, **15**, 269–85.

KING, L. C. (1967) *Morphology of the Earth*, 2nd edn, Oliver & Boyd.

KING, W. B. R. (1954) 'The geological history of the English Channel', *Q. Jl geol. Soc*. **110**, 77–101.

PELTIER, L. C. (1950) 'The geographical cycle in periglacial regions as it is related to climatic geomorphology', *Ann. Ass. Am. Geogr*. **40**, 214–36.

SPARKS, B. W. (1971) *Rocks and Relief*, Longman.

STODDART, D. R. (1969) 'Climatic geomorphology: review and re-assessment', *Progress in Geography*, **1**, 160–222.

TRICART, J. and CAILLEUX, A. (1965) *Introduction à la géomorphologie climatique*, Paris, SEDES. Eng. trans. in preparation, Longman.

WATTS, W. W. (1947) *Geology of the Ancient Rocks of Charnwood Forest, Leicestershire*, Leicester, Backus.

WOOLDRIDGE, S. W. and LINTON, D. L. (1955) *Structure, Surface and Drainage in South-east England*, 2nd edn, London, Philip.

Chapter 11   Landforms in arid and semi-arid climates

Profit may be combined with pleasure in the reading of general accounts of travel, exploration and warfare in deserts. There are many books of this type and the half-dozen listed below are examples of them.

BAGNOLD, R. A. (1935) *Libyan Sands*, Hodder and Stoughton.
DOUGHTY, C. M. (1908) *Wanderings in Arabia*, Duckworth.
FINLAYSON, H. H. (1936) *The Red Centre*, 2nd edn, Angus & Robertson.
LAWRENCE, T. E. (1927) *Revolt in the Desert*, Cape.
STEIN, SIR A. (1933) *On Ancient Central-Asian Tracks*, Macmillan.
THOMAS, B. (1932) *Arabia Felix*, Cape.

In addition the volumes of the *Geographical Journal* contain many articles about deserts, not all of them geographical in the narrow sense, but all containing useful and interesting information.

The following are more specialised references, most of the classic American papers being omitted as they can readily be found through the paper by Tator or from Cotton.

BAGNOLD, R. A. (1941) *The Physics of Blown Sand and Desert Dunes*, Methuen.
BALCHIN, W. G. V. and PYE, N. (1955) 'Pediment profiles in the arid cycle', *Proc. Geol. Ass.* **66**, 167–81.
BOSWORTH, T. O. (1922) *Geology of North-west Peru*, London, Macmillan.
COTTON, C. A. (1942) *Climatic Accidents*, Christchurch, N.Z., Whitcombe and Tombs.
DAVIS, W. M. (1909) *Geographical Essays*, Boston, Ginn (reprinted 1954, Dover).
GAUTIER, E.-F. (1928) *Le Sahara*, Paris, Payot.
GROVE, A. T. (1958) 'The ancient erg of Hausaland, and similar formations on the south side of the Sahara', *Geogrl J.* **124**, 528–33.
HANNA, S. R. (1969) 'The formation of longitudinal sand dunes by large helical eddies in the atmosphere', *J. appl. Met.* **8**, 874–83.
HUME, W. F. (1925) *Geology of Egypt*, Cairo, Government Press, vol. i.
JAMES, P. E. (1959) *Latin America*, 3rd edn, Cassell.
KING, L. C. (1948) 'A theory of bornhardts', *Geogrl J.* **112**, 83–6.
KING, L. C. (1950) 'A study of the world's plainlands', *Q. Jl Geol. Soc.* **106**, 101–32.
KING, L. C. (1951) *South African scenery*, 2nd edn, Oliver & Boyd.
KING, L. C. (1967) *Morphology of the Earth*, 2nd edn, Oliver & Boyd.
LANGBEIN, W. B. and SCHUMM, S. A. (1958) 'Yield of sediment in relation to mean annual precipitation', *Trans. Am. geophys. Union*, **39**, 1076–84.
MABBUTT, J. A. (1952) 'A study of granite relief from South-West Africa', *Geol. Mag.* **89**, 87–96.
MABBUTT, J. A. (1955) 'Pediment landforms in Little Namaqualand', *Geogrl J.* **121**, 77–85.
PEEL, R. F. (1941) 'Denudational landforms in the Central Libyan Desert', *J. Geomorph.* **5**, 3–23.
PEEL, R. F. (1966) 'The landscape in aridity', *Trans. Inst. Br. Geogr.* **38**, 1–23. A very useful general review of some of the main problems.

TATOR, B. A. (1952) 'Pediment characteristics and terminology', *Ann. Ass. Am. Geogr.* **42**, 293–317.
THORNBURY, W. D. (1954) *Principles of Geomorphology*, Wiley.

Chapter 12 Landforms in the humid tropics

BIROT, P. (1960) *Le Cycle d'érosion sous les différent climats*, Universidade do Brasil. Faculdade nacional de Filosofia. Centro de Pesquisas de Geografia do Brasil.
BRYAN, K. (1940) 'The retreat of slopes', *Ann. Ass. Am. Geogr.* **30**, 254–67.
COTTON, C. A. (1961) 'The theory of savanna planation', *Geography*, **46**, 89–101.
COTTON, C. A. (1962) 'Plains and inselbergs of the humid tropics', *Trans. R. Soc. N.Z.: Geology*, **1**, 269–77.
DOUGLAS, I. (1969) 'The efficiency of humid tropical denudation systems', *Trans. Inst. Br. Geogr.* **46**, 1–16.
FAIRBRIDGE, R. H., ed. (1968) *Encyclopedia of Geomorphology*, Reinhold. See especially note and references on etchplains by M. F. Thomas.
KING, L. C. (1967) *Morphology of the Earth*, 2nd edn, Oliver & Boyd.
LAMOTTE, M. and ROUGERIE, G. (1962) 'Les apports allochtones dans la genèse des cuirasses ferrugineuses', *Revue Géomorph. dyn.* **13**, 145–60.
LANGFORD-SMITH, T. and DURY, G. H. (1965) 'Distribution, character and attitude of the duricrust in the northwest of New South Wales and the adjacent areas of Queensland', *Am. J. Sci.* **263**, 170–90.
PRESCOTT, J. A. and PENDLETON, R. L. (1952) 'Laterite and lateritic soils', *Technical Communication, 47, Commonwealth Bureau of Soil Science, Commonwealth Agricultural Bureau.*
RUXTON, B. P. (1967) 'Slopewash under mature primary rainforest in northern Papua', in J. N. Jennings and J. A. Mabbutt, eds, *Landform Studies from Australia and New Guinea*, Cambridge University Press, pp. 85–94.
SIMONETT, D. S. (1967) 'Landslide distribution and earthquakes in the Bewani and Torricelli Mountains, New Guinea, statistical analysis, *ibid.*, pp. 64–84.
SPARKS, B. W. (1971) *Rocks and Relief*, London, Longman.
TRICART, J. (1965) *Traité de géomorphologie, V, Le modelé des régions chaudes, forêts et savanes*, Paris, SEDES. Eng. trans. in preparation, Longman.

Chapter 13 Landforms in glaciated highlands

General accounts of glaciation and very extensive bibliographies are given in:

COTTON, C. A. (1942) *Climatic Accidents*, Christchurch, New Zealand, Whitcombe and Tombs.
DALY, R. A. (1934) *The Changing World in the Ice Age*, Yale University Press.

EMBLETON, C. and KING, C. A. M. (1968) *Glacial and Periglacial Geomorphology*, Arnold.

FLINT, R. F. (1957) *Glacial and Pleistocene Geology*, Wiley.

SISSONS, J. B. (1967) *The Evolution of Scotland's Scenery*, Oliver & Boyd. Although a regional geomorphology, most of the relief of Scotland owes so much to glaciation that this book is a mine of information on Scottish glacial features.

Additional references quoted in this chapter are:

BATTLE, W. R. B. and LEWIS, W. V. (1951) 'Temperature observations in bergschrunds and their relationship to cirque erosion', *J. Geol.* **59**, 537–45.

CAILLEUX, A. (1952) 'Polissage et surcreusement glaciaires dans l'hypothèse de Boyé', *Revue Géomorph. dyn.* **3**, 247–57.

CAROL, H. (1947) 'Formation of roches moutonnées', *J. Glaciol.* **1**, 57–9.

CULLINGFORD, R. A. and SMITH, D. E. (1966) 'Late-glacial shorelines in eastern Fife', *Trans. Inst. Br. Geogr.* **39**, 31–51.

DONNER, J. J. (1970) 'Land sea level changes in Scotland', in D. Walker and R. G. West, eds, *Studies in the Vegetational History of the British Isles*, Cambridge University Press, pp. 23–39.

DURY, G. H. (1953) 'A glacial breach in the North Western Highlands', *Scott. geogr. Mag.* **69**, 106–17.

FILIPPI, F. DE (1932) *Himalaya, Karakoram and Eastern Turkestan*, Arnold.

GARWOOD, E. J. (1910) 'Features of Alpine scenery due to glacial protection', *Geogrl J.* **36**, 310–39.

HAYNES, V. M. (1968) 'The influence of glacial erosion and rock structure on corries in Scotland', *Geogr. Annlr,* **50**, ser. A, 221–34.

LEWIS, W. V. (1938) 'A meltwater hypothesis of cirque formation', *Geol. Mag.* **75**, 249–65.

LEWIS, W. V. (1948) 'Valley steps and glacial valley erosion', *Trans. Inst. Br. Geogr.* **14**, 19–44.

LEWIS, W. V. (1949) 'The function of meltwater in cirque formation: a reply', *Geogrl Rev.* **39**, 110–28.

LEWIS, W. V. (1953) 'Tunnel through a glacier', *The Times Science Review*, no. 9, 10–13.

LEWIS, W. V. (1954) 'Pressure release and glacial erosion', *J. Glaciol.* **2**, 417–22.

LEWIS, W. V., ed. (1960) 'Norwegian cirque glaciers', *R.G.S. Res. Ser.*, no. 4.

LINTON, D. L. (1949, 1951) 'Some Scottish river captures re-examined', *Scott. geogr. Mag.* **65**, 123–31, and **67**, 31–44.

MCCALL, J. G. (1952) 'The internal structure of a cirque glacier', *J. Glaciol.* **2**, 122–30.

MCCANN, S. B. (1966) 'The main Postglacial raised shoreline of western Scotland from the Firth of Lorne to Loch Broom', *Trans. Inst. Br. Geogr.* **39**, 87–99.

MANLEY, G. (1952) *Climate and the British Scene*, Collins.

NYE, J. F. (1952) 'The mechanics of glacier flow', *J. Glaciol.* **2**, 82–93.

PERUTZ, M. F. (1950) 'Direct measurement of velocity distribution in a vertical profile through a glacier', *J. Glaciol.* **1**, 382–3.

SELIGMAN, G. (1947) 'Extrusion flow in glaciers', *J. Glaciol.* **1**, 12–22.

SHARP, R. P. (1953) 'Deformation of a vertical bore hole in a piedmont glacier', *J. Glaciol.* **2**, 182–3.

SUGDEN, D. E. (1968) 'The selectivity of glacial erosion in the Cairngorm Mountains, Scotland', *Trans. Inst. Br. Geogr.* **45**, 79–92.

SYMPOSIUM (1944) 'The mechanics of glacier flow. A discussion', *J. Glaciol.* **2**, 339–41.

WEST, R. G. (1968) *Pleistocene Geology and Biology*, Longman.

WRIGHT, W. B. (1937) *The Quaternary Ice Age*, 2nd edn, London, Macmillan.

Chapter 14 Landforms in glaciated lowlands

Excellent accounts of the features of lowland glaciation will be found in:

EMBLETON, C. and KING, C. A. M. (1968) *Glacial and Periglacial Geomorphology*, Arnold.

FLINT, R. F. (1957) *Glacial and Pleistocene Geology*, Wiley.

SISSONS, J. B. (1967) *The Evolution of Scotland's Scenery*, Oliver & Boyd.

The fuller details of some of the features, especially the glacial diversions, described in this chapter may be found in:

BADEN-POWELL, D. F. W. (1948) 'The chalky boulder clays of Norfolk and Suffolk', *Geol. Mag.* **85**, 279–96.

BOULTON, G. S. (1970a) 'On the origin and transport of englacial debris in Svalbard glaciers', *J. Glaciol.* **9**, 213–29.

BOULTON, G. S. (1970b) 'On the deposition of subglacial and melt-out tills at the margins of certain Svalbard glaciers', *ibid.* **9**, 231–45.

CHARLESWORTH, J. K. (1953) *The Geology of Ireland*, Oliver & Boyd

CHARLESWORTH, J. K. (1957) *The Quaternary Era*, Arnold.

CHORLEY, R. J. (1959) 'The shape of drumlins', *J. Glaciol.* **3**, 339–44.

DURY, G. H. (1951) 'A 400-foot bench in south-eastern Warwickshire', *Proc. Geol. Ass.* **62**, 167–73.

GREGORY, K. J. (1965) 'Proglacial lake Eskdale after sixty years', *Trans. Inst. Br. Geogr.* **36**, 149–62.

KENDALL, P. F. and WROOT, H. E. (1924) *Geology of Yorkshire*, privately published.

PEEL, R. F. (1949) 'A study of two Northumbrian spillways', *Trans. Inst. Br. Geogr.* **15**, 75–89.

SHOTTON, F. W. (1953) 'The Pleistocene deposits of the area between Coventry, Rugby and Leamington and their bearing on the topographic development of the Midlands', *Phil. Trans. R. Soc.*, B, **237**, 209–60.

SPARKS, B. W. and WEST, R. G. (1964) 'The drift landforms around Holt, Norfolk', *Trans. Inst. Br. Geogr.* **35**, 27–35.

WEST, R. G. (1968) *Pleistocene Geology and Biology*, Longman.

WEST, R. G. and DONNER, J. J. (1956) 'The glaciations of East Anglia and the East Midlands', *Q. Jl geol. Soc.* **112**, 69–92.

WHITEHEAD, T. H. *et al.* (1928) 'The country between Wolverhampton and Oakengates', *Mem. Geol. Survey, U.K.*

WILLS, L. J. (1912) 'Late glacial and post-glacial changes in the lower Dee valley', *Q. Jl geol. Soc.* **68**, 180–98.

WOOLDRIDGE, S. W. and LINTON, D. L. (1955) *Structure, Surface and Drainage in South-eastern England,* 2nd edn, London, Philip.

WOOLDRIDGE, S. W. and HENDERSON, H. C. K. (1955) 'Some aspects of the physiography of the eastern part of the London basin', *Trans. Inst. Br. Geogr.* **21**, 19–32.

Chapter 15   Periglacial landforms

General accounts of periglacial forms and extensive bibliographies are given in:

CAILLEUX, A. and TAYLOR, G. (1954) *Cryopédologie*, Paris, Hermann.

EMBLETON, C. and KING, C. A. M. (1968) *Glacial and Periglacial Geomorphology*, Arnold.

TRICART, J. (1967) *Traité de géomorphologie, Tome II: le modelé des régions périglaciaires*, Paris, P.U.F.

WEST, R. G. (1968) *Pleistocene Geology and Biology*, Longman.

The more detailed references mentioned in the chapter are:

HOPKINS, D. M. (1949) 'Thaw lakes and thaw sinks in the Imaruk area, Seward peninsula, Alaska', *J. Geol.* **57**, 119–31.

JAHN, A. (1961) 'Quantitative analysis of some periglacial processes in Spitzbergen', *Uniwersytet Wrocławski, Nauka o Ziemi*, II, 1–34.

MAARLEVELD, G. C. (1965) 'Frost mounds', *Mededelingen van de Geologische Stichting, Nieuwe Serie*, **17**, 3–16.

MACKAY, J. R. (1963) 'The Mackenzie delta area, N.W.T.', *Memoir No. 8, Geographical Branch, Mines and Technical Surveys, Ottawa*.

MÜLLER, F. (1959) 'Beobachtungen über Pingos', *Meddelelser om Grønland*, **153**, 1–127.

PELTIER, L. C. (1950) 'The geographical cycle in periglacial regions as it is related to climatic geomorphology', *Ann. Ass. Am. Geogr.* **40**, 214–36.

PÉWÉ, T. L., CHURCH, R. E. and ANDRESEN, M. J. (1969) 'Origin and palaeoclimatic significance of large-scale patterned ground in the Donnelly Dome area, Alaska', *Geol. Soc. Am., Special Paper*, 103.

SHUMSKII, P. A. (1964) *Principles of Structural Glaciology*, New York, Dover.

SPARKS, B. W., WILLIAMS, R. B. G. and BELL, F. G. (1972) 'Presumed ground ice depressions in East Anglia' (in the press).

SUSLOV, S. P. (1961) *Physical Geography of Asiatic Russia*, San Francisco, Freeman.

SVENSSON, H. (1969) 'A type of circular lakes in northernmost Norway', *Geogr. Annlr*, A, **51**, 1–12.

WASHBURN, A. L. (1956) 'Classification of patterned ground and review of suggested origins', *Bull. geol. Soc. Am.* **67**, 823–66.

WATT, A. S., PERRIN, R. M. S. and WEST, R. G. (1966) 'Patterned ground in Breckland: structure and composition', *J. Ecol.* **54**, 239–58.

WHITE, H. J. O. (1924) 'The geology of the country near Brighton and Worthing', *Mem. Geol. Survey, U.K.*

WILLIAMS, R. B. G. (1964) 'Fossil patterned ground in eastern England', *Biuletyn Peryglacjalny*, **14**, 337–49.

WILLIAMS, R. B. G. (1968) 'Some estimates of periglacial erosion in southern and eastern England', *ibid.* **17**, 311–35.

Chapter 16   Erosion surfaces and their interpretation

BAULIG, H. (1928) *Le plateau central de la France*, Paris, Colin.

BAULIG, H. (1952) 'Surfaces d'aplanissement', *Annls Géogr.* **61**, 161–83, and 245–62.

BURY, H. (1910) 'On the denudation of the western end of the Weald', *Q. Jl geol. Soc.* **66**, 640–92.

COLEMAN, A. (1952) 'Some aspects of the development of the lower Stour, Kent', *Proc. Geol. Ass.* **63**, 63–86.

CRICKMAY, C. H. (1933) 'The later stages of the cycle of erosion', *Geol. Mag.* **70**, 337–47.

DAVIS, W. M. (1895) 'The development of certain English rivers', *Geogrl J.* **5**, 128–46.

EVERARD, C. E. (1954) 'The Solent river', *Trans. Inst. Br. Geogr.* **20**, 41–58.

HACK, J. T. and GOODLETT, J. C. (1960) 'Geomorphology and forest ecology of a mountain region in the central Appalachians', *Prof. Pap. U.S. geol. Survey*, 347.

HOBBS, W. H. (1922) *Characteristics of Existing Glaciers*, New York, Macmillan.

KING, L. C. (1950) 'The study of the world's plainlands', *Q. Jl geol. Soc.* **106**, 101–31.

KING, L. C. (1951) *South African Scenery*, 2nd edn, Oliver & Boyd.

KING, L. C. (1955) 'Pediplanation and isostasy: an example from South Africa', *Q. Jl geol. Soc.* **111**, 353–9.

KING, L. C. (1967) *Geomorphology of the Earth*, 2nd edn, Oliver & Boyd.

PELTIER, L. C. (1950) 'The geographic cycle in periglacial regions', *Ann. Ass. Am. Geogr.* **40**, 214–36.

SPARKS, B. W. (1949) 'The denudation chronology of the dip-slope of the South Downs', *Proc. Geol. Ass.* **60**, 165–215.

SPARKS, B. W. and WEST, R. G. (1965) 'The relief and drift deposits', pp. 18–40 in J. A. Steers, ed., *The Cambridge Region 1965*, British Association for the Advancement of Science.

SWEETING, M. M. (1950) 'Erosion cycles and limestone caverns in the Ingleborough district', *Geogrl J.* **115**, 63–78.

WOOLDRIDGE, S. W. and LINTON, D. L. (1955) *Structure, Surface and Drainage in South-east England*, 2nd edn, London, Philip.

Abbreviations used in the references

Am. J. Sci.	American Journal of Science
Ann. Ass. Am. Geogr.	Annals of the Association of American Geographers
Annls Géogr.	Annales de Géographie
Bull. geol. Soc. Am.	Bulletin of the Geological Society of America
Bull. Soc. belge Etud. géogr.	Bulletin de la Société Belge d'Etudes Géographiques
Bull. Soc. belge Géol.	Bulletin de la Société Belge de Géologie
Geogrl J.	Geographical Journal
Geogrl Rev.	Geographical Review
Geol. Mag.	Geological Magazine
Geogr. Annlr	Geografiska Annaler
J. appl. Met.	Journal of Applied Meteorology
J. Ecol.	Journal of Ecology
J. Geol.	Journal of Geology
J. geol. Educ.	Journal of Geological Education
J. Geomorph.	Journal of Geomorphology
J. Glaciol.	Journal of Glaciology
J. Soil Sci.	Journal of Soil Science
Mem. Geol. Survey, U.K.	Memoirs of the Geological Survey, United Kingdom
Phil. Trans. R. Soc.	Philosophical Transactions of the Royal Society
Proc. Dorset nat. Hist. archaeol. Soc.	Proceedings of the Dorset Natural History and Archaeological Society
Proc. Geol. Ass.	Proceedings of the Geologists' Association
Proc. K. ned. Akad. Wet.	Proceedings of the Koninklijke Nederlandsche Akademie van Wetenschappen
Proc. Trans. Croydon nat. Hist. scient. Soc.	Proceedings and Transactions of the Croydon Natural History and Scientific Society
Prof. Pap. U.S. geol. Survey	Professional Papers of the United States Geological Survey
Q. Jl geol. Soc.	Quarterly Journal of the Geological Society
R.G.S. Res. Ser.	Royal Geographical Society Research Series

Res. Ser. Am. geogr. Soc.	Research Series of the American Geographical Society
Revue belge Géog.	Revue Belge de Géographie
Revue Géomorph. dyn.	Revue de Géomorphologie Dynamique
Scott. geogr. Mag.	Scottish Geographical Magazine
Trans. Am. geophys. Union	Transactions of the American Geophysical Union
Trans. Inst. Br. Geogr.	Transactions of the Institute of British Geographers
Trans. R. Soc. N.Z.: Geol.	Transactions of the Royal Society of New Zealand: Geology
Vjschr. naturf. Ges. Zürich	Vierteljahsschrift der Naturfroschende Gesellschaft in Zürich
Z. Geomorph.	Zeitschrift für Geomorphologie

Index

A

Limousin, 471, 472, 473
linear erosion, 351
Linton, D. L., 39, 137, 141, 145, 152–3, 290, 294, 317, 397–8, 430–2, 475–9, 482
lithology, effect on relief, 168–79, 352–3
Lithothamnion, 267
Little Chesterford, 440
Little Loch Broom, 404
Littlehampton, 151, 224
Littlehampton anticline, 151–2, 166
Littorina Sea, 365
Livradois, 474
Llangollen, 173, 432
Llangranog, 239
lobes of delta, 277
local base level, 110
local superimposition of drainage, 145
Loch Dee, 178
Loch Eil, 398–9
Loch Erricht, 395
Loch Linnhe, 369, 395
Loch Lochy, 395
Loch Lomond, 367
Loch Morar, 394, 396
Loch Ness, 394, 395
Loch Shiel, 361, 395, 399
Loch Torridon, 395
lodgment till, 408–9
Loe Pool, 260
loess, 323, 442
Loire, river, 105, 186, 276, 303, 472, 474
London Basin, 89, 153, 430, 475–9
Long Hills, 249
Long Melford, 419
long profiles, 96–7, 109, 308–9
longitudinal coast, 279
longitudinal consequent, 127
longitudinal stream, 131, 153
longshore current, 221, 240
longshore drift, 225, 240–3, 247, 256, 276
Lop Nor, 400
Lough Neagh, 413
Lower Chalk, 168, 212
Lower Gravel Train, 430
Lower Greensand, 180, 430, 456, 476
Ludlow Shales, 173–4

Luxor, 332
Lynher, river, 176
Lynmouth, 107

M

Maarleveld, G. C., 445
Mabbutt, J. A., 339
Mackay, J. R., 445
Mackenzie, river, 279, 445–6
Mackin, J. H., 99, 101, 104, 110, 113, 114
Macmillan, D. H., 216, 217, 230, 233, 242, 243
macrotidal, 222
Madagascar, 344
magma, 183–6
Magnesian Limestone, 28
magnesium sulphate, 28, 30
magnitude-frequency concept, 107
Main Dolomite, 201
Malaya, 316
Maldive Islands, 271
Malham Tarn, 194
Mallaig, 361, 394, 398
Malvern Hills, 168
Mancos Shale, 77, 82
Manley, G., 358–9
marble, 48
marine terrace, 237, 289–94
Marlborough Downs, 209, 212, 458
marram grass, 251
marshes, 246, 253–5
Marsupites Chalk, 182–3
mass movements, mechanisms of, 60–6
mass movements, rapid flow, 53–4
mass movements, sliding, 54–5
mass movements, slow flow, 51–3
Matterhorn, 376
Matthes, G. H., 122–3
maturity, 8, 16
McCall, J. G., 376, 382
McCann, S. B., 369
meander, 12, 120–6, 145
meander belt, 12, 121–2, 300, 465
meander core, 302
meander cut-off, 12, 121, 300, 465
meanders, incised, 300

521

Index

Medina, 318
Mediterranean, 218, 223, 474
Mediterranean terraces, 293, 474
Medway, river, 145, 476
meio laranja hills, 352
Melbourn Rock, 117, 170, 182
Melton, M. A., 49, 50, 82, 87–8, 89–90,
    91, 116–17, 118
meltwater, 213, 371, 372, 382, 406, 413,
    418, 435
meltwater erosion, 207–8
meltwater streams, 111, 410, 418, 420
Mendips, 200
Merlin's Cave, 226
Merrick, 178
mesa, 189
mesotidal, 222
metamorphic rocks, 47–8
Meuse, river, 126
mica, 35–6, 47, 343–4
Mickleham, 213
micropiracy, 95
microtidal, 222
Middle Chalk, 137, 182, 212
Midhurst, 134
Midlands, 361, 406
Milford, 256
Miller, A. A., 302, 308–9
Millstone Grit, 171, 200
Mimms depression, 206, 430
minerals, rock-forming, 33–5
minerals, susceptibility to weathering,
    35–6
Mio-Pliocene peneplain, 478
Mississippi, river, 100, 107, 121, 122–3,
    276, 277, 285, 303
mobile coasts, 280
mogote, 353
Mole gap, 172
Mole, river, 204
mollisol, 438
Moluccas, 223
monadnock, 17, 464
Mongolia, 323, 442
monoclinal stream, 131
Monroe, W. H., 201
Monteith readvance, 367
montmorillonite, 206, 345
Moray Firth, 418

moraines, lateral, 388, 416
moraines, push, 410
moraines, terminal, 388, 410, 419
Moreton-in-the-Marsh, 428
Morfa Dyffryn, 225
Morisawa, M., 99–100
Moselle, river, 126
moulin kame, 417
Mount Caburn, 132–4
Mount Etna, 319
Mount Olga, 318
Mount Ruapehu, 106
Mount Sinai, 332
Mozambique, 481
mudflow, 50, 54, 103
Müller, F., 445
Mullwharchar, 178
muscovite, 33, 36
Musgrave range, 318

N

naled, 448
Nar Valley Clay, 448
Natal, 337, 481
Nazareth Bank, 271
neap tide, 215
Neath, river, 10, 194, 410
needle ice, 443
Needles, 226, 289
negative feedback, 61
negative movement of base level, 17–18,
    282, 289–311
Nene, river, 428, 483
Netherlands, 440
Netley Heath, 477
neutral shorelines, 280
névé, 360
New Guinea, 272
New Jersey, 444
New Mexico, 87
New Red Sandstone, 316
Newer Drift, 404–5, 410, 425
Newport, 419
Newtondale, 426–7, 434
nick points, 298–300, 308–9
Nile, river, 275–6, 332
nip, 235, 239
Nith valley, 418